Locusts of Power

In this highly original environmental history, Samuel Dolbee sheds new light on borders and state formation by following locusts and revealing how they shaped both the environment and people's imaginations from the late Ottoman Empire to the eve of the Second World War. Drawing on a wide range of archival research in multiple languages, Dolbee details environmental, political, and spatial transformations in the region's history by tracing the movements of locusts and their intimate relationship to people in motion, including Arab and Kurdish nomads, Armenian deportees, and Assyrian refugees, as well as states of the region. With locusts and moving people at center stage, surprising continuities and ruptures appear in the Jazira, the borderlands of today's Iraq, Syria, and Turkey. Transcending approaches focused on the collapse of the Ottoman Empire or the creation of nation-states, Dolbee provides a new perspective on the modern Middle East grounded in environmental change, state violence, and popular resistance.

Samuel Dolbee is Assistant Professor of History at Vanderbilt University.

"It turns out that the locust can speak and Samuel Dolbee tells us how. Grounded firmly in the fertile intersection of Ottoman, post-Ottoman and environmental studies, this book demonstrates how human-pesticide relationships shaped the transformation of Ottoman Jazira: from "desert" to agricultural land; from a place for the sedenterization of nomads to one for the forced nomadization of sedentary populations for genocidal purposes, and how the same land then became parts of post-Ottoman Turkey, Iraq, and Syria. It will become a classic and deservedly so."

Lerna Ekmekcioglu, Massachusetts Institute of Technology

"In this deeply empirical and eloquent book, Samuel Dolbee offers a history of a part of the Middle East that scholars have missed (or ignored)—the Jazira. Following Dolbee following locusts across this landscape opens up modes of political and environmental analyses that point the way for future studies."

Alan Mikhail, Yale University

Studies in Environment and History

Other Books in the Series

Andy Bruno *Tunguska: A Siberian Mystery and Its Environmental Legacy*
Lionel Frost et al. *Cities in a Sunburnt Country: Water and the Making of Urban Australia*
Adam Sundberg *Natural Disaster at the Closing of the Dutch Golden Age: Floods, Worms, and Cattle Plague*
Germán Vergara *Fueling Mexico: Energy and Environment, 1850–1950*
Peder Anker *The Power of the Periphery: How Norway Became an Environmental Pioneer for the World*
David Moon *The American Steppes: The Unexpected Russian Roots of Great Plains Agriculture, 1870s–1930s*
James L. A. Webb, Jr. *The Guts of the Matter: A Global Environmental History of Human Waste and Infectious Intestinal Disease*
Maya K. Peterson *Pipe Dreams: Water and Empire in Central Asia's Aral Sea Basin*
Thomas M. Wickman *Snowshoe Country: An Environmental and Cultural History of Winter in the Early American Northeast*
Debjani Bhattacharyya *Empire and Ecology in the Bengal Delta: The Making of Calcutta*
Chris Courtney *The Nature of Disaster in China: The 1931 Yangzi River Flood*
Dagomar Degroot *The Frigid Golden Age: Climate Change, the Little Ice Age, and the Dutch Republic, 1560–1720*
Edmund Russell *Greyhound Nation: A Coevolutionary History of England, 1200–1900*
Timothy J. LeCain *The Matter of History: How Things Create the Past*
Ling Zhang *The River, the Plain, and the State: An Environmental Drama in Northern Song China, 1048–1128*
Abraham H. Gibson *Feral Animals in the American South: An Evolutionary History*
Andy Bruno *The Nature of Soviet Power: An Arctic Environmental History*
David A. Bello *Across Forest, Steppe, and Mountain: Environment, Identity, and Empire in Qing China's Borderlands*
Erik Loomis *Empire of Timber: Labor Unions and the Pacific Northwest Forests*
Peter Thorsheim *Waste into Weapons: Recycling in Britain during the Second World War*
Kieko Matteson *Forests in Revolutionary France: Conservation, Community, and Conflict, 1669–1848*
Micah S. Muscolino *The Ecology of War in China: Henan Province, the Yellow River, and Beyond, 1938–1950*

George Colpitts *Pemmican Empire: Food, Trade, and the Last Bison Hunts in the North American Plains, 1780–1882*
John L. Brooke *Climate Change and the Course of Global History: A Rough Journey*
Paul Josephson et al. *An Environmental History of Russia*
Emmanuel Kreike *Environmental Infrastructure in African History: Examining the Myth of Natural Resource Management*
Gregory T. Cushman *Guano and the Opening of the Pacific World: A Global Ecological History*
Sam White *The Climate of Rebellion in the Early Modern Ottoman Empire*
Edmund Russell *Evolutionary History: Uniting History and Biology to Understand Life on Earth*
Alan Mikhail *Nature and Empire in Ottoman Egypt: An Environmental History*
Richard W. Judd *The Untilled Garden: Natural History and the Spirit of Conservation in America, 1740–1840*
James L. A. Webb, Jr. *Humanity's Burden: A Global History of Malaria*
Myrna I. Santiago *The Ecology of Oil: Environment, Labor, and the Mexican Revolution, 1900–1938*
Frank Uekoetter *The Green and the Brown: A History of Conservation in Nazi Germany*
Matthew D. Evenden *Fish versus Power: An Environmental History of the Fraser River*
Alfred W. Crosby *Ecological Imperialism: The Biological Expansion of Europe, 900–1900, second edition*
Nancy J. Jacobs *Environment, Power, and Injustice: A South African History*
Edmund Russell *War and Nature: Fighting Humans and Insects with Chemicals from World War I to Silent Spring*
Adam Rome *The Bulldozer in the Countryside: Suburban Sprawl and the Rise of American Environmentalism*
Judith Shapiro *Mao's War against Nature: Politics and the Environment in Revolutionary China*
Andrew Isenberg *The Destruction of the Bison: An Environmental History*
Thomas Dunlap *Nature and the English Diaspora*
Robert B. Marks *Tigers, Rice, Silk, and Silt: Environment and Economy in Late Imperial South China*
Mark Elvin and Tsui'jung Liu *Sediments of Time: Environment and Society in Chinese History*
Richard H. Grove *Green Imperialism: Colonial Expansion, Tropical Island Edens and the Origins of Environmentalism, 1600–1860*
Thorkild Kjærgaard *The Danish Revolution, 1500–1800: An Ecohistorical Interpretation*
Donald Worster *Nature's Economy: A History of Ecological Ideas, second edition*
Elinor G. K. Melville *A Plague of Sheep: Environmental Consequences of the Conquest of Mexico*
J. R. McNeill *The Mountains of the Mediterranean World: An Environmental History*
Theodore Steinberg *Nature Incorporated: Industrialization and the Waters of New England*
Timothy Silver *A New Face on the Countryside: Indians, Colonists, and Slaves in the South Atlantic Forests, 1500–1800*
Michael Williams *Americans and Their Forests: A Historical Geography*
Donald Worster *The Ends of the Earth: Perspectives on Modern Environmental History*
Robert Harms *Games against Nature: An Eco-Cultural History of the Nunu of Equatorial Africa*

Warren Dean *Brazil and the Struggle for Rubber: A Study in Environmental History*
Samuel P. Hays *Beauty, Health, and Permanence: Environmental Politics in the United States, 1955–1985*
Arthur F. McEvoy *The Fisherman's Problem: Ecology and Law in the California Fisheries, 1850–1980*
Kenneth F. Kiple *The Caribbean Slave: A Biological History*

Locusts of Power

Borders, Empire, and Environment in the Modern Middle East

SAMUEL DOLBEE
Vanderbilt University

CAMBRIDGE
UNIVERSITY PRESS

CAMBRIDGE
UNIVERSITY PRESS

Shaftesbury Road, Cambridge CB2 8EA, United Kingdom

One Liberty Plaza, 20th Floor, New York, NY 10006, USA

477 Williamstown Road, Port Melbourne, VIC 3207, Australia

314–321, 3rd Floor, Plot 3, Splendor Forum, Jasola District Centre, New Delhi – 110025, India

103 Penang Road, #05–06/07, Visioncrest Commercial, Singapore 238467

Cambridge University Press is part of Cambridge University Press & Assessment, a department of the University of Cambridge.

We share the University's mission to contribute to society through the pursuit of education, learning and research at the highest international levels of excellence.

www.cambridge.org
Information on this title: www.cambridge.org/9781009200318
DOI: 10.1017/9781009200301

First published 2023

A catalogue record for this publication is available from the British Library.

ISBN 978-1-009-20031-8 Hardback
ISBN 978-1-009-20035-6 Paperback

Contents

List of Figures and Maps

Preface

Locusts of Power explains how locusts and moving people made the modern Middle East. As one of the first environmental histories of the late Ottoman Empire, it focuses on the region known as the Jazira, the borderland area today stretching across Iraq, Syria, and Turkey that was recently unified under ISIS and has variously been envisioned as part of Kurdish or Assyrian homelands. The book uncovers the environmental coherence of this politically embattled region from the late Ottoman to the immediate post-Ottoman period (1858–1939) by following the locust, the insect that over those decades repeatedly took advantage of the region's political ecology. Not only was the region a mix of cultivated and uncultivated arid lands, it was also divided between Ottoman provinces and, subsequently, British Iraq, French Syria, and republican Turkey. What this meant was that locusts used political divisions to thwart various human dreams of transforming the Jazira into a breadbasket by emerging from the edges of cultivation and feasting. With each chapter centered on one group of moving people who were compared to locusts – among them Arabic- and Kurdish-speaking nomads, Armenian deportees, and Assyrian refugees – the book unearths state efforts to use borders to control these groups, and the ways that these groups managed to resist. The insects were profoundly connected to the motion of people on both a material and metaphorical level. They were organisms that forced people to move and starve, but they were also organisms upon which humans projected their own anxieties, hopes, and fears. *Locusts of Power* presents this view based on archival materials, fiction, memoirs, and newspapers in Arabic, English, French, German, Modern Turkish, and Ottoman.

The book thus offers an integrative account of state formation in the modern Middle East, while also connecting it to global transformations of the human relationship with the environment. The history suggests how the Ottoman world was far from a "borderless" space and examines how locusts and people alike moved across Ottoman provincial borders and left confused and quarreling officials in their wake. Rather than ignoring borders in the Ottoman period, people such as nomads actively used them for their own purposes. Moreover, the book exposes how motion, agrarian development, and state violence came together in the Jazira during one of the world-historical events of the century: the Armenian genocide. After failing to develop the Jazira for decades, Ottoman officials turned the environment into a weapon by sending hundreds of thousands of Ottoman Armenians there. Yet even amidst the horrific violence, people managed to resist, with thousands of Armenians – mostly children – surviving by living with the Jazira's nomads. Finally, the book contends that post-Ottoman borders had a different significance than is conventionally believed. The borders took meaning not just from colonial decrees or national declarations in faraway capitals; the new borders significantly emerged in the context of an Ottoman legacy of provincial borders. They also were underpinned by a host of technologies and interventions of agrarian development. As new chemical insecticides largely wiped out the region's distinctive locusts, the mobile political economy that had characterized the Jazira for so many decades unraveled. It was from these clouds of pesticides used on all sides of the Jazira that Iraqi, Syrian, and Turkish national projects emerged. The book thus offers an account of regional transformation, tracing how an area envisioned as a site of nomadic settlement became the site of the Armenian genocide, then the most agriculturally productive lands of Syria, and then the heartland of ISIS. Attention to the environmental rhythms of all of these events connects the particularities of the end of the Ottoman Empire with global transformations of the human relationship with the nonhuman world.

Acknowledgments

I must begin by acknowledging the people and places that are the subject of this book. Without the mobile people whose traces somehow made it into state archives, the photographs of the fiercely defiant stares of Armenian survivors, and the chassis-rattling road between Harran and Viranşehir, this book would not be.

Formulating these thoughts has required time and space, and I am grateful to a number of institutions for granting these comforts to me. The work began as a dissertation at New York University, under the gentle and sage care of Zachary Lockman. Whether through affably delivered tough questions, quizzically smiling squints, or drafts red with Track Changes, Zach radiated a scholarly generosity that taught me how to think with historical rigor while also never losing sight of the political stakes of the work in the present. Karl Appuhn has always believed in me more than I have believed in myself, and I wrote much of this book in the hope it would earn one of his highest terms of academic praise: that it is "badass." Manu Goswami, Dina Rizk Khoury, and Sam White offered generous readings of the work at that early stage. The research and writing benefited from grants from the American Academic Research Institute in Iraq, the Social Science Research Council, the American Council of Learned Societies, and New York University. My work as a teacher during this time was a joy, not only because of my students at NYU, but also because of protections afforded by GSOC-UAW Local 2110, and I thank the generations of graduate student organizers who made it possible. I received postdoctoral support and funding from the Crown Center for Middle East Studies at Brandeis University, the Mahindra Humanities Center at Harvard University, and the Program in Agrarian

Studies at Yale University, where I wrote under the watchful gaze of a poster of a chicken. I thank James Scott and Kalyanakrishnan Sivaramakrishnan for their good-humored guidance. I thank Shivi as well as Jonathan Wyrtzen, Alan Mikhail, Meltem Toksöz, and Thomas Wickman for providing incisive comments during a virtual book-manuscript workshop in the terrifying early months of the pandemic. Further work has taken place in the basement of the Barker Center while teaching in Harvard's History & Literature program, where I am grateful for the leadership of Lauren Kaminsky and Phil Deloria, and the bright-eyed enthusiasm of my students. Special thanks to Briana Smith, Duncan White, and all those who workshopped the book's introduction and offered characteristically thoughtful advice. During my five years in the borderlands of Cambridge-Somerville, Harvard's Political Anthropology/Ecology Working Group grounded me, and I'm thankful for the patient anthropologists who allowed the obtuse questions of a historian. I completed work on the book in my new scholarly home in the History Department and Program in Climate and Environmental Studies at Vanderbilt University. I thank Emily Greble, David Hess, Julia Phillips Cohen, Tasha Rijke-Epstein, Ruth Rogaski, Samira Sheikh, Eddie Wright-Rios, Meng Zhang, and all of my new colleagues for making me feel so welcome in Nashville.

This book would also not exist without the labor of librarians, archivists, and archival workers in too many places to name. I am particularly indebted to the staff at Istanbul's Ottoman Archives for the documents and tea. I am also grateful to the librarians and student workers of Widener Library's Philips Room for fielding many requests on long summer afternoons (and also reminding me when closing time was approaching). Thank you to Reza Nazir for his help in the German archives. Thank you to İsmail Pamuk for serendipitously and generously helping me with photo permissions at Istanbul University. And thank you to Meredith Sadler for making the lovely maps and cover.

I am grateful to the Studies in Environment and History editors Ling Zhang, for her advice on the project, and J. R. McNeill, who first introduced me to environmental history when I was a confused MA student and interloper in the Georgetown History Department basketball game many years ago. I appreciate the patience and diligence of Cambridge University Press, particularly acquisitions editor Lucy Rhymer, content manager Hannah Weber, and editorial assistant Emily Plater. I thank Kathleen Fearn for her meticulous copyediting, and Vidya Ashwin for her management oversight. I also thank the two anonymous reviewers for their careful readings, which both improved and encouraged the book. Portions of the book are based on the articles "The Desert at the End of Empire: An Environmental History of the Armenian Genocide," published in the May 2020 issue of *Past & Present*, and "Empire on the Edge: Desert, Nomads, and the Making of an Ottoman Provincial Border," published in the March 2022 issue of *American Historical Review*. I thank the editors for permission to use amended versions of these articles.

I am lucky to have smart and kind friends, colleagues, and mentors who have read drafts, provided sources, or told me funny things while I have worked on this project. They include Victoria Abrahamyan, Oscar Aguirre-Mandujano, Taylan Akyıldırım, Seda Altuğ, Ahmad Amara, Gabriel de Avilez Rocha, Reem Bailony, Nora Barakat, Alice Baumgartner, Nadim Bawalsa, Steven Biehl, Edna Bonhomme, Guy Burak, Lâle Can, Talha Çiçek, Camille Cole, Vikrant Dadawala, Sarah Dadouch, Emre Can Dağlıoğlu, Nicholas Danforth, Seyward Darby, Muriam Haleh Davis, Mark Drury, David Fedman, Michael Ferguson, Susanna Ferguson, Ella Fratantuono, Matthew Ghazarian, Benan Grams, Zoe Griffith, Yeter Can Gümüş, Vladimir Hamed-Troyansky, Shireen Hamza, Stevan Harrell, Shay Hazkani, Jesse Howell, Faisal Husain, Nurçin İleri, Aaron Jakes, Cemal Kafadar, Ümit Selim Kurt, Ekin Kurtiç, Pauline Lewis, Zhou Hau Liew, Michael Christopher Low, Pascal Menoret, Marijana Mišević, Taylor Moore, Sato Moughalian, Khatchig Mouradian, Golnar Nikpour, Michael O'Sullivan, Ramazan Hakkı Öztan, Uğur Zekeriya Peçe, Zozan Pehlivan, Peter Perdue, Graham Auman Pitts, China Sajadian, Caterina Scaramelli, Nir Shafir, Sarah Shields, Yan Slobodkin, Corey Sobel, Laura Stocker, Ajantha Subramanian, Michael Talbot, Jordi Tejel, Khachig Tololyan, Abdullah Uğur, Harvey Weiss, Patrick Whitmarsh, Elizabeth Williams, Alex Winder, Dilan Yıldırım, Murat Cihan Yıldız, Seçil Yılmaz, Adrien Zakar, and many others not named here. I thank Fredrik Meiton for reading and listening to so many versions of these arguments (and improving them), whether while walking over bridges in New York or hiking in New Hampshire. I also thank Chris Gratien for many conversations and archival trips related to this project, as well as that drive from Harran to the Öğretmenevi of Viranşehir that I did not think we would complete. Any mistakes are of course my responsibility, and are probably a result of not listening closely enough to their advice.

Thank you to Tayga Hatice Urus for sending me so many direct messages on Instagram, and also for praying for me. Thank you to Joey Dolbee for making me laugh. Thank you to Arianne Sedef Urus for going to yoga with me a decade ago and also for reading the entire manuscript (I am still reconsidering my prepositions). And finally, thank you to my parents, Susan Peecher and William Dolbee, who can be blamed for my obsession with space thanks to the many gloriously long car rides they took me on when I was a child. It is to them – and their decades of labor as teachers – that I dedicate this book.

Note on Spelling and Units of Measurement

Deciding on how to spell the names of places and people in a polyglot contested space is fraught and can convey a certainty or one-sidedness that was in actuality contingent and dynamic. Should I write Ras al-Ayn (as it is typically spelled in English), Raʾs al-ʿAyn (the Arabic transliteration), Resulayn (the Turkish spelling), Serê Kaniyê (the Kurdish version), or Rēš Aynā (the Syriac version)? For the sake of simplicity, I have opted for the preferred English spelling whenever possible and also acknowledge the flaws of this approach (see, for example, the panoply of renderings of Deir ez-Zor). When transliterating Ottoman Turkish, I have used the modern Turkish equivalents, and when transliterating Arabic, I have followed the standards of the *International Journal of Middle East Studies* and omitted diacritical marks except for ʿayn and hamza. Unless otherwise noted, all translations are my own.

There are a few spellings from Turkish that require explanation for those unfamiliar:

"c" in Cemal or Cevdet is pronounced like a "j," as in jaguar
"ş" in Viranşehir is pronounced like a "sh," as in shark
"ç" in Suruç is pronounced like a "ch," as in cheetah
"ğ" in Karacadağ is silent and elongates the preceding vowel

For measurements of weight and distance, I have used conversions given by Redhouse:

1 kiyye = 1 okka = 1.28 kilograms = 2.83 pounds
1 dunham = 0.9393 square kilometers = 0.58 square miles

Introduction

"The desert journey continues very boringly," wrote a reporter for the Istanbul newspaper *Akşam* in the summer of 1928.[1] The train was headed eastward from the outskirts of the Syrian city of Aleppo (see Figure 1 for a map of the route). "To pass for hours in the middle of a brown expanse amid suffocating heat in a train car that is always shaking is," the reporter complained, "unpleasant." It would get worse. Suddenly, a droning insect flew into the train car. And then another. They were locusts. Someone closed the windows. But the insects continued to collide into the side of the train "incessantly." In their percussive onslaught, the reporter might have heard the rhythm of the region's recent history. After all, it was these creatures that had helped make a landscape that witnessed nomadic sedentarization campaigns, the Armenian genocide, and interwar refugee resettlement. The train hurtled onward.

The most common locust in the region was the Moroccan locust (*Dociostaurus marocannus*; Turkish: *Fas çekirgesi*; Arabic: *al-jarad al-marrakishi*; Figure 2).[2] The name derived from where a Swedish entomologist first "discovered" the creature. In reality, the insects lived in a wide range of places, from Morocco to Central Asia. In

[1] Y. M., "Şark Mektupları: Harran Ovasından Mardin'e kadar . . . ," *Cumhuriyet*, June 17, 1928.

[2] Latchininsky, "Moroccan Locust," 167. The region was also occasionally afflicted by the migratory desert locust (*Schistocerca gregaria*, also known as the Najdi or Sudan locust) – and, in fact, there was a swarm of this species in 1928 – but the Moroccan locust was native to the region. Natural History Museum Archive (NHMA), Syria and Lebanon 6211 – 157, Rapport sur la lutte anti-acridienne dans les Pays du Levant adhérent à l'Accord International du 20 Mai 1926, 1927–1928, p. 3.

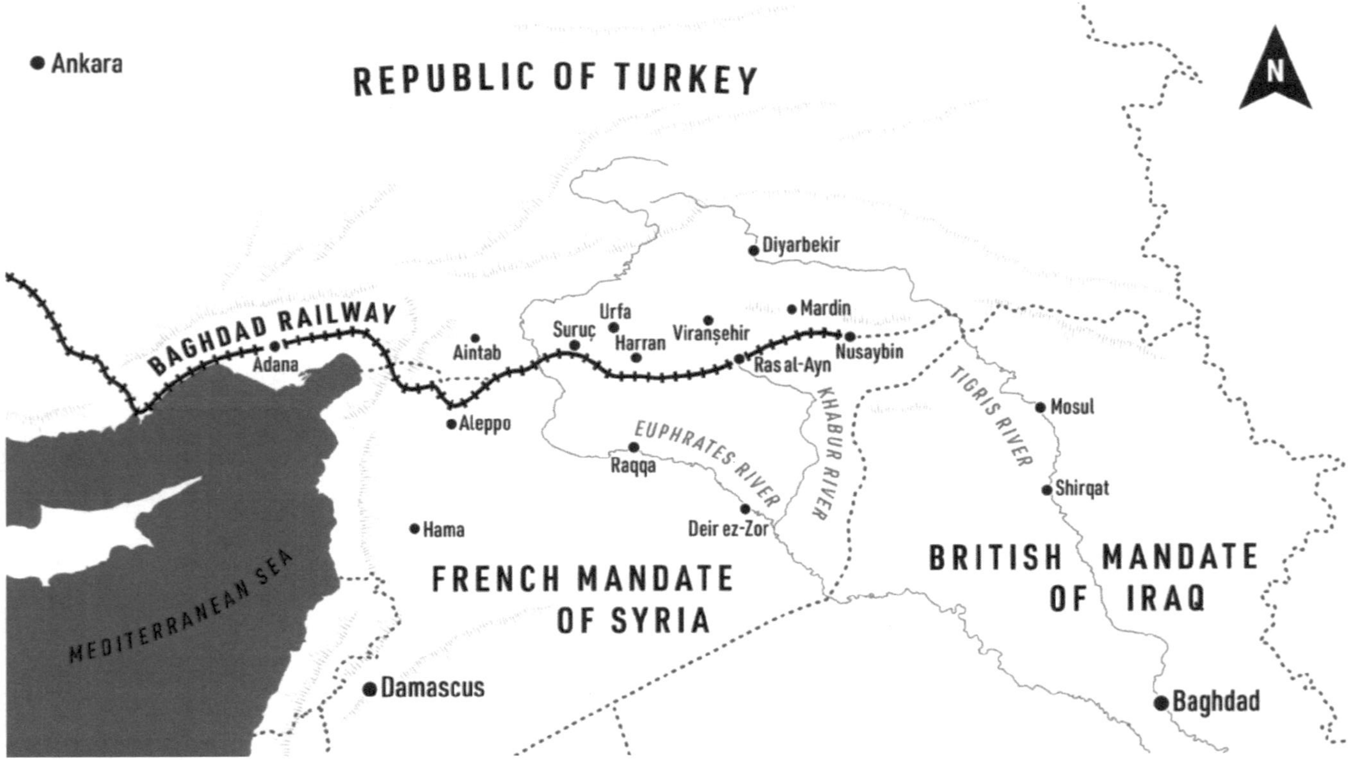

FIGURE 1 Baghdad Railway and post-Ottoman borders of the Jazira

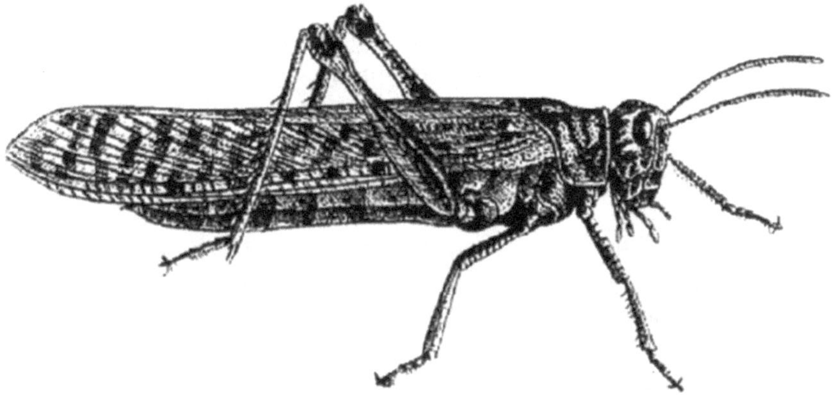

FIGURE 2 The Moroccan locust (*Dociostaurus marocannus*). Maurin, *Invasion des sauterelles*

most years, they remained harmless grasshoppers, but sometimes – because of a mix of precipitation, population density, and weather – they accelerated into what is known as their "gregarious" phase. Their population exploded, and their physiology changed. They swarmed and ranged up to 200 kilometers (124 miles). They blotted out the sun and consumed nearly everything in their path. In the words of one observer, they left nothing behind but "black stumps and their own excreta."[3] They were particularly destructive in zones of expanding cultivation, where planted fields existed alongside their preferred desert and steppe egg-laying grounds. In fact, the insects seemed so connected to human cultivation that elsewhere they were referred to in Arabic as the "human locust" (*al-jarad al-adami*).[4]

Unbeknownst to the bored reporter on the train and overshadowed by infamous figures such as Sykes and Picot, the locusts on the railway in 1928 were in their own way etching borders. In their flight, destruction, and perhaps even excrement, they mapped out an agroecology known as the Jazira – now largely forgotten to those who live outside of it – that stretched from the Tigris to the Euphrates at the foot of the Anatolian plateau. Extending between the cities of Aleppo, Diyarbekir, and Mosul, the Jazira was arid yet fertile, straddling the line where rain-fed agriculture was possible. For centuries, the Jazira functioned as an administrative unit. But when the Ottoman Empire worked to transform the region in the mid-nineteenth century, it attempted to do so through

[3] Lyon, *Kurds, Arabs and Britons*, 174. [4] Latchininsky, "Moroccan Locust," 171.

provinces that divided the connected landscape. When locusts moved across the Jazira, they did so beyond the bounds of provincial borders and often beyond the control of state officials.

Border-crossing movement persisted after the end of the Ottoman Empire, when the insects – and the railway – ensured that Syria and Turkey were curiously linked. The railway had been built in the waning years of the Ottoman Empire with the aim of connecting Berlin and Baghdad. After the Ottoman Empire dissolved in late 1922, a number of successor states emerged in its place, including the Republic of Turkey and the French Mandate of Syria under the neocolonial League of Nations. In the process of dividing once-unified imperial holdings, French and Turkish officials sought a borderline to separate Turkey and Syria and the Ottoman past from the post-Ottoman present. They found such a demarcation in the railway. As the railway moved east of Aleppo, it became the actual border between the countries, Syria to the south of the line and Turkey to the north. Officials thus transformed an infrastructural project intended to rejuvenate the Ottoman Empire into the actual dividing line between post-Ottoman states. Locusts paid little heed to these divisions. In the key of one map depicting the insects' cross-border range (Figure 3), officials had replaced the symbol denoting the railroad. In lieu of the Ottoman infrastructure-turned-cleaver of post-Ottoman states was the thick blue line denoting the range of the locusts. What the map characterized as "the border of the winged" extended from Syria into southeast Turkey.[5]

As locusts moved across Ottoman and post-Ottoman borders in the late nineteenth and early twentieth centuries (Figures 4 and 5), they animated a mobile ecology entangled with people, including Arab and Kurdish nomads, Armenian deportees, and Assyrian refugees. The groups occupied different relationships to states, with some objects of reform and others targets of destruction. Yet in the Jazira, all of these people encountered locusts, which they variously fled, feared, and, in some cases, ate. They were also connected culturally. In fact, all of these groups found themselves compared to the insects at one point or another. Locusts of power, then, refers to the way that locusts shaped not only the "everlasting nature-imposed condition of human existence," as Marx put it, but also the imagination of people in the Jazira.[6] The play on the phrase

[5] NHMA, Turkey 6217, Map of Locust Invasion of 1928–1929 in Turkey.
[6] Marx, *Capital*, vol. 1, 290.

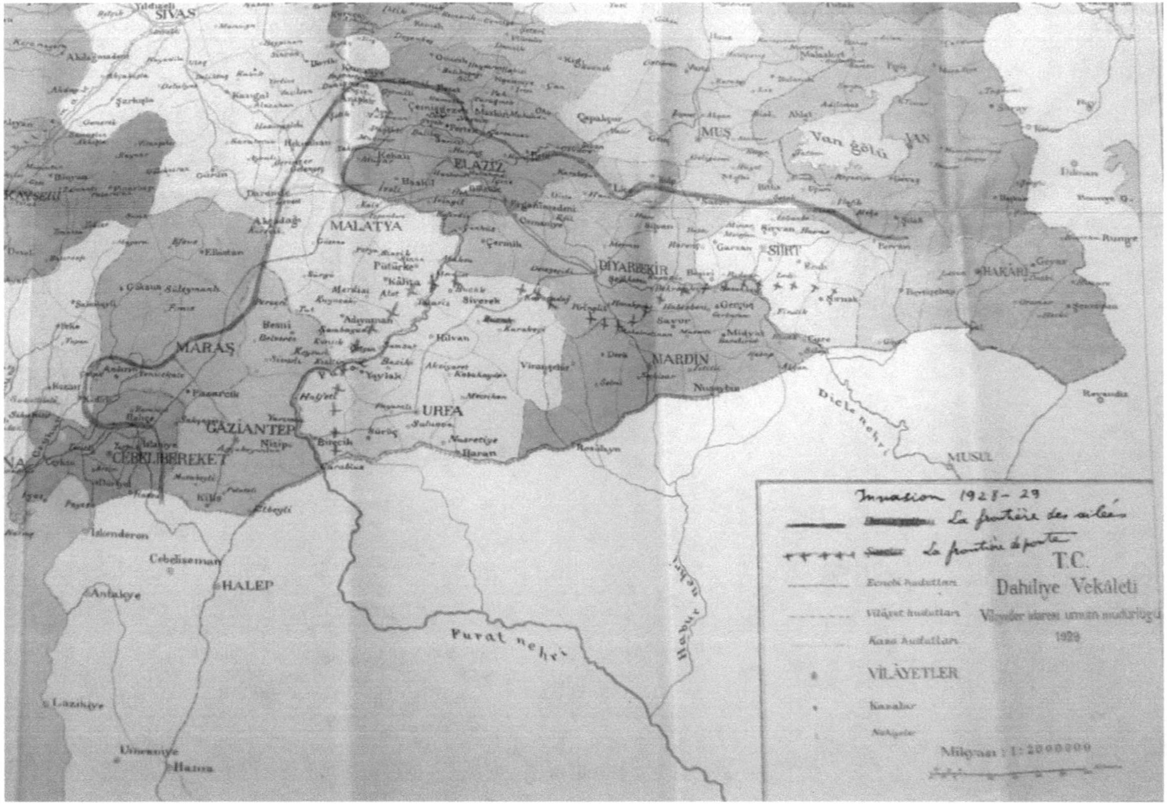

FIGURE 3 "Border of the winged," 1928–1929. NHMA, Turkey 6217. Courtesy of Trustees of the Natural History Museum, London

FIGURE 4 Range of Moroccan locusts. Based on Uvarov, "Ecology of the Moroccan Locust"

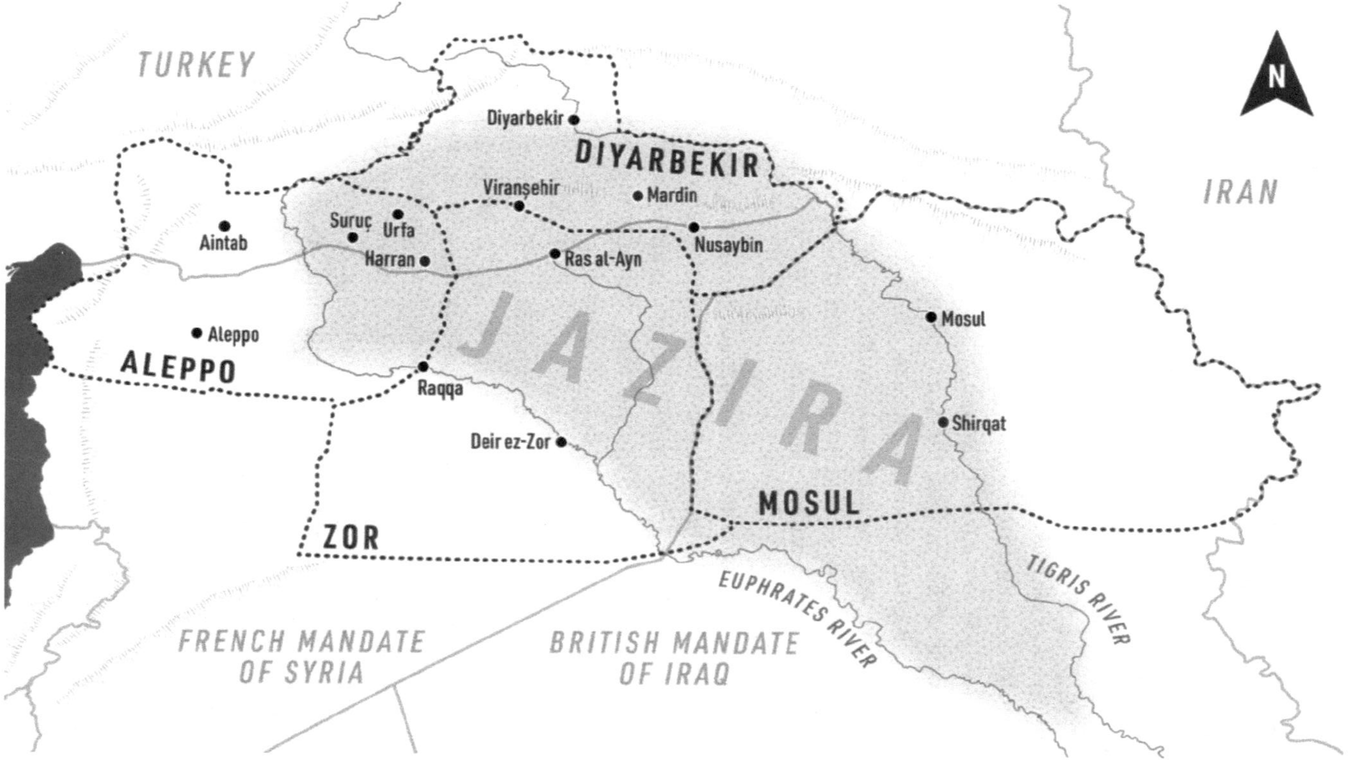

FIGURE 5 The Jazira, Ottoman provincial borders, and post-Ottoman borders. Provincial borders based on Cuinet, *La Turquie d'Asie*

"locus of power," moreover, alludes to the significance of the Jazira's place on political and environmental margins for this kind of power.

Indeed, in moving on the edge, both locusts and these groups of mobile people sometimes managed to evade state officials. As a result, locusts and the many moving people of the Jazira might seem marginal in the sense of being unimportant. But in fact, they were marginal in the literal sense of being on the edge of desert and nondesert, one province or nation-state and another. And this place gave them power, similar to what Stephanie Camp, borrowing from Edward Said, has termed a "rival geography."[7] While this position allowed them to carve out some measure of autonomy, these forms of resistance or agency did not exist separately from the structures of power against which they were articulated.[8] Like the Pacific coast of Colombia, the wintertime snows of New England, the small plots of postrevolutionary rural Haiti, or the floating coast of Beringia, the Jazira was a space that simultaneously protected and limited its people.[9] Its landscapes could be used as a weapon, but the mix of arid ecology and political borders also made the Jazira a place into which people might escape.[10]

But it would not remain so. If locusts made the region seem a wasteland to outsiders, the Jazira's status also invited violent efforts at demographic engineering. In 1858, Ottoman officials could do little against locusts but compel peasants to collect the insects' eggs and pray that a Sufi-blessed holy water might attract the insectivorous starling. By 1939, people all across the Jazira could realistically imagine a world without locusts thanks to chemical insecticides and expanded cultivation. Across this same time period, the Jazira shifted from being the site of nomadic sedentarization campaigns to the killing fields of the Armenian genocide to the location of interwar refugee resettlement. With the virtual eradication of locusts, the region known for its verdant grasses and flocks of sheep became some of the region's most productive cotton- and wheat-growing lands in the twentieth century. At the same time, its people also became defined and targeted in relation to nationalist projects in new ways. Monocrop agriculture and minefields fortified the border that locusts – and people – had once easily crossed. Nevertheless, the Jazira and the

[7] Camp, *Closer to Freedom*, 7; Katz and Smith, "An Interview with Edward Said"; Said, *Culture and Imperialism*, xx.

[8] Johnson, *River of Dark Dreams*, 214.

[9] Demuth, *Floating Coast*; Gonzalez, *Maroon Nation*; Leal, *Landscapes of Freedom*; Wickman, *Snowshoe Country*.

[10] De León, *Land of Open Graves*, 8, 43; Truett, *Fugitive Landscapes*, 9.

power of its place on the edge would not be gone forever. Its particular political ecology has been the scene for various imaginings of Armenian, Assyrian, or Kurdish national homes, and in 2014 even became the heartland of the so-called Islamic State. Though often presented as outside of history, these recent events are connected to the region's legacy of agrarian development, state violence, and popular resistance.

THE JAZIRA AND BORDERS OF HISTORIOGRAPHY

The history of the Jazira is to some extent, in Kate Brown's iconic words, "a biography of no place."[11] Though revealed by the territorial extent of locust swarms and violent state-making efforts, the region is largely absent from people's consciousness. It is also at odds with the spatial bounds of much historical scholarship, which more often accept Ottoman provinces or post-Ottoman nation-states as the frames of analysis, rather than treating them as objects of analysis in their own right. But while sharing much with the borderland suddenly made visible by Chernobyl that Brown refers to as the *kresy*, the Jazira is also significantly different.[12] After all, the Jazira – unlike the *kresy* – had a history as an administrative unit for centuries before the twentieth.[13] The Jazira's history is not of no place, but rather of place undone, its disappearance from maps – and collective consciousness – a rather recent development.

The region was renowned in the ancient world, and for centuries it had appeared in geographies and as part of polities. It was there that the ancient epic of settlement *Gilgamesh* emerged, there that Alexander the Great marched along essentially the same path that the Turkish newspaper reporter took on the Baghdad Railway in 1928, and there that the Romans and Sasanians fought over the province of Mesopotamia.[14] The Umayyads formed a Jazira province, and even made their capitals within it at Raqqa and Harran.[15] The Jazira moreover appeared regularly in geographies alongside more recognizable toponyms such as Egypt, Syria, Hijaz, Yemen, and Iraq.[16] A mix of cultivation and pastoralism made life possible in the region, though urban writers often decried nomads as

[11] Brown, *A Biography of No Place*, 1. [12] Ibid., 2. [13] Ibid., 1.

[14] Foster, *The Epic of Gilgamesh*; Kriwaczek, *Babylon*, 212; Bosworth, *Conquest and Empire*, 79; Edwell, *Between Rome and Persia*, 28.

[15] Sarmani, *Tarikh Iqlim al-Jazira al-Furatiyya khilal al-'asr al-Marwani al-Umawi*, 7; Robinson, *Empire and Elites after the Muslim Conquest*, 28, 33.

[16] Antrim, *Routes and Realms*, 97.

evidence of a retreat of human civilization and expansion of the desert environment.[17]

With the emergence of Ottoman rule in the region in the sixteenth century, the Jazira's status changed. Though districts with capitals in Urfa or Raqqa intersected with the Jazira, they were not referred to as such.[18] And the Jazira increasingly functioned as the hinterlands of the Mediterranean entrepôt of Aleppo.[19] The political economic transformations of the world economy rerouted commerce away from these parts, and it seems that the Little Ice Age also contributed to a general movement away from the plains and toward hillsides for cultivation.[20] In the perhaps exaggerated words of Evliya Çelebi, the Jazira town of Harran – once the capital of the Umayyad polity – had become "ruins, the houses dirt."[21] As throughout the empire, so too in the Jazira – officials responded to these dilemmas by trying to incorporate nomadic pastoralists as local notables in ways that defied straightforward understandings of centralization.[22]

In these cracks of empire and the world economy, verdant vegetation grew.[23] Some observers offered specific descriptions of the Jazira's bountiful pastures. According to oral history, "a saddlebag of wilted flowers of buttercups, chamomile, milk thistles, haloxylon, and milfoil" had attracted the Arabic-speaking Shammar tribe to migrate from the Najd in central Arabia to the Jazira in the early nineteenth century.[24] Others gestured to the

[17] Eger, *The Islamic–Byzantine Frontier*, 155. For examples of geographers complaining about nomads and the desert, see Ibn Hawqal, *Kitab Surat al-Ard*, 211; Ibn Jubayr, *Rihlat Ibn Jubayr*, 214–220; Ibn Batuta, *Rihlat Ibn Batuta*, 236–237; Mustawfi, *Nuzhat al-Qulub*, 106.

[18] İnalcık, *An Economic and Social History of the Ottoman Empire*, xxxvi.

[19] Masters, *The Origins of Western Economic Dominance in the Middle East*, 8–9.

[20] White, *The Climate of Rebellion*, 236–237; Hütteroth, "Settlement Destruction in the Gezira Between the 16th and 19th Century," 286; Tabak, *The Waning of the Mediterranean*, 220.

[21] Evliya Çelebi, *Evliya Çelebi Seyahatnamesi*, vol. 3, 161.

[22] Abou-El-Haj, *Formation of the Modern State*, 154; Husain, *Rivers of the Sultan*, 136; Kasaba, *A Moveable Empire*, 54–55, 70, 72, 75–76, 83; Khoury, *State and Provincial Society in the Ottoman Empire*, 9; Masters, *The Origins of Western Economic Dominance in the Middle East*, 116; Salzmann, *Tocqueville in the Ottoman Empire*, 11; Thomas, *A Study of Naima*, 5–6, 55; Winter, "Alep et l'émirat du desert (*çöl beyliği*)"; Yaycıoğlu, *Partners of the Empire*.

[23] Bettina Stoetzer uses the term "ruderal" to refer to this dynamic in Berlin as a way of characterizing "the communities that emerge spontaneously in disturbed environments usually considered hostile to life: the cracks of sidewalks, the spaces alongside train tracks and roads, industrial sites, waste disposal areas, or rubble fields." "Ruderal Ecologies," 297.

[24] al-ʿUjayli, *Ahadith al-ʿAshiyat*, 33.

riches of the Jazira more crudely. For example, Mark Sykes – the British official whose name is forever associated with the modern state system of the Middle East thanks to his role in the Sykes–Picot Agreement – admitted that he was "sadly ignorant of botany."[25] Thus, instead of naming the flowers of the Jazira that he observed in the early twentieth century, he could only attest to the fact that "yellow, blue, and purple" dappled his sight lines, their "honey-scent" so overpowering as to be "oppressive." Like many others, Sykes also observed that amidst so many colors, there were few people. The land, he said, seemed "almost uninhabited."

The fraught term used for this allegedly empty landscape by both Europeans and Ottomans alike was "desert" or "steppe" (Ottoman: *çöl*; Arabic: *badiya*; Kurdish: *berrî*).[26] The Jazira region sits astride the lands where rain-fed agriculture was possible. It was stiflingly hot in the summer, sometimes approaching 50° Celsius (122° Fahrenheit) in the - shade.[27] In such weather, there was nothing to do, observed one French missionary, "but wash oneself with fresh water and as much as possible not move."[28] But of course many people could do other things in such heat, in part because the category of desert is also as much cultural as material. Rather than a neutral description of ecological reality, the term often functioned as what scholar Diana Davis has called "environmental Orientalism."[29] This is to say that when European travelers or officials deployed the loaded term, it was an accusation. For them, "desert" meant that someone had made it that way, and more often than not they blamed local people or nomads for this state of affairs.[30] While it is difficult to speak of any monolithic Ottoman discourse about the desert, on this question a figure such as the nineteenth century's foremost Ottoman reformer Midhat Pasha sounded a lot like his American contemporary George Perkins Marsh, the renowned ungulate hater and former American ambassador to the Ottoman Empire, who suggested that were it not for goats and camels, then forests would cover "Arabian and African deserts" (Marsh also blamed deforestation for causing locusts).[31] Midhat Pasha similarly suggested that nomads and their animals had transformed the Jazira from a bountiful breadbasket of

[25] Sykes, *The Caliph's Last Heritage*, 328.
[26] Kıran, *Kürt Milan Aşiret Konfederasyonu*, 36.
[27] Archives de la Province dominicaine de France (APDF), Haute Djezireh 15, April 16, 1945.
[28] APDF, Haute Djezireh 56, Rudder, July 30, 1940. [29] Davis, *The Arid Lands*, 2.
[30] Ibid., 82.
[31] Ibid., 93; Marsh, *Man and Nature*, 291; Mikhail, "Foreword: Ottoman and Nature," vii.

world renown to little more than a dusty migratory ground.[32] But as Sykes's simple descriptions of flowers as well as the more eloquent observations of others suggest, the Jazira's desert did not conceal its potential. With the empire engaged in reforms known as the Tanzimat and attempting to incorporate large numbers of refugees fleeing ethnic cleansing on the empire's borders in the Balkans and with the Russian Empire, the Jazira – as a desert to be redeemed – became a space to solve the empire's problems, even as the region did not exist as a province or a district on its own.

Piecing the Jazira back together offers a different vantage on the end of empire and the making of the modern Middle East. In op-ed pages and survey courses alike, the 1916 Sykes–Picot Agreement often functions as shorthand for the artificial division of the region by colonial powers.[33] A historiography often grounded in one Ottoman province or one post-Ottoman nation-state can unintentionally support this historical trajectory.[34] Yet these accounts do little to explain the Jazira's locusts and the people who moved in relation to them, and it is thus difficult to appreciate what changing political structures meant for people and animals on the edge.[35] By being situated on both the political and environmental margins, these groups not only escaped state officials; they also largely evaded the gaze of historians, who have both reproduced the blind spots of state officials and introduced some of their own.

[32] Midhat Paşa, *Tabsıra-ı İbret*, 83.

[33] For critiques of this approach, see Bâli, "Sykes–Picot and 'Artificial' States"; Nicholas Danforth, "Could Different Borders Have Saved the Middle East?" *New York Times*, May 14, 2016; Daniel Neep, "Focus: The Middle East, Hallucination, and the Cartographic Imagination"; *Discover Society*, January 3, 2015; David Siddhartha Patel, "Repartitioning the Sykes–Picot Middle East?" *Middle East Brief* 103 (2016); Pursley, "'Lines Drawn on an Empty Map.'"

[34] With respect to works that take nation-states (and, perhaps, capital cities) or provinces as frames, a few that have been significant to this work are Gelvin, *Divided Loyalties*; Jongerden and Verheij, ed. *Social Relations in Ottoman Diyarbekir*; Khoury, *State and Provincial Society in the Ottoman Empire*; Marcus, *The Middle East on the Eve of Modernity*; Neep, *Occupying Syria under the French Mandate*; Shields, *Mosul before Iraq*.

[35] As for the Jazira more particularly, scholars treating the people who found refuge there in the interwar period have most often taken the space seriously: Altuğ, "Sectarianism in the Syrian Jazira"; Aydın, Emiroğlu, Özel, and Ünsal, *Mardin – Aşiret – Cemaat – Devlet*; Barut, *Al-Takawwun al-Tarikhi al-Hadith lil-Jazira al-Suriyya*; Fuccaro, *The Other Kurds*; Kayalı, *Imperial Resilience*; Khidr, *Tarikh al-Muhammad al-Jarba wa Qabilat al-Shammar al-ʿArabiyya fi Iqlim Najd wa al-Jazira*; Tachjian, *La France en Cilicie et en Haute-Mésopotamie*; Tejel Gorgas, "Un territoire de marge en haute Djézireh syrienne (1921–1940)"; Velud, "Une expérience d'administration régionale en Syrie durant le mandat français."

Regional history has witnessed a wealth of work concerned with borders and transnational frames, which altogether has restored space to a status beyond what David Harvey has critically described as "a simple and immutable container in which social processes occur."[36] Yet it still remains difficult to trace the impact of borders because of how few works bridge the Ottoman and post-Ottoman eras. The siloed nature of the historiography helps to leave rupture intact as a narrative device, even though it remains supposed more than it is substantiated. Even recent attempts to transcend the divide have presented late Ottoman space as "borderless" or described the Jazira as having had "no borders" between Ottoman conquest in the sixteenth century and its post-Ottoman division.[37] There is, of course, some truth to these characterizations, especially in contrast to the kinds of territorial limits emerging in the post-Ottoman period.

Yet many – ranging from nomads to Ottoman officials to locusts – would have been surprised to learn that the world they lived in was "borderless."[38] Indeed, both locusts and people in the Jazira repeatedly manipulated the mismatch between borders and environment. If the Ottoman state might be thought of as an "empire by nature" or "imperial ecology," as early works of Ottoman environmental history have suggested, moving groups in the Jazira located and manipulated the fissures in these terms.[39] They did not rely exclusively on either political borders or

[36] Harvey, *Spaces of Global Capitalism*, 77. On borders themselves: Altuğ and White, "Frontières et pouvoir d'état"; Ateş, *The Ottoman–Iranian Borderlands*; Ellis, *Desert Borderland*; Meiton, *Electrical Palestine*; Öztan, "The Great Depression and the Making of Turkish–Syrian Border"; White, *The Emergence of Minorities in the Middle East*. On people crossing borders, see Arsan, *Interlopers of Empire*; Bailony, "From Mandate Borders to Diaspora"; Banko, *The Invention of Palestinian Citizenship*; Bawalsa, *Transnational Palestine*; Berberian, *Armenians and the Connected Revolutions in the Russian, Iranian, and Ottoman Worlds*; Cân, *Spiritual Subjects*; Fahrenthold, *Between the Ottomans and the Entente*; Fuccaro, *The Other Kurds*; Gutman, *The Politics of Armenian Migration to North America*; Khater, *Inventing Home*; Khuri-Makdisi, *The Eastern Mediterranean and the Making of Global Radicalism*; Provence, *The Last Ottoman Generation*; Tejel, *Syria's Kurds*; Turna, *19. YY'den 20. YY'ye Osmanlı Topraklarında Seyahat, Göç ve Asayış Belgeleri*.

[37] Kayalı, *Imperial Resilience*, xvii; Schayegh, *The Middle East and the Making of the Modern World*, 246, 254, 265.

[38] This formulation is inspired by and expands upon existing accounts of Ottoman provincial borders. Abu Manneh, "The Establishment and Dismantling of the Province of Syria, 1865–1888"; Fattah, *The Politics of Regional Trade*; Hanssen, *Fin de Siècle Beirut*, 25–54; Salzmann, *Tocqueville in the Ottoman Empire*; Sluglett, "The Resilience of a Frontier."

[39] Mikhail, *Nature and Empire in Ottoman Egypt*, 36; White, *The Climate of Rebellion*, 17.

the Jazira's arid landscape. Instead, in a region whose borders have been alternately championed and contested with terms like "natural" and "artificial," the Jazira's moving groups exploited the "hybrid" spaces in between.[40] In other words, they took advantage of how borders made the desert and the desert made borders. Even in the waning years of World War I, with the empire devastated and foreign schemes to divide it imminent, Ottoman officials continued to consider how to close the gap between the Jazira's expansive environment and its piecemeal political borders, the better to manage people who wandered on the edge of cultivation "like locusts." Moreover, after the end of empire, the world may well have seemed less rigid than it has been rendered. As the flight of locusts across the railway in 1928 underscored, the environment remained connected in part because its political administration was fractured.

PESTS IN HISTORY

The impact of locusts on human existence has been great. Like other creatures deemed pests, they have caused famine and disease, inspired massive state-directed schemes at control (with their own sometimes-deleterious consequences), and fired the human imagination. Often harmless in the singular, they become something different in a group. Perhaps most disconcertingly, they repurpose human infrastructures in unintended ways as part of their survival. Thus, the term "pest" seems to refer to creatures on their own, but in reality, the status as pest is only possible in connection with the human-built world and its culture; boll weevils require cotton bolls, and house flies require houses.[41] Perhaps because of this in-between state, writing about creatures deemed pests as

[40] For examples of scholarly use of "natural" and artificial," see Chatty, "Bedouin Tribes in Contemporary Syria," 158; Khoury, "The Tribal Shaykh, French Tribal Policy, and the Nationalist Movement in Syria between Two World Wars," 183. For examples of historical use of "natural" and "artificial," see Altuğ, "The Turkish–Syrian Border and Politics of Difference in Turkey and Syria (1921–1939)," 60; Balistreri, "Revisiting *Millî*," 52; Gelvin, *Divided Loyalties*, 152; al-Husri, *Ara' wa-Ahadith fi al-Wataniyya wa-l-Qawmiyya*, 106; Schayegh, *The Middle East and the Making of the Modern World*, 151; Williams, "Cultivating Empires," 137–138. On hybrid landscapes, see Sutter, "The World with Us," 96. See also Fiege, *Irrigated Eden*; Mitman, *Breathing Space*; Nash, *Inescapable Ecologies*; Pritchard, *Confluence*; Walker, *Toxic Archipelago*; White, *The Organic Machine*; and Wilson, *Seeking Refuge*.

[41] In a different context, William Cronon summarized this approach when he wrote that "the pig was not merely a pig but a creature bound among other things to the fence, the dandelion, and a very special definition of property." Cronon, *Changes in the Land*, 10.

historical subjects has proven difficult, caught, as they are, between roles as objects of mundane annoyance and of existential dread.[42] For the most part, they lurk – omnipresent but unremarkable – in the shadows of a human-built world. When they appear in literature, it occurs often in relation to symbolically potent disgust or cataclysm. And then, with the plot device achieved or the symbolic point made, they are gone, and human ability to control the world is restored.[43]

Like pests in literature more broadly, locusts also appear briefly – if sometimes cataclysmically – in regional historiography. Salim Tamari's *Year of the Locust*, for example, foregrounds how the insect swarms that befell the coast of greater Syria during World War I served another purpose, in that their appearance "erased four centuries of a rich and complex Ottoman patrimony."[44] The catastrophe furnished a gruesome and decisive reason, as if more were necessary, for nationalists and the historiographies they wrote to leave behind the Ottoman past. As for the impact of locusts in the Jazira more specifically, historians have typically treated the insects less as actors than as extras. In these accounts, locusts function as a gloss for intermittent bouts of famine, rural depopulation, or disorder in a general sense.[45]

[42] On locusts, see Bello, "Consider the Qing Locust"; Dağyeli, "The Fight against Heaven-Sent Insects"; Few, "Killing Locusts in Colonial Guatemala"; Forestier-Peyrat, "Fighting Locusts Together"; Lockwood, *Locust*; Mohamed, "'The Evils of Locust Bait'"; Peloquin, "Locust Swarms"; Rothschild, "Sovereignty, Virtue, and Disaster Management"; Seitz, "Scientizing the Steppe"; Thistle, *Resettling the Range*; van der Watt, "'To Kill the Locusts, but Not Destroy the Farmers.'" On pests more generally, see Biehler, *Pests in the City*; Cronon, *Changes in the Land*, 142–145, 153–155; Crosby, *Ecological Imperialism*; Giesen, *Boll Weevil Blues*; Fissell, "Imagining Vermin"; Jones, "Becoming a Pest"; Mavhunga, "Vermin Beings"; McNeill, *Mosquito Empires*; McWilliams, *American Pests*; Melillo, *The Butterfly Effect*; Raffles, *Insectopedia*; Russell, *War and Nature*; Ticktin, "Invasive Others"; Walker, *The Lost Wolves of Japan*; Walker, *Toxic Archipelago*, 22–70; Wang, *Mad Dogs and Other New Yorkers*.

[43] Whether the plagues in the Bible, the rat in the opening scene of Richard Wright's *Native Son*, or the locusts in Pearl Buck's *The Good Earth*, they only make brief – albeit memorable – appearances. There are, of course, exceptions beyond Gregor Samsa, including Tewfik al-Hakim's "Fate of a Cockroach," which is set in "a spacious courtyard – as viewed of course by the cockroaches" but is in reality "the bathroom floor in an ordinary flat." Buck, *The Good Earth*, 220–222; Wright, *Native Son*; al-Hakim, *Fate of a Cockroach and Other Plays*, 1; Kafka, "The Metamorphosis." On anthropocentrism and views of vermin in the early modern Ottoman Empire, see Çelik, "Humans in *Animalscapes*."

[44] Tamari, *Year of the Locust*, 5. On the invasion itself, see Foster, "The 1915 Locust Attack." For research squarely focused on locusts beyond the Jazira in the late Ottoman and Republican periods, see Çavuş, "Osmanlı Devleti'nde Çekirglere Karşı Mücadele," Özbilge, *Çekirgeler, Kürtler ve Devlet*; Özer, *Anadolu'da Görülen Çekirge İstilaları*.

[45] Khoury, *State and Provincial Society in the Ottoman Empire*, 42; Marcus, *The Middle East on the Eve of Modernity*, 132; Lewis, *Nomads and Settlers in Syria and Jordan*, 40; Shields, *Mosul Before Iraq*, 131.

Locusts of Power challenges the narrative marginalization of pests writ large and locusts in particular by connecting their swarms to shifts in the region's political economy. Indeed, locusts were distinct from other creatures deemed pests in how their existence and threat were connected to seasonal agrarian rhythms. For the Jazira, there was no single "year of the locust." Their springtime swarms afflicted the Jazira 1859 to 1870, 1882 to 1894, 1899 to 1911, 1913 to 1917, 1920 to 1924, and 1927 to 1933. Their presence dated to antiquity. In the words of James C. Scott, monocrop agriculture in the ancient Jazira offered "a permanent feedlot for insect pests," locusts included.[46] It seems to have been at this time too that sedentary agriculturalists began to deride moving people and moving insects in similar ways, with some evidence that the ancient Assyrian word for "steppe nomad" derived from "locust."[47] Over many centuries, locusts helped create an environment in which nomadic pastoralism made sense. For example, nomadic pastoralists' camels alone could feed on thistles left in the wake of devastating locust swarms.[48] Animal husbandry provided a flexible – though of course not impervious – strategy of gathering resources in the Jazira. In the event of locusts, one could move a flock of sheep but not a field of cotton.

Standard scientific accounts of the locust's life cycle – especially in the simplified diagrammatic form – often overlook the role of human influence. But the insects were in many ways creatures of humans, just as humans were creatures of locusts.[49] Locusts typically appeared in the Jazira some time between March and May, emerging from eggs at a length of five millimeters and growing to a length of about thirty millimeters with wings within about forty days of birth. They also became capable of mating, which they did for periods of six to twenty-four hours. Afterward, the males died, and the females laid pockets filled with around forty eggs in small holes that they spent hours digging. Shortly afterward, they died, too, with the next generation emerging in about ten-and-a-half months' time. The locusts' springtime arrival occurred at a delicate time for the Jazira's agroecology and various groups attempting to extract value from it. Located just on the edge of where rain-fed agriculture was possible, the region had long been home to a mix of cultivation and

[46] Scott, *Against the Grain*, 110. [47] Webb, *Imagining the Arabs*, 25–26.

[48] The National Archives-United Kingdom (TNA-UK), AIR 23/281, SSO Mosul to Air Headquarters, Baghdad, August 31, 1925.

[49] This approach overlaps with Richard White's pithy aim "to look for the natural in the dams and the unnatural in salmon." White, *The Organic Machine*, xi.

pastoralism. Wintertime rains nourished grains such as wheat and barley, typically harvested in April or May. The same rains enabled verdant springtime pastures. Nomads fed their flocks of sheep and camels with these grasses and brought their flocks close to cities such as Aleppo and Baghdad in March and April to sell their butter and wool to local merchants as well as foreign commercial firms. It was also around this same time that cultivators would plant summer crops, most notably sesame and cotton. In Nusaybin, there was rice; in Diyarbekir, there were famously large watermelons; and anywhere along river valleys, there were pomegranates. And it was in these springtime months too that state officials assessed and collected taxes. Into this intricate choreography of value, swarms of locusts entered again and again, seemingly out of nowhere. To humans, they could be a surprise. But for the locusts, expansion of cultivation offered an inviting buffet of cereals and likely expanded their power. Like so many creatures deemed pests, they repurposed human designs and siphoned off value in proportions at odds with their small size.

Any response to a locust variation of Timothy Mitchell's famous paraphrase of Gayatri Spivak – "Can the mosquito speak?" – must be in the affirmative.[50] When Ottoman officials and cultivators joined many around the world in trying to capitalize on the global cotton shortage of the American Civil War, locusts repeatedly devastated the Jazira's newly planted cotton crops, forestalling any regional transformation abetted by the white gold. When the Ottoman state hoped to settle Chechen refugees in the Jazira in 1866, locusts swooped in to make conditions much more difficult. After the Constitutional Revolution in the Ottoman Empire in 1908, the failure to destroy locusts provided a pretext for toppling a reformist Ottoman governor. In the midst of the devastation of World War I, locusts would be identified by the Ottomans as more of a threat than the armies they were fighting. And the insects' persistence after the Ottoman Empire – as underscored by their flight across the railway in 1928 – would challenge states to control the nonhuman inhabitants within their borders. The virtual eradication of locusts in the region in the mid-twentieth century, meanwhile, helped to provide the foundation necessary for transforming the Jazira into the region's most productive agricultural lands. Thus locusts can speak, in the sense that they had a material impact on the world despite the tendency of some social sciences to ignore nonhuman forces and their complicated interconnections with humans.

[50] Mitchell, "Can the Mosquito Speak?" in *Rule of Experts*, 19–53.

In addition, locusts have also been repeatedly spoken for. This is to say that they were not simply "biological agents," but rather also operated on a cultural level, as they shaped the language and metaphors with which people have understood their world.[51] Braudel compared nomads to the insects, and Foucault used "swarming" or "teeming mass" (*grouillement*) to refer to the unruly object of modern governance.[52] In the Jazira more specifically, the creatures have appealed to many, somewhat like the parasite in other contexts.[53] As a symbol for power, motion, and difference, locusts were attractive to figures as far apart on the political spectrum as the Turkish writer of Kurdish descent Yaşar Kemal and the ideologue of Turkish nationalism Ziya Gökalp. In the eyes of Kemal, Gökalp, and many others, the insects signified a staggering array of groups: Arabic-speaking nomads in general, the Shammar and ʿAnaza tribes more specifically, Armenian deportees, Assyrian refugees, camels, Chechen refugees, disease, droughts, Kurdish rebels, Kurdish refugees, Iraqi locust-control officers, Ottoman soldiers, tax collectors, the will of God, and war profiteers. Of course, there is a long history of insect metaphors alongside mass violence – Jews as lice, Hutus as cockroaches, Japanese as flies – to say nothing of what Fanon more generally described as colonial deployment of "zoological language" to dehumanize colonized populations.[54] These dynamics were certainly present in the Jazira, but they were not the only uses for the ubiquitous comparison. The reference to locusts conveyed different things, from omnipotence to powerlessness, denigration to pity.

The clearest example of where human dreams and locust realities came together were in the ancient ruins of bygone empires that dotted the Jazira. As dirt had accumulated upon them over the years, the ruins transformed into hills known as *tall*. In the nineteenth century, these traces of past cities attracted European archaeologists.[55] Often pointed in the direction of ruins by groups such as the Shammar, the diggers brought their finds to museums and universities in Europe and the United States, where the exhibits then fertilized the imagination of future colonial officers.[56] But

[51] Walker, *Toxic Archipelago*, 54.

[52] Braudel, *La Méditerranée*, 170; Foucault, *Surveillir et punir*, 145.

[53] Bein, "The Jewish Parasite," 9, 13, 19; Césaire, *The Original 1939 Notebook of a Return to the Native Land*, 46–47; Lafargue, *The Right to Be Lazy*, 45; Serres, *Le Parasite*.

[54] Copeland, *Cockroach*, 48; Fanon, *Les damnés de la terre*, 45; Raffles, *Insectopedia*, 146; Russell, *War and Nature*, 97–100.

[55] Bernhardsson, *Reclaiming a Plundered Past*, 37, 41.

[56] Layard, *A Popular Account of Discoveries at Nineveh*, 268; Leick, *Mesopotamia*, 195. For accounts of travelers or officials whose experience of the region was filtered through the lens of the Assyrian friezes at the British Museum, see Bell, *Amurath to Amurath*, 71;

for Ottoman officials, the ruins meant something else. They were a sign that the deserts of the Jazira might be made to bloom, so much so that Ottoman warships took names in their honor and intellectuals hailed the ancient glory in connection with modern infrastructural projects such as the Baghdad Railway.[57] The ruins made humans dream.

The locusts, however, had different uses for these hills that inspired fantasies of granaries: procreation. Uncultivated and yet in proximity to cultivated lands, the *tall* proved ideal for both locust mating and egg-laying. In other words, the same geographic features that attracted European and American archaeologists and the same humble hills that sparked Ottoman hopes of agrarian transformation also provided locusts with forward operating bases through which to reproduce and consume the cereals of Ottoman reform. Far from superhuman afflictions emerging by chance, locusts, instead, were organisms emerging in conjunction with newly planted fields of grain adjacent to vast swaths of uncultivated land with low population density and limited state control.

Just as infrastructures of long-dead humans affected the lives of locusts, the death of locusts and the methods of killing them would also affect the lives of humans in the Jazira, and beyond. Between the mid-nineteenth century and mid-twentieth century, the means of killing insects changed all around the world. In the late-nineteenth-century Jazira, state-of-the-art locust killing techniques involved locust egg collection through forced labor, or Sufi-blessed holy water believed to attract the insectivorous starling.[58] But gradually chemical insecticides crowded out these methods, and especially in the wake of World War I, corporations that had sold chemicals with which to kill humans sought new markets.[59] Locusts had often been referred to as "plagues." With these new substances, the entomological and the epidemiological further collapsed; in both Turkish and Arabic the words used for pesticide (*ilaç* and *dawa'*) were the same ones used for "medicine."[60] The substances were aimed at

Jebb, *By Desert Ways to Baghdad*, 196; Lukach, *The Fringe of the East*, 242, 243; Nogales, *Four Years Beneath the Crescent*, 174.

[57] Makdisi, "Ottoman Orientalism," 790; *Halep Salnamesi* 1315, 169; *Musul Salnamesi* 1330, 53; *Servet-i Fünun*, 7 Kanunusani 1325 (January 20, 1910); Nazif, *El-Cezire Mektupları*, 8; Muhammad Kurd ʿAli, "Sikkat Hadid Baghdad," *Al-Muqtabas* 3.2 (1908): 143–146.

[58] Encouragement of natural predators or crop diversification was common at the time. McWilliams, *American Pests*, 109–110.

[59] Russell, *War and Nature*.

[60] Rogers, "Germs with Legs, Disease, and the New Public Health," 605; Roy, *Malarial Subjects*, 249.

insects, but arsenic compounds lodged in soft tissues and organophosphates interfered with the nervous systems in the bodies of people the world over.[61] It was no accident whom these chemicals hurt. In the Jazira, nomads in particular suffered from the clouds of pesticides sprayed on all sides. And so it came to be that they grappled with the chemical changes to their world decades before Fanon invoked the seeming omnipotence of DDT as part of colonialism, and decades, too, before Rachel Carson decried the same substance's carcinogenic potency.[62] The region and its locusts are part of the broader story of capitalism, agriculture, and the environment, culminating in soaring cancer rates, mass insect die-offs, and concerns for collapse of many different kinds.

SOURCES AND STRUCTURE

Reconstructing a region that haunted everyday life yet does not exist as a state administrative unit required consulting a broad range of sources in a number of languages, including state and missionary archival sources as well as memoirs, newspapers, novels, poetry, and scientific literature in Arabic, English, French, German, Modern Turkish, and Ottoman Turkish. It also required reading against archival spatial logics that can structure scholarly vision in subtle yet important ways. Finding the Jazira involved reading about its component parts. After all, locusts or any number of humans might appear minor according to government communiqués or consular records from, for example, Aleppo alone. But broader patterns became evident from connecting those swarms or migrations with simultaneous events in Diyarbekir, Mosul, and Baghdad. The same holds true for the post-Ottoman period when the space became fractured between Iraq, Syria, and Turkey. State officials were largely aware of the Jazira, even if they administered it in a piecemeal way. But their correspondence was compartmentalized archivally in ways that belied the region's coherence. Even if sources invoked the Jazira, archival catalogues or finding aids rarely did, both because it was largely not a unit of state administration and because it is not a widely recognizable unit of space today.[63]

[61] Nash, *Inescapable Ecologies*.

[62] Carson, *Silent Spring*; Fanon, *Les damnés de la terre*, 45; Nixon, *Slow Violence*, 6–7.

[63] On Cilicia as a coherent space with similar historiographical dilemmas, see Toksöz, *Nomads, Migrants, and Cotton in the Eastern Mediterranean*; Gratien, *The Unsettled Plain*.

I eventually came to see the incomprehensibility of the basic terms of the project as evidence of the depth of the political and environmental transformations that I was trying to unearth. My invocation of the Jazira often produced blank stares or questions about whether I meant the Arabian Peninsula (*al-Jazira al-ʿArabiyya*) or the southeastern Anatolian city of Cizre (also known as Jazirat Ibn ʿUmar). My attempt to obviate these responses with unwieldy yet accurate terms like "the area between Aleppo, Diyarbekir, and Mosul" rarely helped. In one case it was even "locust" that invited such confusion. An archival worker first suspected I meant the "Locust" neighborhood of the Turkish city of Bursa, and, when I clarified that I meant "locust locusts," asked what a locust was. When people did recognize the terms of the project, they often revealed something about their own background or political commitments. There was the bookshop owner who nonchalantly asked if I would be writing about the Armenian genocide, and the acquaintances who immediately understood the geography to be the area known as Rojava, both of which were far from politically neutral topics in Turkey while I was conducting research. *Locusts of Power* explains both the forgetting and the recognition of the Jazira through a history of borders, empire, and environment. The chapters proceed in a chronological fashion. Each centers one group of humans and their material and metaphorical entanglements with the Jazira's insect overlords, from the Arabic-speaking Shammar nomads to the largely Kurdish-speaking Millî to Armenian deportees and, finally, to Assyrian refugees.

Chapter 1 focuses on locusts and the Arabic-speaking Shammar nomadic group between 1858 and 1890. It explains how locusts foiled Ottoman attempts to transform the Jazira into a cotton-growing heartland in the midst of the American Civil War. As locusts challenged the designs of certain humans, they also ensured that the Jazira landscape remained productive, depending on how one moved within it. It was in part the landscape created by locusts that undermined Ottoman attempts to forcibly settle the Shammar during the 1860s and made far more difficult the settlement of Chechen refugees at Ras al-Ayn. And it was this same landscape of locusts that incubated a revolt in 1871, as the Shammar protested the formation of the special administrative district of Zor, created in an effort to match the desert with administrative borders with the help of the empire's foremost reformers, Cevdet Pasha and Midhat Pasha. The revolt was crushed and ended with different branches of the Shammar attached to separate districts of the Jazira. But it did not end the power of locusts and mobility, and so people continued to imagine how to close the gap between Ottoman provinces and the environment they divided up.

Chapter 2 foregrounds locusts and the seminomadic Kurdish group known as the Millî between 1890 and 1908. In a shift from the policies of previous years, the Ottoman state created light cavalries known as the Hamidiye Brigades. Typically seen as co-opting Kurdish groups in defense against Russian invasion or Armenian rebellion, the Millî under the leadership of Ibrahim Pasha actually pushed south into the Jazira. The chapter traces how provincial borders came to be invoked alongside the environmental border of the desert in clashes between the Millî and the Shammar. All the while, locusts created conditions favorable to pastoralists. Locusts also forced people to move and, in other cases, offered a plausible justification for nomadic encroachment on the landholdings of urban notables. The chapter concludes with the death of Ibrahim Pasha, and the new political era it seemed to portend, even as locusts continued their work on the edge.

Chapter 3 examines Armenian deportees and locusts in the Jazira between 1908 and 1918. It places the Armenian genocide within the longer history of efforts to control the Jazira, as the district created for the settlement of nomads in 1871 transformed into the final destination for many of the empire's Armenian citizens. The chapter exposes the complicated ways the violence affected and was affected by the environment. Indeed, one German locust expert even suggested that the deportations of the genocide coupled with war mobilization to make the locust invasions worse, because so much land was left fallow. But the environment also managed to help some escape, whether children who survived by working as shepherds for pastoralists or the Armenian who, while concealing his identity, worked as the locust-control officer of the Jazira. In a mark of the enduring challenge of the Jazira and its provincial division, Ottoman officials discussed how to draw better borders in the region throughout, from the lead-up to deportations in 1915 all the way to the end of the war in 1918.

In the wake of the war, the Jazira was divided between the Republic of Turkey, French Syria, and British Iraq, and Chapter 4 examines the region's divided connection from 1918 to 1939 through locusts, the war's refugees, and various interwar state-making projects. While all of the states in the Jazira endeavored to differentiate themselves from the Ottomans, they undertook strikingly similar projects based on refugee resettlement and arsenic compounds, the new insecticides that each state wielded to kill locusts. The substance virtually ended the reign of locusts in the region, and it was at roughly the same time that the Jazira very nearly made its way onto the map. The late 1930s witnessed an unprecedented autonomist movement advocating for a French protectorate of the Jazira, and the ensuing conflict encompassed all of the protagonists of *Locusts of Power*, from the Shammar and the Millî to Armenian genocide survivors and Assyrian refugees. The autonomist effort failed, and the Jazira became the foremost space of Syrian agricultural development, Ottoman dreams of prosperity finally coming to fruition.

Altogether, the book traces how borders and agriculture shaped the Jazira and its people as the region shifted from a space of locusts to one where they were largely absent. The figures involved in this transformation include a nomadic leader who revolted against Ottoman provincial borders in 1871, starlings believed to be attracted by a Sufi-blessed holy water that laid waste to locust swarms, an Ottoman Armenian locust-control officer who used his position in the midst of the genocide to save those sentenced to death, and locusts who vaulted over zinc walls intended to stop them by climbing up the Jazira's long blades of grass. By the mid-twentieth century, these worlds would seem to be interred under fields of cotton and wheat. But the Jazira's landscape of revolt and the bones of the past would not remain buried.

I

Sultans of the Open Lands (1858–1890)

In the mid-twentieth century, the Syrian doctor, politician, and writer ʿAbd al-Salam al-ʿUjayli collected oral histories in the Jazira. Some of the stories explained the migration of the Jarba branch of the Arabic-speaking Shammar nomads to the region from the Najd, in what is now Saudi Arabia. According to one story, in the early nineteenth century it was at first only the shaykh of the Shammar, his unnamed wife, and an unnamed enslaved person who made the journey. Attempting to conceal his wealth so as not to rouse the suspicions of locals, the shaykh told those who asked that the enslaved man was in fact his cousin from a Black mother.[1] Convinced of the bounty of the Jazira's grasses, the shaykh dispatched the enslaved man to return to the Najd, carrying a saddlebag filled with "dried out grasses and roots of plants of varied colors and wilted flowers of buttercups, chamomile, milk thistles, haloxylon, and milfoil."[2] When he reached Najd and the shaykh's followers, they asked, "Where is the paradise whose riches these are?" And so, as the story goes, they left Najd behind, bound for the fertile lands at the foot of the Anatolian plateau between the Tigris and the Euphrates. Al-ʿUjayli described the matter ominously, utilizing an insect metaphor: "The crawling locusts of the Shammar … entered the Jazira as ruthlessly acquisitive invaders."[3] Al-ʿUjayli's presentation of the Shammar as locusts was part of a long legacy of dehumanizing descriptions of pastoralists.

Scholars have complicated such disparaging depictions of mobile people. The historian Sarah Shields summarized a significant approach in the study of these groups when she wrote that "the nomads were

[1] Al-ʿUjayli, *Ahadith al-ʿAshiyat*, 30. [2] Ibid., 33. [3] Ibid., 36.

important to Mosul's economy *because they were nomads.*"[4] In other words, she suggested how scholars might inquire about what nomadic pastoralists did, rather than how they diverged from teleologies of civilizational stages or modernization theory, whether used by state officials or by scholars themselves. Subsequently, many have examined nomadic pastoralists on these terms and revealed how they were not simply objects of Ottoman reform but, indeed, active agents of those processes.[5] This literature has further challenged conventional wisdom about divisions between the state and tribe, as well as nomadic pastoralism and cultivation.[6] In many situations – as was the case of the Shammar – people who lived in tents relied on wool, yes, but also agriculture, toll collecting, and state subsidies.

While the comparison of locusts to the Shammar obscures these nuances, it nevertheless calls attention to the perhaps-unexpected ways that both insects and people were enmeshed in the Jazira. In the wake of defeat in the Crimean War (1853–1856) and the 1858 Land Code, the Ottoman state renewed its endeavor to clarify its governing mechanisms and simplify the relationship between subjects and the state. But the mobility of groups like the Shammar and locusts made this task difficult in the Jazira. Indeed, the seasonal motion of both locusts and people was entangled as they similarly crossed various provincial borders across the arid Jazira, leaving quarreling officials in their wake. Locusts in Urfa could be blamed on negligence in Mosul; the Shammar could avoid a strict governor in Diyarbekir by fleeing into Aleppo. The locusts and Shammar also both made out well in the midst of the American Civil War, when the Shammar benefited from booming wool prices, and the locusts again and again consumed the cotton popping up on the outskirts of the region. Yet despite the intersections and even causative links between locusts and nomadic pastoralism, state officials largely viewed the issues separately. With the Shammar, the state attempted to co-opt them or coerce them. They envisioned a number of schemes of these sorts, which included charging a Shammar shaykh with protecting a telegraph line, creating a line of cordon across which nomads could not pass so as to promote settlement by Kurdish tribes, settling Chechen refugees to act as

[4] Shields, "Sheep, Nomads and Merchants in Nineteenth-Century Mosul," 782; emphasis original. See also, Shields, *Mosul Before Iraq*.

[5] Amara, "Civilizational Exceptions"; Barakat, "Marginal Actors?"; Çiçek, *Negotiating Empire in the Middle East*; Minawi, *The Ottoman Scramble for Africa*.

[6] Barakat, "Making 'Tribes' in the Late Ottoman Empire"; Husain, *Rivers of the Sultan*.

a bulwark against the desert, and, finally, forming the special administrative district of Zor in 1871 to bring together desert edges of various provinces that the Shammar exploited. The last measure instigated a violent revolt, which Ottoman officials largely presented in terms of civilizational backwardness rather than any material conditions. Indeed, the insistence on seeing the mobile people of the Jazira as distinct from the buzzing insects or scorching summers derived from the fact that officials had fewer options with the locusts and the weather. Because the insects emerged from vast, sparsely populated regions, extensive control policies were largely out of the question. Prayers for the intervention of the insectivorous starling seemed like the most reliable policy. As locusts persisted, though, so, too, could the ambiguity of the desert and human motion in it, meaning the edge – whether of an environmental or political sort – could be a place of power, even as state officials tried to close the gap between environmental and political borders.

THE SHAMMAR AND THE TANZIMAT

ʿAbd al-Salam al-ʿUjayli may have compared the Shammar to destructive locusts when they entered the Jazira toward the beginning of the nineteenth century, but in the decades that followed, the Shammar gradually became part of the region's fabric and a key part of Ottoman governance. The region had historically been a part of overland trade networks linking the Indian Ocean world with the Mediterranean. But overseas trade and, later, steam travel increasingly cut into the profits afforded by the route. Commerce, of course, did not disappear. And the status of the region as a transit zone was in fact quite harmonious with the way the Shammar utilized the space as an environmental margin useful for the production of sheep and camels. Through both conflict and alliance with local populations, the Shammar came to cement their control over the Jazira. Many groups – including the ʿUbayd, Dulaym, and Jabbur – fled the appearance of the Shammar.[7] With the Arabic-speaking Tayy nomadic group, meanwhile, the Shammar solidified their alliance through the marriage of Shammar leader Sufoq to ʿAmsha, the daughter of the Tayy leader. In the words of British archaeologist and diplomat Austen Henry Layard, ʿAmsha was the "queen of the desert," her body covered in the "tattooed ends of flowers," her nose weighed down by "a prodigious gold ring" so large she had to remove it to eat, and her camel saddle possibly confused

[7] Nieuwenhuis, *Politics and Society in Early Modern Iraq*, 125.

with "some stupendous butterfly skimming slowly over the plain."[8] She would go on to play a crucial role in Shammar political power.

Like many places in the Ottoman Empire at the beginning of the nineteenth century, the Jazira, too, witnessed Ottoman reliance on local intermediaries. The Shammar were perhaps not as famous as the notable families of cities such as Aleppo or Mosul, but they occupied similar places of power. Sufoq had spent several years under arrest in Istanbul with his son Farhan, and upon returning to the Jazira, he ensured that the Shammar backed Ottoman governors in military campaigns on a number of occasions, including against the rebellious governor of Egypt, Mehmed Ali Pasha.[9] The Shammar also acted as government functionaries in some towns.[10] The Ottomans had long relied on the office of "Lord of the Desert" – historically held by a member of the Mawali confederation – to project power in the region.[11] It seems the Shammar took up this title, if not the office itself. Their leader was hailed in the region as "Sultan of the Open Lands" (*sultan al-barr*), which Layard understood as "the King of the Desert."[12] As they collected taxes on their lands, the realm of the desert may even have expanded, as cultivators in some cases abandoned their lands and fled to cities.[13] Of course, Shammar power in the margins did not mean that they remained there or were somehow not connected to cities. Rather, they seasonally migrated to the edges of cities, with a moving population that rivaled the population of the very same cities. In many cases, they took care of animals owned by urban residents. In addition to relying on their vast flocks of sheep, the Shammar also collected taxes on villages and goods in transit through the region, though periodically they would suspend their collection of taxes in return for a salary from the Ottoman government, which they disparagingly referred to as "the Roumi," meaning those from Anatolia – but literally "the Romans."[14]

8 Layard, *A Popular Account of Discoveries at Nineveh*, 69–70, 72.

9 Ceylan, *The Ottoman Origins of Modern Iraq*, 40, 45; Nieuwenhuis, *Politics and Society in Early Modern Iraq*, 126; Williamson, "A Political History of the Shammar Jarba Tribe," 49, 63.

10 Nieuwenhuis, *Politics and Society in Early Modern Iraq*, 32–33.

11 Winter, "Alep et l'émirat du desert (*çöl beyliği*) au XVIIe–XVIIIe siècle," 93–98; Masters, *The Arabs of the Ottoman Empire*, 97.

12 Al-Zakariyya, *'Asha'ir al-Sham*, vol. 2, 617; Layard, *A Popular Account of Discoveries at Nineveh*, 66.

13 Çiçek, *Negotiating Empire*, 70–71, 97; Pehlivan, "Abandoned Villages in Diyarbekir Province."

14 Ceylan, *The Ottoman Origins of Modern Iraq*, 139; Centre des archives diplomatiques, Nantes (CADN), 166PO/D7/14, October 24, 1860.

And what of those heirs of Byzantium and, before that, Rome? By the middle of the nineteenth century, the empire was in the midst of a decades-long effort at chipping away at the power of local notables. The power of the notables had allowed the empire to incorporate vastly diverse geographies, but it had also left the Ottoman state struggling to maximize production and tax revenue so as to meet the existential challenges of nineteenth-century nationalist revolutionaries and European imperialists alike.[15] The empire-wide reforms known as the Tanzimat attempted to remedy these challenges in a number of ways. One such measure to circumvent local notables was cancelling tax farming in 1839, only to have to reinstate it due to revenue shortfalls (a practice that would be repeated several times in the nineteenth century). The Ottoman state also worked to settle nomadic groups, so as to clarify their revenue obligations to the state.[16]

The Crimean War of 1853 to 1856 added further complications to an empire working to settle accounts. A conflict ostensibly over Russian custodianship of holy places in Jerusalem, the war quickly embroiled Britain, France, and Sardinia, which fought on the side of the Ottomans for the sake of preserving the territorial integrity of the empire. The war's impact was disastrous. Not only did it hasten the first instances of Ottoman foreign borrowing and, subsequently, onerous debt arrangements, but it also led to nearly 1,000,000 people leaving Crimea and the Caucasus as refugees.[17] To respond to this challenge, the empire established its first Immigration Law (Muhaceret Nizamnamesi) in 1857, which, in addition to codifying existing practices, also offered incentives for expanding cultivation in less developed portions of the empire.[18] Immigrants received tax exemptions for six years on lands in the Balkans and twelve for those in Anatolia. Following this promulgation was the momentous Land Code (Arazi Kanunnamesi) of 1858. The impact of the instrument varied over the years and across the empire. But one of its main aims was to register lands in the empire with title deeds, thereby clarifying tax responsibilities while also promoting cultivation.[19] In an 1859 display of the twin necessities of refugee settlement and expansion of cultivation, the empire formed the Refugee Commission

[15] On the importance of local notables, see Khoury, *State and Provincial Society in the Ottoman Empire*; Yaycıoğlu, *Partners of the Empire*.

[16] Barakat, "Making 'Tribes' in the Late Ottoman Empire," 484.

[17] Badem, *The Ottoman Crimean War*.

[18] Kasaba, *A Moveable Empire*, 110–111.

[19] Quataert, "The Age of Reforms," in *An Economic and Social History of the Ottoman Empire, 1300–1914*, ed. İnalcık, 856–861.

(Muhacirin Komisyonu) to coordinate on these matters. The Jazira would prove to be a key part of this effort at expanding cultivation and state power in marginal areas of the empire. As the minister of religious endowments Suphi Pasha wrote, it was time for a new policy among the "several million Arabs [*urbân*] … in the vast and fertile deserts … between the lands of Damascus, Aleppo, and Iraq."[20]

THE LOCUST AND THE STARLING

By the middle of the nineteenth century, two half-brothers were at the center of Shammar leadership. ʿAbd al-Karim and Farhan were the grandsons of the first Shammar shaykh to decamp to the Jazira, and the sons of Sufoq, "the King of the Desert," who had been killed by Ottoman troops in 1847.[21] With about 50,000 tents and 15,000 horsemen by one estimate, their motion stretched between Urfa in the west and Baghdad in the southeast, with the Jazira as the heart of their power.[22] The brothers had different reputations, ʿAbd al-Karim – the son of ʿAmsha – as a "man of action" and Farhan as a "man of politics," owing, perhaps, to the time he spent in Istanbul during his youth.[23]

At the beginning of the 1860s, the Ottomans saw Farhan and ʿAbd al-Karim as potential allies and hoped to coax them in the direction of "civilization." But such simplistic formulations were often more complicated in practice. For example, in 1860, hostilities broke out between the Shammar and branches of the ʿAnaza, an Arabic-speaking tribe that typically stayed southwest of the Euphrates and thus outside of the Jazira.[24] After clashes near Harran, Ottoman troops from nearby Urfa arrived on the scene. According to consular records, the Shammar insisted that the state forces remain out of the engagement. As the Ottoman forces stood down, the Shammar defeated the ʿAnaza. They bribed the Ottoman officer with several thousand sheep and camels and then headed east into the Jazira. Reports in state newspapers, however, transformed the clash into an unambiguous victory for the Ottoman state. By this account, the victor of the "brilliant raid" was the Ottoman commander, and the vanquished not simply the ʿAnaza but also the Shammar, who were

[20] "Suphi Paşa Hazretleri'nin Layihası'ndan," 21.
[21] Çetinsaya, *Ottoman Administration of Iraq*, 86.
[22] CADN, 166PO/D7/14, Consulate in Baghdad to Marquis de Lavalette, October 24, 1860.
[23] CADN, 166PO/D7/14, Consulate in Baghdad to Marquis de Lavalette, October 24, 1860.
[24] CADN, 166PO/D7/14, Consulate in Baghdad to Marquis de Lavalette, October 24, 1860.

erroneously presented as having fought on the side of the ʿAnaza.[25] Thus reform efforts were not a simple process of state versus tribes, but this narrative nevertheless had power.

Internal Ottoman deliberations, meanwhile, spoke of other motives, particularly with respect to the Shammar chief ʿAbd al-Karim, whom Ottoman officials wished to enlist in their efforts to control the Jazira more effectively. In the coming decade, ʿAbd al-Karim would use the Jazira to challenge Ottoman efforts to divide it up. But in 1860, he appeared redeemable to Ottoman officials. The Ottoman commander of Urfa declared that in the fighting with the ʿAnaza, ʿAbd al-Karim had displayed "good service, manliness, and trustworthiness," which seemed to be a surprise given that "he had never met with a government official before."[26] There was thus hope that ʿAbd al-Karim and the 4,000 to 5,000 tents he led – somewhere between 20,000 and 30,000 people – might be used for the "protection of the desert." In hoping to use ʿAbd al-Karim as a way of expanding Ottoman control into a sometimes-difficult environment, Ottoman officials also hoped to affect a change within ʿAbd al-Karim, from "savagery" (*bedeviyet*) into "civilization" (*medeniyet*). In sum, the Ottoman administration used one nomadic group against another and planned to use ʿAbd al-Karim as a protector of Ottoman interests in the peculiar environment of the desert. In the process, they hoped to transform ʿAbd al-Karim himself.

But transforming ʿAbd al-Karim could only go so far without managing the locusts that regularly ravaged the region. Although the barley had been harvested by the time the insects arrived in Mosul and developed wings in 1860, the locusts did cause "considerable damage to the standing wheat."[27] Ottoman officials – as they did most years – impressed peasants on the edges of the Jazira into forced labor brigades. They collected some 100,000 okka (283,000 lb) of locust eggs before the insects had developed wings, and another 50,000 okka (141,500 lb) of locusts after they began to fly.[28] People then dug holes in the ground in which they snuffed out the insects by burying them. The Ottoman governor of Mosul Veysi Pasha noted that

[25] CADN, 166PO/D7/14, Consulate of France in Baghdad to Marquis de Lavalette, December 19, 1860. See *Journal de Constantinople*, October 17, 1860, pp. 2–3.

[26] Türkiye Cumhuriyeti Cumhurbaşkanlığı Devlet Arşivleri Başkanlığı Osmanlı Arşivi (BOA), İ.DH 462/30764, Mehmed Takiyüddin to the Grand Vizier, 9 Safar 1277 (August 27, 1860).

[27] TNA-UK, FO 195/603, Rassam to Bulwer, May 14, 1860.

[28] BOA, MVL 756/36, Governor of Mosul Veysi Pasha to the Grand Vizier, 9 Şevval 1276 (April 30, 1860).

government efforts to include "tribes" in the effort to collect locusts and locust eggs contributed "incalculable" (*hesapsız*) numbers to the cause, too.

In case human labor would not be enough, Veysi Pasha had utilized a different kind of relationship with the environment. After all, even if he took precautions within the territory under his control, locusts could always appear, as they had in previous years, "from the direction of the desert."[29] For this dilemma, there was one prescription: "the famous locust water [*meşhur olan çekirge suyu*] of Konya province." Many believed that hanging containers of the substance from mosques would attract birds and then instigate "a war." The birds attracted by the substance – also known by its Arabic name of *ma' al-jarrad* – were starlings (Turkish: *sığırcık*, Arabic: *samarmar*; Figure 6).[30] The locust-eating birds were described as

FIGURE 6 The starling. D'Herculais, *Les Sauterelles*

[29] BOA, A.}MKT.UM 354/56, Governor of Mosul Veysi Pasha, 27 Şevval 1275 (May 30, 1859).

[30] BOA, MVL 756/36, Governor of Mosul Veysi Pasha to Grand Vizier, 9 Şevval 1276 (April 30, 1860).

far back as Pliny the Elder.[31] The belief in a holy water attracting them was widespread, though stories differed. According to various accounts, those afflicted by locusts in different parts of the Ottoman Empire procured holy water alternately from Iran, Khorasan, or the foot of Mount Ararat.[32] Historian James Grehan has written about this phenomenon, describing the *samarmar* as a "magical bird" owing its power to Sufis intent on "perpetuating its lore" rather than any actual reality or ecological relationship.[33] But the reputation of the starling as both reality and legend persisted into the nineteenth century and beyond.[34] Indeed, in Mosul in 1860, the Tigris flowed with locusts. Veysi Pasha explained that starlings had appeared north of the city and relentlessly slaughtered the swarms of insects. The locusts had no escape from the birds but by "throwing themselves into the water," and so the mighty river flowed with the vanquished insects that wielded such power in the region. Thus as with mobile people, so, too, with mobile locusts. The Ottoman state managed to harness the power of yet another moving group – starlings – to protect itself. Still, Veysi feared that the locusts might return, laying waste to the region's rich summer crops of "cotton, sesame, onions, tobacco, and vegetables."

The connections between nomads and locusts were both discursive and material. The British consul in Aleppo dated Aleppo's decline to the emergence nearly eighty years before – by his estimation – of "swarms" of nomads in the region.[35] Ottoman officials used the adjective of "damaging" (*muzirre*) to describe the Shammar and the annoying insects alike, while various British consular officials similarly termed them "a pest to the country" and "worthless hordes."[36] Beyond language, there were also more direct connections between insects and mobile people. In the spring of 1861, the locusts struck and devastated agriculture. "Devouring every particle of vegetation on their road," they left the land between Nusaybin

[31] Pliny the Elder, *The Natural History*, 39.27.

[32] Jennings, *Christians and Muslims in Ottoman Cyprus and the Mediterranean World*, 178; Parrot, *Journey to Ararat*, 144–145; Niebuhr, *Description de l'Arabie*, 154.

[33] Grehan, "The Legend of the Samarmar," 125.

[34] Göçen, "Sığırcık Suyu Şeyhleri"; Hızlı, "Çekirge İstilasına Çözüm Sığırcık Kuşu."

[35] TNA-UK, FO 195/55, Skene to Bulwer, May 12, 1860.

[36] BOA, MVL 613/25, Council of Kurdistan to Grand Vizier, 9 Zilhicce 1277 (June 18, 1861); TNA-UK, FO 195/676, Rassam to Bulwer, December 9, 1861; FO 78/1607, Taylor, Report on Trade in Diyarbekir Pashalik, December 31, 1861.

and Urfa bare.[37] It was only the "natural obstacles" of the mountains near Mardin and Mount Karaca (also known by its Turkish name of Karacadağ) that stopped the locusts' unprecedented march, it having been many years since they had struck the fertile lands around Diyarbekir. If this was the degree of suffering on the edges of the Jazira, it was even worse in its heartlands. With no pasture left for their animals thanks to the locusts, the Shammar – led by ʿAbd al-Karim – moved northward to the very same region whose elevation had stopped the march of the locusts.[38] The Shammar sought pasture, but also to purchase grain out of concern that prices might increase due to famine.[39] Local government officials called for Ottoman troops to maintain order, while foreign onlookers predicted that soldiers in the field "will destroy what little cultivation the locusts have left."[40] In other words, it was imperial troops that may well have seemed like locusts for their destructive capacities in this case.

The volatile situation gave way to conflict. Disputes between the Shammar and seminomadic Kurdish groups in the region – among them the Kiki and the Millî – escalated. Reports suggested that the Shammar had used the crops of these groups as "pasture for their animals."[41] In response, the Kurdish groups struck the Shammar, and allegedly took many of their camels.[42] Overlaying the tension with respect to land usage was ethnic difference, as Mount Karaca functioned not only to stop locusts and signify a changing geography, but also, albeit in a somewhat blurry way, a sense of changing ethnicity. While no clear dividing line existed, to the north of Mount Karaca, Kurdish became more commonly spoken, while to the south Arabic was more commonly used. The interconnected motion of locusts and nomads structured all of these clashes and interactions. The moving people and animals also had an environmental impact. According

[37] TNA-UK, FO 78/1607, Taylor, Report on Commercial Conditions of Diyarbekir Pachalik, May 29, 1861.

[38] BOA, MVL 613/25, Council of Kurdistan to Grand Vizier, 9 Zilhicce 1277 (June 18, 1861).

[39] TNA-UK, FO 78/1607, Taylor, Report on Commercial Conditions of Diyarbekir Pachalik, May 29, 1861.

[40] BOA, MVL 613/25, Council of Kurdistan to Grand Vizier, 9 Zilhicce 1277 (June 18, 1861).

[41] Centre des archives diplomatiques, La Corneuve (CADC), 60CPC/3, Aleppo to Ministry of Foreign Affairs, September 14, 1861.

[42] BOA, A.}MKT.UM 498/35, Governor of Harput to Grand Vizier, 9 Safar 1278 (August 16, 1861).

to the British consul in Diyarbekir, the locusts and nomads together had turned the region into "literally desert."[43]

The depredations of both also invited speculation on how the Ottoman state might resolve the tensions between different nomadic groups. British observers emphasized several kinds of lines. The consul in Aleppo called for the "formation of a line" of settled nomads to end these circuits of motion.[44] He subsequently added another evocative detail to his plans for nomadic control. He called for the placement of "light guns" on "the numerous artificial mounds" dotting the Jazira.[45] In other words, the distinctive ruins of the Jazira's past civilizations would be used as a base for state violence in the effort to resurrect these civilizations. The British consul in Diyarbekir took a different view, complaining instead of the "destructive and antagonistical effects" of nomads moving between various provinces. He thus proposed aligning the particular environment of the Jazira with the political borders through which it was managed. More specifically, he called for Baghdad province to be extended such that it include "the desert limits" of the districts of Urfa, Diyarbekir, and Mardin and extend all the way south to the Euphrates.[46] He also saw an ethnic logic at work in such an arrangement, noting that Mount Karaca ended the Jazira's plains and also served as a "natural barrier between the Arab and the Turks and Koord." Drawing a neat border around the desert, in his view, would enable control of all of these different currents.

Ottoman officials would take the same view in coming years, but they were not ready in 1861, when they preferred to co-opt the nomadic motion of ʿAbd al-Karim's brother Farhan for their own purposes. Urfa district governor Mehmed Takiyüddin – he who had taken credit for the Shammar defeat of the ʿAnaza in 1860 – restated his hopes for "gradual transition from savagery to civilization [*bedeviyetten medeniyete*]."[47] Indeed, at the same time as many complained of the depredations of nomads, the Ottoman state also entrusted to nomads the protection of what was arguably the empire's most important technological project: the telegraph line. Having reasoned that "no body of horsemen" from the government "would suffice to protect the line," the Ottomans entrusted security of the line to the Shammar chief Farhan, ʿAbd al-Karim's

[43] TNA-UK, FO 195/676, Taylor to Bulwer, August 5, 1861.
[44] TNA-UK, FO 195/55, Skene to Bulwer, May 12, 1860.
[45] TNA-UK, FO 195/55, Skene to Bulwer, December 11, 1860.
[46] TNA-UK, FO 195/676, Taylor to Bulwer, December 12, 1861.
[47] BOA, A.}MKT.UM 505/13, Mehmed Takiyüddin to Grand Vizier, 9 Rabiaülevvel 1278 (September 14, 1861).

brother.[48] Farhan had often shown a greater willingness to negotiate with Ottoman officials than his brother, and this pattern would continue in years to come. In the meantime, the collaboration bespoke the mutual reliance of Ottoman officials and mobile populations in marginal environments. It was the protection of Farhan and the Shammar, then, that enabled the first telegraphic communication between Istanbul and Baghdad in late June of 1861, which carried news of the death of Sultan Abdülmecid and the ascension of his brother Abdülaziz.[49]

LOCUSTS, SHAMMAR, AND THE AMERICAN CIVIL WAR

Locusts did not catalyze the same questions about provincial borders as the Shammar did. But they could have. In late May of 1862, the insects arrived again. The French consul in Baghdad wrote, "I was told that in the memory of man no such thing had been seen at Baghdad."[50] The insects "devastated pasture" and destroyed all vegetables except for "onions and some roots." They did not stop there, spreading to Mosul, and then so far west that they crossed the Euphrates by late May. In Aleppo, so many locust bodies accumulated in the city's stores of drinking water that residents had to rely on well water, as if they were under siege.[51] Clearly, the locusts moved with little regard for provincial borders. But provincial borders were the administrative units through which Ottoman officials managed the vast region of the Jazira. The Ottoman district governor of Urfa explained that all of the ills of locusts in 1862 could be attributed to provinces to the east.[52] Though local officials had gone to great lengths to destroy locust eggs left over from the previous year's invasion in Baghdad and Mosul, the district governor explained that nevertheless "winged" locusts appeared in great quantities from the direction of Mosul, Baghdad, and Diyarbekir. Their "winged" nature was especially significant, because before the insects developed wings about forty days after hatching, they could be destroyed much more easily. But after the insects developed wings, they were essentially impossible to stop. As they flew, they devastated crops and laid more eggs, which would then

[48] TNA-UK, FO 195/676, Kemball to Bulwer, April 15, 1861.
[49] CADC, 23CCC/12, Consul General in Baghdad to Ministry of Foreign Affairs, July 3, 1861.
[50] CADN, 166PO/D7/14, June 18, 1862.
[51] TNA-UK, FO 195/717, Rassam to Bulwer, May 26, 1862; CADC, 4CCC/32, Consul General in Aleppo to Ministry of Foreign Affairs, July 10, 1862.
[52] BOA, MVL 764/15, Governor of Urfa, 5 Eylül 1278 (September 17, 1862).

become their forward operating bases for the next year. And they did so while crossing provincial borders, blurring the neat division of the administrative responsibility through which the Jazira was to be controlled.

Like locusts, nomads also presented spatial challenges in 1862 for Ottoman officials by taking advantage of the discrepancy between the environment and Ottoman political borders. But the Ottoman campaign of that year actually began with government officials working to use the Jazira – so long the environmental refuge of the Shammar – for their own ends. In response to Shammar raids and failure to pay taxes, Ottoman officials planned on "surrounding" the Shammar "in the desert of Mesopotamia" in the springtime when snowmelt and rains ensured high water levels that would make it difficult for the Shammar and their animals to cross the Euphrates.[53] The plan also attempted to transcend the Ottoman division of the region, by sending military detachments from all of the districts that divided up the Jazira from east to west, an arc stretching from Baghdad to Kirkuk, Mosul, Diyarbekir, Urfa, and, finally, Aleppo. It was this crescent-like formation that would enclose the nomads in the desert and, they hoped, restore the Fertile Crescent.

Yet Ottoman hopes were dashed as the environment – and Shammar movement within it – once again defied official efforts at control. After reportedly stealing some 8,000 sheep near Mount Karaca, the Shammar fled into the desert. It was then that the torture of the military commenced.[54] In the sparsely inhabited area, the troops struggled to find supplies. They "spent entire days deprived of food," and rumors abounded that they "fed on the flesh of camels that had died in the famine."[55] They pursued the Shammar all the way to the Khabur River as part of orders to capture Farhan, but, as the French consul in Baghdad intoned, locating the "rebel chief" was difficult given his "wandering nature [*humeur ambulante*]." By April, the troops returned to their bases for fear of the onset of summertime heat.[56] The British consul in Mosul called the expedition a "total failure."[57] The Ottoman troops had, he wrote, "cut a bad figure before the wild sons of the desert." A campaign that began with plans to enlist the environment against the Shammar had resulted in the opposite occurring, a reminder that even when Ottoman

[53] TNA-UK, FO 195/717, Rassam to Constantinople, March 3, 1862.
[54] TNA-UK, FO 195/717, Taylor in Diyarbekir, March 11, 1862.
[55] CADC, 62CPC/5, Baghdad to Chouvenel, MAE, May 7, 1862.
[56] TNA-UK, FO 195/717, Rassam to Bulwer, April 28, 1862.
[57] TNA-UK, FO 195/717, Rassam to Bulwer, April 28, 1862.

administrations transcended provincial division, the Shammar could still use the environment as a weapon against the state.

The unruly spatial bounds of locusts and nomads had especially harsh consequences given the way they undermined the Ottoman effort to expand cotton cultivation amidst the global shortage caused by the American Civil War. As elsewhere around the world, in the Ottoman Empire, too, opportunistic landowners increased cultivation of the distinctive fiber (or at least hoped to) everywhere from Thessaly and Izmir on the Aegean to the Nile Delta.[58] In the Jazira, however, locusts intervened again and again to thwart these attempts. In Baghdad, locusts ravaged the cotton crops, of which British trading houses had encouraged cultivation.[59] In Mosul, locusts forced delays in the planting of the crop, which then put the cotton at risk of maturing so late in the fall that rain might destroy it.[60] North of Aleppo in the cotton fields of Kilis, the French consul described how "in twenty-four hours all of the plantations of this valuable plant were devoured."[61] In Aleppo, the British consul explicitly blamed the insects, noting that Britain could not rely on Aleppo to compensate for the global cotton shortfall "on account of damage caused by locusts."[62] But the insects were not alone. Nomads also played a role. The British consul in Aleppo claimed that nomads were the "great obstacle" to cultivation of the crop because of the way they would "turn their flocks into cotton fields."[63] Indeed, just as locusts were a seasonal affliction on cotton, so, too, were nomads. It was precisely when cotton began to emerge from the ground that nomads and their hungry flocks arrived close to cities.

Despite the deleterious impact of locusts and nomads on cotton, the promise of the crop still held an allure for cultivators. Typical cotton production in Aleppo province was about 1,000 bales, 500 of which were exported. But in 1862 the amount exported from Aleppo's port on the

[58] On the Ottoman Empire and the Middle East, see Kurmuş, "The Cotton Famine and Its Effects on the Ottoman Empire," 160–169; Owen, *Cotton and the Egyptian Economy*, 89; Rodrigue, *French Jews, Turkish Jews*, 49; Toksöz, *Nomads, Migrants and Cotton in the Eastern Mediterranean*, 12; Kasaba, *The Ottoman Empire and the World Economy*, 64. On the rest of the world, see Beckert, *Empire of Cotton*, 253–257.

[59] CADC, 23CCC/12, Baghdad to Ministry of Foreign Affairs, June 18, 1862.

[60] TNA-UK, FO 195/717, Rassam to Bulwer, May 26, 1862.

[61] CADC, 4CCC/32, Consul of France in Aleppo to Minister of Foreign Affairs, July 10, 1862.

[62] TNA-UK, FO 78/1689, Skene to Bulwer, July 31, 1862.

[63] TNA-UK, FO 195/761, Skene, Report on the Trade of Aleppo during the Year 1862, December 31, 1862.

Mediterranean was 29,000 bales.[64] In the coming years, cotton cultivation would remain afflicted by volatility. The 1863 cotton crop in Aleppo was "completely destroyed" by locusts.[65] Other problems could emerge, too, such as lack of labor to collect the crop when it was finally ready for harvest, as was reported in Diyarbekir in 1863.[66] Officials worried about similar problems in the next year, given the way that even settled populations fled locust invasions. As one Muslim merchant of Diyarbekir allegedly predicted, "If the locusts do come this year the pasha will find himself alone in the Province."[67] Still, the benefits seemed to outweigh the costs for many. In Mosul, cultivators would respond to the destruction of locusts by replanting cotton "up to two and even three times in the span of three months."[68] In 1864 in Aleppo, some five times the usual amount of land was devoted to cotton.[69] Despite these efforts, the cotton-growing center of the late Ottoman Empire would end up near Adana, a region that was largely not afflicted by the Moroccan locust.[70]

Rising prices for fibers extended not only to cotton but also to wool, one of the key products of groups such as the Shammar. Numerous observers remarked on this dynamic. The French consul in Mosul called wool "the principal business" of the city in 1863, thanks in no small part to high prices instigated by the American Civil War.[71] The British consul in Diyarbekir went further. Of wool, he wrote, "Former prices and yields were nothing in comparison to those in the present day."[72] Nomads evidently recognized the opportunity. According to the British consul, they previously "had no idea of the real value of their produce," but with the infusion of money, pastoralists began to "fight for the last farthing." The effects were felt in Aleppo, too, where people who had been "starving a few years ago" were allegedly "in possession of L1,000 each by the sale of cotton and wool."[73]

[64] TNA-UK, FO 195/761, Skene, Report on the Trade of Aleppo during the Year 1862, December 31, 1862.
[65] TNA-UK, FO 195/761, Skene to Bulwer, June 30, 1863.
[66] TNA-UK, FO 195/799, Report on Trade and Agriculture in Diyarbekir in 1863, March 31, 1864.
[67] TNA-UK, FO 195/799, Taylor to Bulwer, January 11, 1864.
[68] CADC, 214CCC/1, Mosul to Ministry of Foreign Affairs, July 16, 1866.
[69] TNA-UK, FO 195/800, Report on the Trade of Aleppo during the Year 1864.
[70] See Toksöz, *Nomads, Migrants and Cotton in the Eastern Mediterranean*, 12.
[71] CADC, 214CCC/1, Mosul to Ministry of Foreign Affairs, Rapport Commercial sur l'année 1862, November 20, 1863.
[72] TNA-UK, FO 195/799, Report on Trade and Agriculture in Diyarbekir in 1863, March 31, 1864.
[73] TNA-UK, FO 78/1828, Skene to Bulwer, June 30, 1864.

While cotton and wool could both mean riches for some, the geography of wool production fit much more easily with the choreography of locusts and nomads that defined the Jazira. The sheep of the Jazira essentially constituted movable farms with which the Shammar could respond to environmental variation. There were of course limits to this flexibility, especially in the instance of epizootics or cold winters that killed off herds, as historian Zozan Pehlivan has detailed.[74] But if locusts arrived, nomads could find pasture elsewhere; cotton cultivators had no choice but to attempt to plant the crop again. And if cotton cultivation occurred on the edges of the Jazira, wool production derived from the space in between the region's great cities, the seemingly empty lands whose "luxuriant pasturage" offered value to those who knew how to move.[75] Typically, merchants from a place like Aleppo or Mosul would spread out into the Jazira beginning in February.[76] At that point, they would pay for fleeces. Of course, great risk was involved in fronting large sums of money, and low profit margins prevailed for exporters, even in boom years. But part of the appeal was how they paid for the fleeces. In addition to using currency, they also used "European cotton goods" as advances.[77] In fact, it was this latter connection in particular that often made wool worth it.[78] The real payoff was in the opening up of markets among nomads for European consumer goods.

But there were also ways of expanding the profit margin, and these patterns undermine many of the common remarks of denigration made toward nomads. To cut down on transport costs, export houses tried to lure the nomads and their flocks as close to cities as possible prior to the shearing of fleeces in April.[79] The purchasers would then export the wool onward to other parts of the Ottoman Empire or to Europe. It was at these times that conflicts between cultivators and nomads would often occur, occasioning categorical statements such as one by the French consul in Aleppo that "nomads ... constitute the principal, and until now insurmountable, obstacle to the current of civilization."[80] Yet the types of

[74] Pehlivan, "El Niño and the Nomads."
[75] TNA-UK, FO 195/799, Report on Trade and Agriculture in Diyarbekir in 1863, March 31, 1864.
[76] CADC, 60PC/3, Mosul to Ministry of Foreign Affairs, June 8, 1862.
[77] TNA-UK, FO 195/799, Report on Trade and Agriculture in Diyarbekir in 1863, March 31, 1864.
[78] TNA-UK, FO 195/800, Report on the Trade of Aleppo during the Year 1864.
[79] TNA-UK, FO 195/799, Report on Trade and Agriculture in Diyarbekir in 1863, March 31, 1864.
[80] CADC, 4CCC/33, Consul in Aleppo to Ministry of Foreign Affairs, May 21, 1863.

conflicts that prompted him to make this comment actually occurred because of the nomads' integration into long-distance trade networks. They may have appeared to be enemies of civilization when they destroyed cotton crops in 1862, for example, but they were also doing something else: destroying their economic competition.

The Shammar in particular seized the opportunity to maximize their economic power. They had occupied an ambiguous place with respect to Ottoman authorities. On the one hand, officials decried their lack of civilization and led campaigns to discipline them. But at the same time, state officials relied on them, not only for their economically productive capacities, but also for their ability to offer security in marginal environments. When the French consul in Baghdad ventured toward the Jazira in 1863, for example, he encountered Farhan, the very same chief who had been charged with protecting the telegraph in 1861 and who had been the target of the 1862 campaign. Of his interaction with the Shammar chief, the consul recalled, "I took the opportunity to compliment him on the immense quantity of sheep, camels, and horses that I had encountered since my departure from Baghdad."[81] So great were the flocks and so rich the animal products they sustained that the French consul called the region the "Normandy of Asia." In response, Farhan laughed and told the consul, "You have not seen anything. All of Mesopotamia is covered with sheep and camels belonging to my people. That is their only wealth." By October of 1863, Farhan even journeyed to Baghdad, where he brought the consul of France "a sample of camel wool."[82] The consul wrote that Farhan "assured me ... that it would be possible to export a great quantity, all the more since this wool has never before been demanded by commerce." In other words, Farhan was marketing a new product. It is unclear if anything came of this connection. But the consul's report evidently caught the attention of one of his higher-ups – perhaps in Paris – as someone had scrawled in the margin "where is the camel sample?"

THE LINE OF CORDON

In response to these interconnected dynamics, Ottoman officials announced a new plan in 1864 to control pastoralist motion and to spur agrarian improvement. A March memo summarized many of the swirling discussions

[81] CADC, 62CPC/5, Delaporte to Drouyn de Lhuys, January 21, 1863.
[82] CADC, 23CCC/12, Delaporte to Ministry of Foreign Affairs, December 30, 1863.

of environment, mobility, and ethnicity. Much like European observers, the Ottomans saw the region as one of "ruins," and an "empty desert that had not seen a trace of development for a long time."[83] With the Ottomans as with the Europeans, blame fell on the effect of nomads like the Shammar. Seminomadic Kurdish tribes had once been in the process of "civilization and settlement" (*temeddün ve tavattun*), but thanks to the depredations of the Shammar "the taste of settlement and civilization disappeared from their minds." The Kurds, the Ottomans explained, would cultivate in the summer. After the harvest, they would move to the Khabur River with their flocks, perched "on the edge of the desert." From this vantage, the desert was the locus of disorder. It was in the desert in particular where the language of mobility and ethnicity blurred. "With the taking of horses, camels, sheep, and cattle," explained the memo, the Kurds "became accustomed to wandering like Arabs [*arap gibi*]." The term "Arab" in this context referred not to the language that they spoke but rather to their practices of nomadism. Indeed, in the same context, Kurds became described as "seminomadic" (*mütearrib*), which literally meant "Arabized" because of the prevailing understanding that "Arab" and "nomad" were synonymous. It was the land to which the seminomadic or Arabized Kurds moved in the winters that constituted the Shammar's "old home" (*vatan-i kadim*), tantamount to "an independent . . . wandering ground."

To convey their control of motion, Ottoman officials used the language of public health. They envisioned their plan on a map, which almost seemed to look southeast from Anatolia (Figure 7). The map depicted a number of forts and villages across the region at Mount ʿAbd al-ʿAziz, Ras al-Ayn, and Viranşehir. It also depicted a "line of cordon" across which the Shammar could not cross.[84] Thus, to address the Shammar's contagious example for the seminomadic groups on the edge of the desert, Ottoman officials proposed spatial controls fit for disease. The sanitary cordon stood as the technology par excellence for fighting epidemics in this period of interconnection, so much so that numerous port-city neighborhoods across the empire took this name, everywhere from Izmir to Salonica.[85] The sanitary cordons of port cities were intended to stop diseases like plague. The sanitary cordon of the Jazira aimed to stop

[83] BOA, İ.MVL 520/23021, Memorandum from Governor of Kurdistan Mustafa, 27 Şubat 1279 (March 10, 1864).
[84] BOA, İ.MVL 520/23021, Memorandum from Governor of Kurdistan Mustafa, 27 Şubat 1279 (March 10, 1864).
[85] Zandi-Sayek, *Ottoman Izmir*, 27.

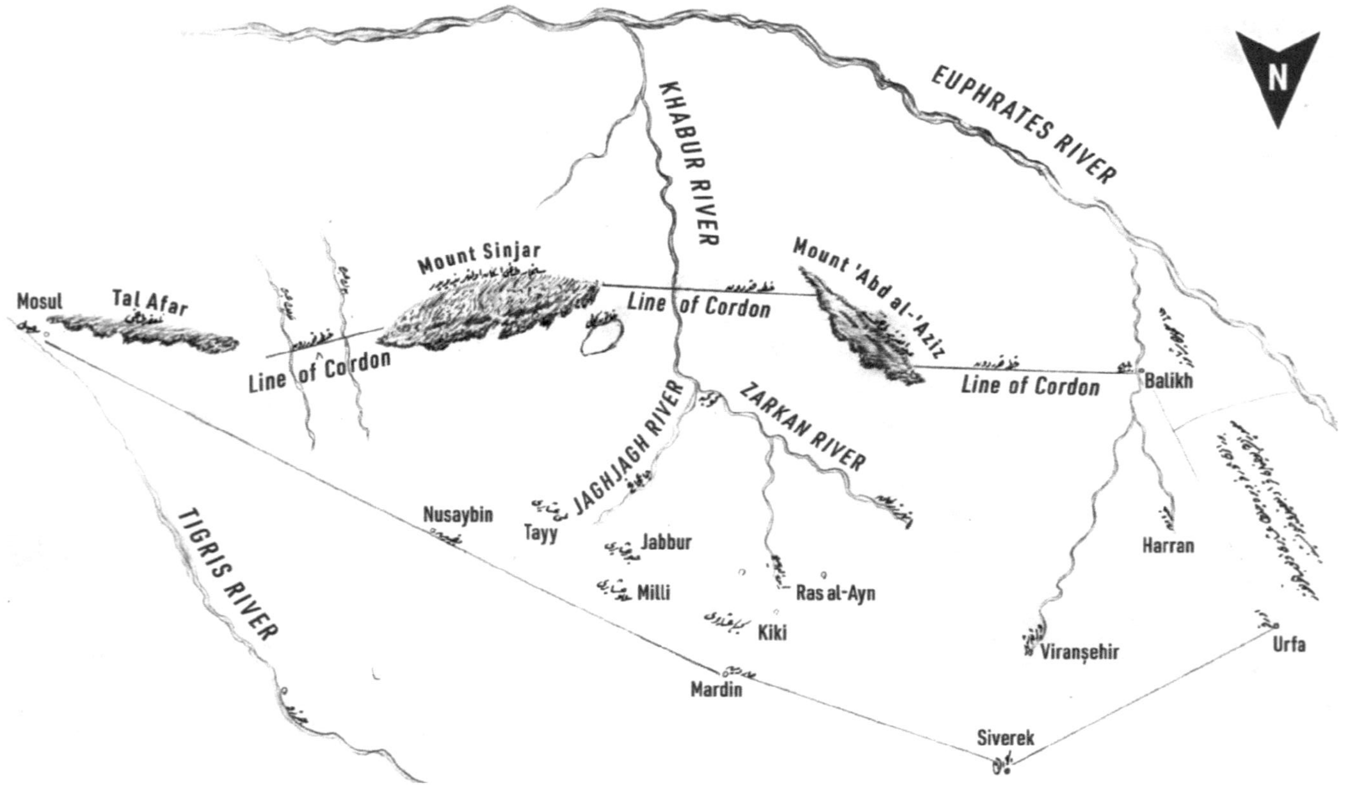

FIGURE 7 Line of cordon. BOA, İ.MVL 520/23021

nomadism by depriving the Shammar of their "summer pastures."[86] Unable to cross into the Jazira, and unable to move further south for fear of their long-time enemies the ʿAnaza, the Shammar would consequently choose to "abandon savagery and choose settlement." And thus, the Shammar's nearly 11,000 tents (as estimated by Ottoman officials) would become houses that rivaled in sum the population of cities like Diyarbekir and Mosul. Or at least so Ottoman officials hoped.

The region's various nomadic and seminomadic leaders were far from passive in these proceedings, and their correspondence with the Ottoman government in fact offered divergent accounts for their interest in settlement. Whereas Ottoman officials emphasized the singular role that the Shammar played in the destruction of the region, the chiefs of seminomadic groups like the Millî blamed nonhuman forces. They explained that "the locusts had attacked the edges of the desert for the past five or six years" and left their lands in "a miserable state."[87] Consular reports told the same story. The British consul of Diyarbekir had traveled in the region and noted that he could not obtain rice or wheat bread in much of the province.[88] In its place, locals used a flour made from millet or acorns, which, the consul grumbled, "produces obstinate constipation." The Kurdish chiefs added that the harsh winter of the previous year and lack of rain in the current year had left their animals with little forage, and many had died off.[89] In other words, the seminomadic chiefs presented their problems as not solely stemming from the Shammar, but rather from the environmental dynamics that ensured that it made sense to be a nomad in the first place.

Similarly distinctive explanations surfaced in correspondence between Eyup Bey of the seminomadic Kurdish Kara Keçe tribe and the Ottoman state. In an Arabic-language letter, Eyup Bey gave thanks that "our domain" (*dayratuna*) – as he termed it – "is safe from bandits."[90] In its Ottoman translation, Eyup Bey's expression of territorial possession became something else, as "our domain" transformed into "the cultivated and desert areas" (*mamur ve çöl havalisi*). In other words, a phrase that in

[86] BOA, İ.MVL 520/23021, Memorandum from Governor of Kurdistan Mustafa, 27 Şubat 1279 (March 10, 1864).
[87] BOA, İ.MVL 510/23021, Letter from heads of Aliyan, Dakura, Kiki, and Millî, 12 Zilkade 1280 (April 19, 1864).
[88] TNA-UK, FO 195/799, Taylor to Bulwer, January 11, 1864.
[89] BOA, İ.MVL 510/23021, Letter from heads of Aliyan, Dakura, Kiki, and Millî, 12 Zilkade 1280 (April 19, 1864).
[90] BOA, İ.MVL 510/23021, Letter from Eyüp Bey, 10 Şevval 1280 (March 19, 1864).

Arabic conveyed a sense of power over place became anodyne environmental description in Ottoman translation. When Kurdistan governor Mustafa carried news of Eyup Bey's potential settlement near Viranşehir, he returned to the monocausal explanation that the Ottoman state had previously emphasized, suggesting that the cause of upheaval in the region could solely be attributed to "the noxious Arab tribes" (*aşair-ı muzirre urbân*).[91] As in 1860, when a Shammar defeat of the ʿAnaza became a triumph of the Ottoman state over recalcitrant tribes, lost in the bureaucratic chain of communication in 1864 were the more complicated causes – among them, locusts – and vocabularies – "our domain" as opposed to "the cultivated and desert areas" – that characterized the Jazira on the ground.

While Ottoman officials cleaned up messy complexity communicated by local actors, local actors also seem to have used language in line with Ottoman discourse as a way of consolidating their own power. In May of 1864, a chief of the Shammar by the name of Jazaʿa wrote a letter in Arabic to the governor of Kurdistan, in which he emphasized Shammar power on the margins while also praising the region's ambitious governor.[92] "We are the people of the tent in this desert, and we have been moving in this vast desert since long ago." Perhaps hoping flattery might be a pathway to power, he wrote, "We have never seen an official like you." In the otherwise Arabic-language letter, Jazaʿa used the Ottoman word *çöl* for desert – a word that includes a letter not commonly used in Arabic – signifying how the Ottoman term for the Jazira's environment made its way into the other languages of the region.[93] The Shammar chief continued using this Ottoman word when he threw his support behind what he perceived to be the Ottoman plan for the region, amounting to "improvement of the desert" (*ʿamarat al-chul*). Clearly, Jazaʿa had a plan, too, namely, to use his praise of Ottoman plans to gain a kind of control within the Shammar like that of Farhan and ʿAbd al-Karim. To achieve this aim, he vowed to protect the region from "the people of the tribes," including "the Shammar Arabs." The response of Mustafa, the governor of Kurdistan, contained both denigration and pragmatic estimation. On the one hand, Jazaʿa had never seen a city before, having spent his life in the desert.[94] His lifestyle of "savagery" was rather like being an "animal," in Mustafa's words. But the

[91] BOA, İ.MVL 510/23021, Memorandum from Kurdistan Governor Mustafa, 19 Mart 1280 (March 31, 1864).

[92] BOA, MVL 678/88, Letter from Shammar Chief Jazaʿa, 2 Zilhicce 1280 (May 9, 1864).

[93] The usage apparently prevailed into the post-Ottoman period. Lange, "Shawāyā," 108.

[94] BOA, MVL 678/88, Letter from Shammar Chief Jazaʿa, 5 Muharrem 1281 (June 10, 1864).

chief might also be of use to the Ottoman state. Whatever the plans for stopping Shammar motion, it was always handy to have someone who could exert power in the environment that Mustafa described as "the mouth of the desert," which stretched from Derik to Nusaybin. This was precisely the environment that repeated locust invasions of the previous years had revealed to be essentially outside of state – or perhaps even human – control.

COFFEEHOUSE ON THE KHABUR: THE CHECHEN SETTLEMENT AT RAS AL-AYN

The political ecology of the Jazira in the 1860s revealed a connected relationship. Locusts and nomads empowered each other. They both came from places without many people. Locusts destroyed cotton planted to capitalize on the commodity's global shortage, while nomads sold wool at high prices thanks, in part, to the cotton shortage. Ottoman officials largely ignored how the movement of people, commodities, and insects fit together as a flexible response to environmental and economic imperatives. Instead, they viewed motion as a reflection of virtue, and a factor in destroying the environment. In the coming years, yet another group of people would be classified in these disparaging terms. Expelled from the Russian Empire, Chechens were just one group of many Muslims – amounting to hundreds of thousands in total – who fled conflict and ethnic cleansing on the edge of the Ottoman Empire to seek safety and security within the empire over the late nineteenth and early twentieth centuries. They were referred to in Ottoman as *muhacir*, which might be translated as "refugee," though of course the meaning would change in the twentieth century in accord with international law. The Ottoman state worked to manage them while also using them as catalysts for the expansion of cultivation everywhere from Libya to Amman to Ankara.[95] Such was to be the case with the Chechens who ended up in the Jazira.

But very early on, it became clear that the Chechen refugees would not act as Ottoman officials hoped. In fact, local officials described the refugees in many of the same ways they had described the nomads of the Jazira. When several thousand households of refugees arrived in Muş and

[95] Blumi, *Ottoman Refugees*; Fratantuono, "Producing Ottomans"; Gratien, "Ottoman Quagmire"; Hamed-Troyansky, "Circassian Refugees and the Making of Amman"; Lorenz, "The 'Second Egypt'"; Rogan, *Frontiers of the State*.

Erzurum in late 1865, the local council of Diyarbekir dismissed them as hailing from "savage nations."[96] As in the case of the nomads of the Jazira, the refugees' agile responses to difficult conditions became coded as lack of civilization. The council thus decried the disorder caused by the group as they sought safety. Some attempted to return home, while others "went among the tribes" in order to survive the "severity of the winter." The Ottomans hoped to move the Chechens from more densely populated regions and toward the edge, using them to bolster the seminomadic settlements envisioned in 1864. In this way, Ottoman officials believed that they would better take advantage of the uninhabited lands near Ras al-Ayn. On the edge of the provinces of Aleppo and Kurdistan, the land had previously been targeted as part of the plan for a cordon of settlement. At the source of the Khabur River, the site boasted ruins of ancient civilization and sulfurous springs.[97] One military officer described the land as "extremely fertile and productive," terms that would become a formulaic refrain over the coming years.[98] Left out of these rosy pronouncements was how Ras al-Ayn was also part of the strip of territory that had been devastated by locust invasions in the previous years.

The idea was to place the Chechens as bulwarks of sedentary cultivation in the midst of a geography of motion. But different visions of the space still persisted, and they would affect the settlement's prospects. One encapsulation of this dilemma appeared in correspondence between Ottoman officials and the Shammar chief ʿAbd al-Karim, the leader eyed by Ottoman officials in Urfa as being ripe for civilization in 1860. In a letter written in Arabic from the nomadic leader to the governor of Kurdistan, ʿAbd al-Karim declared of the Shammar, "We are the shaykhs of the Jazira from Aleppo to the gates of Baghdad."[99] In doing so, ʿAbd al-Karim explicitly invoked the regional bounds of the Jazira that extended beyond Ottoman provinces like Aleppo and Baghdad. It was this discrepancy between the provincial map and the broader environment that in part underwrote Shammar mobility and power. The translation of the letter into Ottoman, however, revealed a different description of the space, one that left out the term "Jazira" altogether. Instead, the text rendered the space solely in terms of Ottoman provinces, merely

[96] BOA, İ.DH 546/38018, 21 Kanunuevvel 1281 (January 2, 1866).

[97] Taylor, "Journal of a Tour in Armenia, Kurdistan, and Upper Mesopotamia," 349.

[98] BOA, İ.DH 546/38018, Derviş Paşa to the Fifth Army Command, 19 Kanunusani 1281 (January 31, 1866).

[99] BOA, MVL 723/41, Shammar Chief ʿAbd al-Karim to Governor of Kurdistan Mustafa, undated.

describing it as "from Aleppo to Baghdad."[100] The omission in translation underscored the tensions at work in the region, between a mobile geography and an administrative infrastructure that split this space up. The discrepancy made it seem at times that the term "Jazira" did not even translate into the language of the Ottoman bureaucracy.

Mobile forces defying jurisdiction would haunt Ottoman efforts at reform even as new steps were taken. To this end, the famed Ottoman reformer Cevdet Pasha arrived in Aleppo in April of 1866. Born in Lovech (in today's Bulgaria), he became a scholar, statesman, and military officer. Cevdet wrote his own magisterial history of the Ottoman Empire and translated Ibn Khaldun's *Muqaddimat* into Ottoman Turkish, both of which exhibited a vision of progress in line with Ottoman notions of civilizational uplift at work in the Jazira.[101] He also helped to write the Land Code of 1858.[102] And prior to his arrival in Aleppo, Cevdet served as the military commander of forces charged with subduing local notables, settling nomads, and making these reforms into a reality in the Adana region.[103] They were so committed that they renamed a local city – Islahiye – after the Turkish word for reform.[104] Yet when this formidable figure arrived in Aleppo to enact change in April of 1866, he had to deal with first things first. Cevdet Pasha, famed Ottoman reformer, estimable intellectual, had to grapple with "an invasion of locusts . . . threatening the harvest with complete destruction."[105] The great Ottoman statesman's initial actions in Aleppo were thus against the small insect that had bedeviled Ottoman efforts to transform the region. In fighting the locusts, it is unclear if Cevdet mimicked the actions of a previous ruler of the region – Mehmed Ali's son Ibrahim Pasha – who met the locusts in the 1830s by catching them in his fez.[106] But once again the connected nature of the region and the fractured nature of its governance meant that Cevdet, and the residents of Aleppo, could not breathe easy even after

[100] BOA, MVL 723/41, Translation of Shammar Chief ʿAbd al-Karim to Governor of Kurdistan Mustafa, 14 Mayıs 1282 (May 26, 1866).

[101] Aydın, *The Politics of Anti-Westernism in Asia*, 23; Thomas, *A Study of Naima*, 2; Tekin, "Reforming Categories of Science and Religion in the Late Ottoman Empire," 131–158.

[102] Karpat, "The Land Regime, Social Structure, and Modernization in the Ottoman Empire," 87.

[103] On the devastating epidemics that occurred as a result of and in the wake of these campaigns, see Gratien, *The Unsettled Plain*, 56–93.

[104] Ibid., 71.

[105] CADN, 166PO/D/1/62, Consul General in Aleppo to Marquis de Moustier, April 12, 1866.

[106] CADC, 4CCC/33, Aleppo to de Lhongs, April 25, 1863.

their efforts ended. After all, as the French consul suggested, it "remained still to be seen ... if the locusts had also been destroyed in surrounding provinces."

The location of the site of the Chechen settlement at Ras al-Ayn was roughly the center of the landscape that had been devastated by locust invasions in 1861, and insects and nomads had an impact there again. Ottoman officials struggled to locate supplies to provision the Chechen refugees in the wake of three years of "constant destruction of their crops ... by the locusts."[107] Nomads also affected the fate of the Chechen settlement. American missionaries in Mardin reported that the Chechens could not immediately reach Ras al-Ayn in accordance with the plan because nomadic groups had heard "of the intent to locate them there" and hence "came up earlier" to the region than usual.[108] In the words of the missionary, when the nomads reached Ras al-Ayn, they "took possession and keep possession." Whether the nomads had moved to the edge of the desert because of the locusts or because of the threat of settlement on their pastures, it was clear that the Ottoman state would have to contend with complicated dynamics of motion to install the Chechens. The nomadic occupation of the space left "the Pasha and all the troops ... on the edge of the mountains overlooking the plain." In other words, Ottoman administration of the Jazira looked much like the map of 1864, which rendered Ottoman administration looking to the southeast from the Anatolian highlands. The grasslands of the Jazira had offered nomads like the Shammar sustenance for decades, and their ability to make value out of these margins had afforded them a kind of power. It would not be crowded out easily.

The settlement of the Chechens finally began to move forward in late 1866 and 1867, occasioning great optimism for agrarian and civilizational change, at least among some. Perhaps the most enthusiastic booster of the Chechen settlement was the British consul Taylor. Having previously been an agent of the East India Company in Basra, he occupied much of his time with archaeology.[109] Accordingly, he took a long (if not accurate) view of the importance of the settlement. He boasted in early 1867 that "for the first time for several hundred years Northern Mesopotamia is comparatively free from thieving Bedouins, is again traversed by carts, and its

[107] TNA-UK, FO 195/799, Taylor to Lyons, 30 July 1866.
[108] TNA-UK, FO 195/799, 30 July 1866, Taylor to Lyons, Copy of Extract: 9 July 1866, Walker to Taylor.
[109] Sollberger, "Mr. Taylor in Chaldaea."

comparatively virgin soil has been broken down and sown with seed."[110] In short, he viewed the Chechen settlement as world historical, and he blithely deemed the Shammar a relic of the past, predicting that they would "forever be effectively excluded from this part of Mesopotamia, so long a useless desert under their blighting domination." With the Shammar gone, Taylor crowed that the Ottomans would restore "the proverbial fertility of the vast space" that had "been successively attested by all ancient and modern authors down to Marco Polo." In place of the Shammar, he imagined "thousands of acres of rich virgin soil" turned over to cotton cultivation.[111] Lockean notions of utility shaped Taylor's pronouncements, as he deemed improvement to be the most important factor in determining value. Left out of his judgment was the idea that there was actually a logic and value to exploitation of the region by pastoralists.

In reality, the harsh conditions of settlement likely made a life of nomadic pastoralism appealing to the Chechens. As the Chechens ran out of provisions, they had to rely on animals to supply their nutritional needs, and "they devoured every four footed beast that fell into their hands."[112] Typically, people in the region relied on animal manure as fuel, but because the Chechens lacked animals, they also lacked manure to use as fuel, and so they chopped down some of the area's few trees, including fruit trees that lined the Khabur River.[113] Meanwhile, for housing, many spent the winter living in what Taylor described as "a species of covered holes they have burrowed in the ground."[114] Boosters had envisioned the Chechens as catalyzing a transformation of the land, turning pastures considered wasteland into ordered rows of wheat and cotton fields. Instead, the Chechens entered the ground itself. The American missionary in Mardin, Augustus Walker, conveyed the impact of the Chechen refugees in a perhaps-overdetermined way, given their mobile nature and impact on the region's resources. He wrote that they were "worse than [a] locust invasion."[115]

In the years to come, the hopes attached to the Chechens of transforming the nomads gave way to fears that it would be the nomads who would

110 TNA-UK, FO 195/889, Taylor to Lyons, January 15, 1867.
111 TNA-UK, FO 195/799, Taylor to Lyons, December 28, 1866.
112 TNA-UK, FO 195/799, Taylor to Lyons, December 28, 1866.
113 Taylor, "Journal of a Tour in Armenia, Kurdistan, and Upper Mesopotamia," 350.
114 TNA-UK, FO 195/889, Taylor to Lyons, January 15, 1867.
115 *The Missionary Herald, Containing the Proceedings of the American Board of Commissioners for Foreign Missions with a View of Other Benevolent Operations for the Year 1866*, vol. 62 (Cambridge: Riverside Press, 1866), 362.

transform the Chechens. In March of 1867, Kurdistan governor Mustafa decried how the Shammar continued to wander in the desert, making them seem more like "birds and beasts" than human beings.[116] Yet at the same time, Ottoman soldiers could make little impact on them in the difficult environment of the Jazira. While Mustafa did not deploy the metaphor of the locusts, he did describe the impact of the Shammar through figurative language that packed a particular punch. He declared that the Shammar had turned all the lands on the edge of the desert into "disorder," but he conveyed this point by literally saying that they had turned this region into "Arab hair." The racist idiom uses stereotypically curly Arab hair to signify lack of control. In Turkish, the term "Arab" also often meant "Black."[117] Accordingly, the term had a particular connotation with the Shammar, given that a significant number of enslaved and formerly enslaved people of African descent lived among them, as had been highlighted by the prominence of an enslaved person in the story of their initial migration to the region.[118] In the same spring, Cevdet Pasha in Aleppo carried a similar if less specific message of denigration, calling all of the inhabitants of the "exact center of the deserts" at Deir ez-Zor "savage."[119] Environment, mobility, and race came together in these characterizations of the Shammar and other nomadic groups of the Jazira.

Further complicating matters was the threat of the locusts, which forever loomed on the horizon, beyond human settlement and, seemingly, control. In 1867, as in 1860, however, nonhumans intervened to lessen the locust burden once again. The British consul from Diyarbekir Taylor warned that even though rains seemed to augur a good harvest, locusts were "still dreaded."[120] He moreover disparaged Ottoman locust-control efforts, which only consisted – by his account – of "sprinkling a few drops of holy water" from Konya, the same substance believed to have attracted the starlings in Mosul in 1860 that left the Tigris flowing with locust

[116] BOA, MVL 734/59, Governor of Kurdistan Mustafa to Aleppo province, 4 Mart 1283 (March 16, 1867).

[117] Boratav, *100 Soruda Türk Folkloru*, 51, 69, 102; Willoughby, "Opposing a Spectacle of Blackness"; Wingham, "*Arap Bacı'nın Ara Muhaveresi.*"

[118] It was a Black member of the Shammar named Dathan who led Layard to the archaeological site at Hatra in 1846. Layard was surprised by the warm embraces Dathan exchanged with all Shammar along the way. Layard, *A Popular Account*, 72. See also Jwaideh and Cox, "The Black Slaves of Turkish Arabia during the 19th Century," 48.

[119] BOA, İ.MVL 571/25663, Ahmed Cevdet Pasha to Grand Vizier, 2 Mart 1283 (March 14, 1867).

[120] TNA-UK, FO 195/889, Report on Commercial Conditions of Consular District of Kurdistan, 18 April 1867.

carcasses. Yet Taylor also witnessed the power of the starlings west of Mardin just at the foot of the Anatolian plateau.[121] There were no clouds in the sky on the spring day in 1867 when he visited, but suddenly he was met with the "instantaneous obscurity" produced by an "impenetrable swarm" of locusts. The wheat survived, since it was already sufficiently mature that it was "unsuited to the delicate tastes of these insects." Still, the locusts posed a threat to summer crops. It was not government officials that met the locusts, however, but rather a natural enemy, namely the starlings referred to in Arabic as, in Taylor's transliteration, "Sammirmed." The fight between the vaunted birds and the swarming insects could be fierce. Taylor had witnessed insects even bring their avian enemies to the ground, starlings "having been completely nibbled by the locusts." But in this case, the birds prevailed. They appeared to so enjoy killing the locusts that they did not even bother to eat them, instead snapping them in two, and washing their beaks out with water from a nearby stream anytime locust carcasses began to clog them. The murderous starlings' flight in the spring of 1867 pointed to how processes beyond human control persisted in the Jazira.[122]

As enmity prevailed between starlings and locusts, the Ottomans attempted to ensure the same dynamic between refugees and nomads. Some reports emerged suggesting that the Chechens had sided with Shammar chief ʿAbd al-Karim in attacks against the ʿAnaza.[123] Taylor wrote that in this regard the Ottoman state might have more to fear from the Chechens than the Jazira's nomads, given the military training many of the refugees had received during their time in the Russian army. He warned that stationing imperial troops near Nusaybin would do little, seeing as the nomads would simply move elsewhere. Taylor claimed that Ottoman officials handled the situation by encouraging the Chechens to raid the Shammar. Allegedly, Derviş Pasha had given his imprimatur, remarking during a visit to the Chechens, "There is a great deal of sport in the desert," before specifying, "I don't mean gazelles and hares but camels, horses and sheep and you can amuse yourselves as you like." Derviş Pasha clearly referred to the animals of the Shammar that had for so long been the lifeblood of the group's mobility and opportunistic

121 Taylor, "Journal of a Tour in Armenia, Kurdistan, and Upper Mesopotamia," 359.

122 Later British officials were less suspicious of the impact of the starlings, but they were circumspect on whether they should introduce the birds to Cyprus given that "the manner in which [the starling] attacks grapes and other fruit almost counterbalances his use as a locust destroyer." TNA-UK, FO 424/132, Wilson to Earl of Dufferin, February 21, 1882.

123 TNA-UK, FO 195/889, Taylor to Elliot, April 20, 1868.

defiance of Ottoman officials. In May of 1868, reports suggested that the Chechens themselves had pulled off raids against the Shammar, having taken some 14,000 sheep.[124] They also allegedly struck against ʿAmsha, the queen of the desert, whom they "despoiled of all her ornaments and clothes." In protest, she dispatched "her own riding camel housed in black" to the various branches of the Shammar in the hope of uniting them against the Chechen raids.

While tension simmered between Chechen refugees and Shammar nomadic pastoralists, the Jazira came within the sights of a much broader Ottoman reform effort focusing on the relationship between central and local administration. The stationing of Cevdet Pasha in nearby Aleppo – once he got past the locusts, of course – represented one mark of this shift. So, too, did the transfer of Midhat Pasha to the governorship of Baghdad in 1868. On the other edge of the Jazira, Midhat Pasha embodied the dynamics of reform in a similar way. Like Cevdet, Midhat had also spent some of his early years in Lovech.[125] Like Cevdet, too, Midhat had also helped to pen a significant piece of Ottoman legislation in the form of the 1864 Vilayet Law, which clarified the relationship between local administration and the central government, with the aim of preventing foreign intervention.[126] He subsequently served as the governor of Danube province, widely seen as a test run for many of the reforms he championed.[127] Midhat would go on to write the Ottoman constitution and serve as the grand vizier overseeing its implementation, for which he was hailed by many as "the father of the free and the deposer of sultans."[128] But before these events, he found himself governor of Baghdad, forming a pincer movement with Cevdet Pasha surrounding the Jazira.

With less esteem and more infamy than Cevdet and Midhat were several local officials who both took their names from carnivorous predators and derived their power from extracting value from the land in violent ways. Arslan ("Lion") Pasha had distinguished himself as the district governor of Mardin. During this time, he had arrested a number of villagers from the Midyat region because they had failed to pay taxes "owing to losses occasioned by locusts and general poverty consequent

124 TNA-UK, FO 195/889, Taylor to Elliot, May 28, 1868.
125 Chambers, "The Education of a Nineteenth-Century Ottoman Alim, Ahmed Cevdet Pasha," 441; Davison, *Reform in the Ottoman Empire*, 144.
126 Davison, *Reform in the Ottoman Empire*, 146–147.
127 Petrov, "Midhat Paşa and the *Vilayet* of Danube."
128 Saliba, "The Achievements of Midhat Pasha," 307.

upon high prices."[129] After a brief skirmish, Arslan took prisoners and returned them to Mardin, where they "were exposed in the July sun to the attacks of myriads of insects abounding at that time." Another of Arslan's enforcers "bound" men "in the sun" and put "syrup" on their faces "to attract the vermin and flies." Still another killed a woman who had complained of his violent ways and then "quartered the body and hung the pieces up on the trees near the high road." Arslan reportedly tolerated the behavior because the official was good at collecting taxes, pointing to the ways that extracting value from the environment in the form of taxes was also occasioned, sometimes, by the use of the environment itself as a weapon.

Some of the same practices were rumored to be associated with a figure named Ismail Pasha but more widely known as Kurt ("Wolf") Pasha, who, too, used environmental violence.[130] An ally of Cevdet Pasha, the illiterate Kurt Pasha was appointed governor of Diyarbekir in 1868 (it had previously been a part of Kurdistan province, which was broken apart in 1867). When an official in Ras al-Ayn was charged with embezzlement, Kurt Pasha punished him by tying a hungry greyhound to the alleged corrupt official's back and placing a basket of bread around the official's neck. The official was to climb the hill up to Mardin in this state, marching from the fertile soil at its foot to the stony hilltop city while the angry dog repeatedly bit him. Taylor concluded, "The numerous acts attributed to him are so vindictively atrocious that I even having some knowledge of the man would fair believe them exaggerated or the acts of a demented being, but the Pasha's antecedents unhappily are too notorious for such suppositions." The lurid tales of violence coupled with the presence of Tanzimat luminaries such as Cevdet and Midhat underscored the stakes of reform in the Jazira. At issue were questions of government efficiency, while the means of violence was the environment itself.

Alongside the Jazira's leonine and lupine administrators was a more mundane technology aimed at conveying the message of civilizational uplift and news of agrarian transformation: *Diyarbekir Gazetesi*, a state-published newspaper inaugurated in 1869. It trumpeted, for example, the familiar lament of how for many years there had been "six hundred villages at the foot of the mountains from Urfa, Siverek, and Mardin all the way to Nusaybin."[131] The very same geography had been described by

[129] TNA-UK, FO 195/889, Taylor to Lyons, March 6, 1867.
[130] TNA-UK, FO 195/889, Taylor to Elliot, September 30, 1868.
[131] *Diyarbekir Gazetesi*, 28 Ağustos 1285 (September 9, 1869).

consuls as being devastated by locusts throughout the 1860s. Kurdish chiefs had similarly suggested how the locusts had pushed them to seeking government support for their settlements. But as in 1864, officials focused on people to blame for the environment, rather than environmental variations such as locusts or drought. By this familiar argument, the region had turned to "ruins" and "the main cause of the ruins were the Shammar Arabs" (*Şamar urbânı*). Government publications thus made clear who was the target of the reform efforts.

But they also signaled how this campaign was part of a movement around the world, both in time and in space. With respect to time, one article in *Diyarbekir Gazetesi* declared that in "the age of prosperity" (*mamuriyet*) such "ruins" as marred the Jazira had to be improved.[132] With respect to global currents, another article in the publication observed that in America and Europe, there were many places comparable to "our desert," yet, in contrast to the Ottoman case, they had been transformed thanks to government diligence.[133] The implication was that the same could occur in the plains south of Diyarbekir. And the newspaper emphasized signs that were cause for optimism. The organ did not speak of Ras al-Ayn in the way that someone like the British consul Taylor had. But its articles nevertheless noted that the fertile soil beside the Khabur River south of Ras al-Ayn had produced enormous watermelons weighing fifteen to twenty kiyye (forty-two to fifty-six pounds), a sign of "the extent to which the lands were fertile and productive."[134] Perhaps even more impressive than preposterously large cucurbits, the Chechen refugees – who had survived the ethnic cleansing on the empire's borders with Russia to be called savages and compared to locusts in the Ottoman Empire – had a coffeehouse. Its windows overlooked the Khabur River. In honor of such developments, Ras al-Ayn's mosque witnessed a prayer to Sultan Abdülaziz, who, it was declared, "revived the desert that had once been ruins."

FROM WANDERING GROUND TO SPECIAL ADMINISTRATIVE DISTRICT

Locusts had played a key role in developments for much of the previous decade, but environmental factors beyond the insects would shape the region's fate in the coming years. In the 1860s, locusts had devastated

[132] *Diyarbekir Gazetesi*, 28 Ağustos 1285 (September 9, 1869).
[133] *Diyarbekir Gazetesi*, 23 Teşrinievvel 1285 (November 4, 1869).
[134] *Diyarbekir Gazetesi*, 27 Teşrinisani 1285 (December 9, 1869).

agriculture, and cotton cultivation in particular during the global cotton shortage instigated by the American Civil War. In the process, locusts ensured that it made sense to remain a nomadic pastoralist in the Jazira. The insects had also chewed away at provisions necessary for refugee resettlement. They even proved so symbolically powerful that Chechen refugees had been compared to them. In 1870, they made yet another appearance in late February in Mardin.[135] Snowfall and a quick turn from cold to warm weather succeeded in wiping out the insects, a reminder that nonhuman factors figured as much into the absence of locusts as human effort. Yet the absence of locusts did not mean an easy season in the Jazira. Drought had afflicted the region beginning the previous winter, and many villages surrounding Aleppo did not even harvest enough to recover seed for the following year.[136] In Mosul, too, crops failed, with some 130 villages emptied, as their populations fled to the mountains where they worked vineyards and the famine crop of maize.[137] Meanwhile, in the deserts of the Jazira, the Shammar found their pastures withered and received special permission to migrate north into Diyarbekir province and the largely Kurdish region of Mount Karaca, so that they and their animals did not die of "hunger and thirst."[138]

It was in this context of drought and human mobility that Ottoman officials once again reconsidered the geographic frame through which they governed the Jazira. A memorandum described a situation in which the Shammar – like other nomadic groups in the empire – adeptly took advantage of all of the "provinces of Baghdad, Aleppo, and Diyarbekir that surrounded the country of the desert."[139] If the Shammar received punishment from any of these provinces, they would right away head for another province. No one was better at this, the memo noted, than the Shammar chief ʿAbd al-Karim, who had long been eyed by Ottoman officials as both a threat and – if reformed – a possible ally. His movement and that of others threatened the broader Ottoman plan for the region, too. The fact that Ras al-Ayn was directly on the edge of the desert – with the land between it and Nusaybin and Deir ez-Zor "empty of any improvement" – made it especially vulnerable to such depredations.

[135] *Diyarbekir Gazetesi*, 23 Nisan 1286 (5 May 1870), p. 3.

[136] CADC, 4CCC/34, Bertrand to Duc de Gramont, 12 August 1870.

[137] TNA-UK, FO 195/949, Rassam to Elliot, 13 June 1870.

[138] *Diyarbekir Gazetesi*, 2 Temmuz 1286 (July 14, 1870).

[139] BOA, İ.MMS 40/1654, Mehmed Rüşdi, 12 Şeval 1287 (January 5, 1871). On similar dynamics on the "internal Ottoman 'border' between Benghazi and Egypt," see Ellis, *Desert Borderland*, 123–128.

Thus the settlement that had been intended as a "wall" against the "desert" seemed close to transforming into part of the desert. All of these observations prompted Ottoman officials to note that "bringing the desert country under a unified administration would produce many benefits."

The plan offered a variation on the Ottoman translation of ʿAbd al-Karim's letter in 1864. That letter represented the way that Ottoman administration did not or could not take account of broader regional connections beyond Ottoman provinces. But the deliberations of 1871 pointed to the ways that Ottoman officials hoped to ameliorate this problem. The orders of the Ottoman grand vizier Mehmed Emin Ali to Arslan Pasha – he of the voracious tax collectors in Mardin – underscored this awareness.[140] The grand vizier began by explaining the area of administration that he had in mind for reform. The region was referred to as Zor, and because of its capital city of Deir on the Euphrates, it was sometimes referred to as Deir ez-Zor. In Ottoman usage, however, Zor typically referred to the district generally, whereas Deir or Deir ez-Zor referred to the city on the Euphrates. According to the grand vizier, Zor encompassed "the desert country between Baghdad, Diyarbekir, and Aleppo." The region extended "from the places close to Mosul," stretched from there to the Euphrates, and on another axis starting from "the places called Sinjar and Khabur and ending with the place named Ras al-Ayn in Diyarbekir." The extensive district would still not perfectly fit with the environment. Mehmed Emin Ali admitted that the desert went beyond the confines of Zor. But he still hoped that the new arrangement would help to bring about change in a region "empty of agriculture." In other words, the highest state official was convinced that the gap between environmental and political borders had to be closed.

As part of this plan to remedy the situation, the grand vizier called for Zor to be designated with a special status in the Ottoman provincial scheme. Like many before him, the grand vizier blamed the region's nomads for this state of affairs, accusing groups such as the Shammar and the ʿAnaza of "destroying traces of prosperity" and turning the area into a "wandering ground" (*cevelangah*). The way to fix it, Mehmed Emin Ali explained, would be to turn it into a "special administrative district" (*mutasarrıflık*; Figure 8). The possibility of this new administrative category stemmed from Midhat Pasha's work on the Vilayet Law, which offered a special status for some places, such that their governor reported

[140] BOA, ŞD 2434/69, Orders from Mehmed Emin Ali to Arslan Pasha, 3 Mart 1287 (March 15, 1871).

FIGURE 8 The special administrative district of Zor, 1871

directly to Istanbul and could allow for greater oversight on all decisions. Over the years, entities that obtained this status included Mount Lebanon, Jerusalem, Cyprus, and Cyrenaica.

At the same time as officials utilized this instrument to rethink administration of Zor, they also looked to reform administration of the eastern portions of the Shammar domains. There, Midhat Pasha went to work. He echoed the rationale put forward by other Ottoman officials about an unruly space ruined by nomads. In Midhat's reckoning, the region from Aleppo to Baghdad – including places such as Urfa, Siverek, Diyarbekir, Mardin, and Mosul – had once been home to "prosperous villages" but fell into "ruins" when their inhabitants either "fled" or "entered a state of savagery" by joining the nomads.[141] Other officials had described the space in terms of existing Ottoman provinces, but Midhat Pasha saw the region that crossed all of these lines, referring to the space in question as "the Jazira."

To achieve reform where others had failed, Midhat hoped to exploit the division of power between the half-brother leaders of the Shammar, ʿAbd al-Karim and Farhan. The Shammar may have frequently used Ottoman borders for their own interests, but Midhat observed that the half-brothers "do not cross the borders of each other."[142] While Farhan wintered near Baghdad and summered near Mosul, ʿAbd al-Karim moved around the Khabur and Zor more commonly. The fraternal borders posed an opportunity. So, too, did the hardship that the nomads had experienced over the previous year. The drought of 1870 had given way to the harsh winter of 1871, which had wiped out many of their animals and left the Shammar in a state of "despair." Midhat believed they might be enticed to settle. He proposed forming a district named "Shammar" near Mosul and allowed that in the future it might extend all the way to the Khabur River, which fell in the district of Zor in the lands of ʿAbd al-Karim to the west. An Arabic proclamation on the matter specified that the state would cede all uncultivated land on the right bank of the Tigris between Tikrit and Mosul to the Shammar.[143]

A REVOLT AGAINST BORDERS

With the Shammar in such a condition Ottoman officials hoped to align the borders of the state with those of the arid environment. In another memo to Arslan Pasha from the grand vizier, the goal was once again articulated as

[141] BOA, İ.DH 630/43847, Midhat Pasha to Grand Vizier, 3 Mart 1287 (March 15, 1871).
[142] BOA, İ.DH 630/43847, Midhat Pasha to Grand Vizier, 3 Mart 1287 (March 15, 1871).
[143] BOA, İ.DH 630/43847, To Farhan, 19 Zilhicce 1287 (March 12, 1871).

"an administration of the desert."[144] Zor had always been "by virtue of its place" the most sensible "center of a desert administration." But in addition to the already existing district of Zor (which had previously been part of Aleppo), officials added Ras al-Ayn and Nusaybin (previously in Diyarbekir), and Sinjar (previously in Baghdad). The proposed district was also to include a number of tribes previously affiliated with Aleppo. Ottoman officials clearly believed that borders were at the heart of their effort to transform the Jazira, to make its administrative borders look more like human mobility in its environment and, thus, easier to govern.

The Shammar understood the plans to be a threat. In the summer of 1871, a portion of the Shammar revolted. There is some dispute on what immediately precipitated the uprising, whether it was an insult by an official from Nusaybin of ʿAbd al-Karim or something else.[145] But as historian Oktay Karaman has suggested, the broader issue was ʿAbd al-Karim's anger at the "linking of Ras al-Ayn and Nusaybin to the district of Zor," and, he believed, the plot behind it to force him and other nomads of "the Jazira to build houses and work the land."[146] Such plans meant an end to the motion that had allowed ʿAbd al-Karim to carve out a space of autonomy on the edges of Ottoman administration.

As the revolt began, likely exaggerated reports of violence gave Ottoman officials the pretext they needed to paint the Shammar as the kind of savages they had long been presented as. Consular officials described widespread destruction in cultivated areas around Mardin and Nusaybin.[147] They took as much "wheat, barley, and objects of all sorts" that "their camels" could carry.[148] Midhat Pasha likened ʿAbd al-Karim to "a savage dog."[149] In his memoir, Midhat referred to the Shammar chief as Genghis Khan.[150] Whether an animal or a destructive scourge, ʿAbd al-Karim was – by Midhat's description – not only an enemy of the

[144] BOA, ŞD 2434/69, Grand Vizier to Arslan Pasha, 29 Haziran 1287 (July 11, 1871).

[145] TNA-UK, FO 195/939, Taylor to Elliot, August 6, 1871; CADN 166PO/D/1/66, Bertrand to Comte de Vogüe, July 29, 1871; 166PO/D/7/15, French Consul in Baghdad, July 19, 1871.

[146] Karaman, "Diyarbakır Valisi Hatunoğlu Kurt Ismail Paşa'nın Diyarbakır'daki Aşiretleri Islah ve İskan Çalışması (1868–1875)," 242; CADN, 166PO/D/54/5, Consul in Mosul to Comte de Montebello, July 29, 1871.

[147] CADN, 166PO/D/1/66, Bertrand to Comte de Vogüe, July 29, 1871; CADN, 166PO/D/54/5, Consul in Mosul to Comte de Montebello, July 29, 1871.

[148] CADN, 166PO/D/54/5, Consul in Mosul to Comte de Montebello, July 29, 1871.

[149] BOA, ŞD 2148/36, Governor of Baghdad Midhat Pasha to Grand Vizier, 8 Eylül 1287 (September 20, 1871).

[150] Midhat Paşa, *Tabsıra-ı İbret*, 111.

state but also an enemy of agriculture. Midhat claimed to have intercepted correspondence between ʿAbd al-Karim and other nomads attempting to incite a broader revolt against the Ottoman administration, which, the Shammar chief insisted, was intent on "demolishing the tent [*bayt al-shaʿr*]" that was symbolic of their nomadic lifestyle.[151] Midhat also evidently wanted to use the conflict to ensure that "no hope remained" for the Shammar "to live in the country of the Jazira."[152]

Agriculture and the Jazira were central to Midhat Pasha's actions, as they were to ʿAbd al-Karim's brother Farhan's decision to stand down during the revolt. Midhat offered amnesty to ʿAbd al-Karim's followers, on the condition that they join Farhan and till the earth.[153] British consular reports suggested that Farhan apologized for his brother's revolt and sought state support for his own control of the Shammar "and the country they inhabit," specifically outside the jurisdiction of the provinces of Baghdad and Diyarbekir.[154] Though the Ottoman side of this correspondence is unclear, Farhan's request would not have been unreasonable, not only given the previous arrangement for the Shammar to settle along the Tigris but also in light of various local actors' – nomadic or otherwise – ability to turn imperial reform in the direction of their particular interests.[155] Moreover, Farhan's proposal pointed to a more complicated use of borders by the Shammar than simply as a portent of oppression or means of escape. For Farhan, the changing borders of the Tanzimat presented an opportunity of a different kind: becoming an Ottoman official.

While presented as illogical avatars of violence and destruction by European observers and Ottoman state officials alike, ʿAbd al-Karim and the Shammar actually employed much the same tactics that they had in the past. They utilized both state borders and the Jazira environment as part of their challenge to the Ottoman effort to make these units align. According to the French consul in Baghdad, the Shammar had actually taken care to attack only the lands of Mosul and Baghdad rather than Urfa and Diyarbekir so as to "show that it is not the authority of the Sultan against

151 BOA, ŞD 2148/36, Governor of Baghdad Midhat Pasha to Grand Vizier, 8 Eylül 1287 (September 20, 1871).

152 Ibid.

153 BOA, ŞD 2148/36, Arabic language proclamation of Governorate of Baghdad to the People, undated.

154 TNA-UK, FO 195/939, Taylor to Elliot, August 6, 1871.

155 Ceylan, *The Ottoman Origins of Modern Iraq*, 142; Fattah and Badem, "The Sultan and the Rebel," 680; Petrov, "Everyday Forms of Compliance"; Rogan, *Frontiers of the State in the Late Ottoman Empire*, 78–81; Saraçoğlu, *Nineteenth-Century Local Governance in Ottoman Bulgaria*, 56.

whom they make war, but against Midhat Pasha."[156] In addition to being conscious of Ottoman borders, the Shammar also adeptly used the environment. Initially, the Ottoman governor of Diyarbekir – the famed Kurt ("Wolf") Pasha – had pursued ʿAbd al-Karim, but the Shammar slipped away into the desert.[157] Left with a disappearing enemy in a harsh environment, Kurt Pasha attacked any nomads he encountered and duly reported his strikes against unaffiliated groups as victories against the Shammar. In reality, however, his troops not only failed to defeat the Shammar in battle, but also even failed to face them. As of early September, according to one report, Kurt Pasha's troops remained sequestered south of Mosul where they were "suffering seriously from deprivations and illnesses" thanks to the late-summer heat.[158] Meanwhile, ʿAbd al-Karim remained to the west between Sinjar and the Khabur, safe in the desert.

But the desert only provided a temporary shelter. The Shammar may have known how to survive in the Jazira more effectively than government troops, but their life in the Jazira had never been one of isolation. The Shammar and Midhat Pasha alike both called upon clear divisions between nomadic and settled life as part of the conflict. But such distinctions had always obfuscated the ways that both groups were interconnected. The so-called civilized cities and villages relied on nomads for animal products, while the so-called uncivilized nomads relied on cities and villages for wheat, barley, and other provisions. ʿAbd al-Karim brought his forces out of the desert and to the southeast toward Baghdad in search of supplies. Then, seeking environmental refuge once again, he crossed the Euphrates and headed for the desert and what he assumed to be the safe confines of the Muntafiq nomadic group. He was mistaken. The Muntafiq promptly captured him and handed the Shammar chief to Midhat Pasha, further evidence of how the civilizational conflict presented between nomads and the state was not so simple. It was the reliance of the Shammar on the world outside of the desert for supplies that prompted ʿAbd al-Karim to leave more familiar territory, and it was connections between the Muntafiq nomads and Midhat Pasha that ensured ʿAbd al-Karim's capture.

A Baghdad court sentenced him to death, and he was transported to Mosul.[159] He had built his power by taking advantage of the disjuncture

[156] CADN, 166PO/D/7/15, July 19, 1871.
[157] CADC, 4CCC/35, Bertrand to Favre, August 19, 1871.
[158] CADC, 4CCC/35, Bertrand to Favre, September 5, 1871.
[159] Williamson, "A Political History of the Shammar," 115.

between the environment of the Jazira and the provincial borders that divided it up. He had been called a "savage," his actions had been said to turn the land into something like "Arab hair." His movement required cordons to stop, as if it were a disease, and, ultimately, a special administrative district with borders meant to encompass the entirety of the desert, denigrated by Ottoman officials as wasteland. But ʿAbd al-Karim knew better. The formation of Zor carried with it a plan to make him settle. And wasteland was a misnomer, if only one knew where to look for the fresh green grasses of springtime, or the white chamomile flowers of June. If he looked west from Mosul, he would have seen the ruins of Nineveh, the immense mounds of the city of antiquity whose excavation prompted the Ottomans to imagine transforming the Jazira. Maybe in the distance beyond, ʿAbd al-Karim could have seen the grasses of the Jazira expanses that in stretching across provincial boundaries, had for so long offered him opportunities for refuge as one of the "Sultans of the Open Lands." But these grasses would not save him that autumn. He was executed on the bridge over the Tigris.

Although the revolt involved a blurring of distinctions between nomadic and settled, state and savage, Midhat Pasha and his allies celebrated the execution of ʿAbd al-Karim as an unambiguous victory of settled agriculture and state control over an unruly environment. Conflicts between states and nomadic groups all around the world at this time took on a distinctly agrarian character. In the United States, seizure of Native American land was used to seed the endowments of the country's land-grant universities, whose research on agriculture would radically transform the land of which Indigenous nations had been dispossessed.[160] A similar dynamic occurred in the Jazira, though on a different scale. Midhat sold ʿAbd al-Karim's camels and purchased "agricultural implements" with the proceeds.[161] The Diyarbekir provincial yearbook carried a similar message of stark contrast between nomadic and settled, suggesting that most of the Shammar shaykhs had never even seen a house before, and now their "savageness had been annihilated."[162] Midhat also took the opportunity to display power across provincial borders. He boasted that the Shammar had underestimated Ottoman

[160] Robert Lee and Tristan Ahtone, "Land-Grab Universities," *High Country News*, March 30, 2020.
[161] BOA, İ.DH 639/44468, Baghdad Governor Midhat Pasha to Grand Vizier, 20 Eylül 1287 (October 2, 1871).
[162] *Diyarbekir Salnamesi* 1288, 187.

forces, believing that – as in the past – provincial borders and climate would act as their ally. Not so in 1871, Midhat insisted, noting that not only had Kurt Pasha of Diyarbekir pursued ʿAbd al-Karim across "the border of Baghdad," but so too had troops in Mosul chased the nomads "into the desert ... in the month of July."[163] Kurt Pasha happily contributed to this narrative of environmental mastery. In direct contradiction of reports of his soldiers' struggles, he declared that troops all across "the desert region" had united against the Shammar and had surmounted not only the nomadic threat but also the environmental challenge of "poisonous air."[164]

For years, the Shammar had taken advantage of the Jazira's status stretching across Ottoman provinces, but Midhat declared that the region would no longer provide refuge. In a rejoinder to his promises of a Shammar district, Midhat crowed that "the name of the Shammar would not remain" and "there was no hope" for "the bandit" ʿAbd al-Karim "to live ... in the Jazira."[165] Nor would any nomads live in the Jazira, Midhat declared. If the nomads needed land, they could be content with the Shamiyya desert, southwest of the Euphrates. In that way he hoped the "country of the Jazira would obtain the previous degree of prosperity" that it had once possessed.

But Midhat Pasha's promises of agricultural reawakening would come up short in the wake of the conflict. Farhan Pasha – the brother of ʿAbd al-Karim, who did not revolt – was remunerated handsomely for standing down. Midhat arranged to allocate to the chief a salary of 5,000 kuruş and another 20,000 at the disposal of an administrative council.[166] Ottoman authorities even set up the Shammar chief in a castle at Shirqat on the Tigris, with some 170 water pumps operating within the first year.[167] Yet from the very beginning, there were cracks in the edifice of the agricultural foundation Midhat attempted to construct. Almost all of those who had settled with Farhan at Shirqat were gone by January 1872 because their camels "needed a peculiar pasture" that they could not find in the

[163] BOA, ŞD 2148/36, Baghdad Governor Midhat Pasha to Grand Vizier, 8 Eylül 1287 (September 20, 1871).
[164] BOA, İ.DH 642/44656, Diyarbekir Governor İsmail Hakkı, 12 Teşrinievvel 1287 (October 24, 1871).
[165] BOA, ŞD 2148/36, Baghdad Governor Midhat Pasha to Grand Vizier, 8 Eylül 1287 (September 20, 1871).
[166] BOA, İ.ŞD 24/1057, Council of State Decree, 18 Cemazeyilahir 1289 (August 22, 1872).
[167] Midhat Pasha, *Tabsıra*, 113; TNA-UK, FO 195/1479, Howden to Richards, June 15, 1884.

immediate vicinity.[168] Midhat Pasha envisioned Shirqat as a way of transforming the lives of the Shammar morally and materially. But they treated the settlement of Shirqat not unlike their other seasonal sites of residence, convenient only insofar as it offered the resources that their animal property required to survive. Moreover, in a mark of the itinerant nature of Ottoman officials as well as the nomads they were charged with managing, Midhat Pasha would not stay either. He was bound for bigger things. In 1872, he took up the position of grand vizier of the empire.

Although the Shammar of Farhan might not have remained in one place in exactly the way authorities wished, Ottoman provincial officials happily hailed the new borders of the Jazira that had prompted ʿAbd al-Karim's revolt. Arslan Pasha had died of poisoning shortly after clashes elsewhere in Zor with the ʿAnaza.[169] His successor Ömer Şevki described how the absence of the "chief bandit" ʿAbd al-Karim had changed the region, giving the administration the ability to entice other nomads to settle, as had occurred in the case of some 2,000 households of the Arabic-speaking Baggara on the Euphrates.[170] He would later boast that lands that had once been home to nothing but "birds and beasts" on the Khabur were becoming home to nomads keen on taking advantage of "fertile and bountiful" soil on the riverbanks.[171] Ömer Şevki even called for Zor to be transformed from a special administrative district (*mutasarrıflık*) to a full province (*vilayet*) in its own right.[172] Officials in Nusaybin and Ras al-Ayn – districts that had been lopped off of other provinces in order to form Zor – also praised their connection to the district formed in 1871.[173] After nearly a decade of considering how to reform the region of the Jazira, it seemed that Ottoman officials had achieved something.[174] They had aligned the scales through which they governed with the environment, or at least made them closer to one another. Provincial reorganization provided fertile ground

168 TNA-UK, FO 195/949, Rassam to Elliot, January 4, 1872.

169 Çiçek, *Negotiating Empire*, 118.

170 BOA, ŞD 2213/19, Military Detachment Commander and District Governor of Zor Ömer Şevki, 11 Kanunusani 1287 (January 23, 1872).

171 BOA, ŞD 2213/29, Ömer Şevki to Grand Vizier, 13 Recep 1289 (September 16, 1872).

172 BOA, ŞD 2149/17, Ömer Şevki to Grand Vizier, 4 Şaban 1289 (October 7, 1872).

173 BOA, ŞD 2213/38, Nusaybin District Report, 13 Kanunusani 1288 (January 25, 1873); Ras al-Ayn District Report, 24 Zilkade 1289 (January 23, 1873).

174 Not everyone agreed. Damascus governor Suphi Pasha, for example, complained that no one had ever settled in Zor and, echoing many, added that when "soldiers went to the east side of the district, [nomads] went to the west side, and when [soldiers] headed to the west side [nomads] went to the east." He also called for Zor to be attached to Damascus province because of nomadic pastoralists from Zor settling around Hama. BOA, ŞD 2270/15, Abdüllatif Suphi to Grand Vizier, 5 Recep 1289 (September 8, 1872).

for high hopes. In 1873, the administrative council of Zor wrote that their region once could not be "crossed from any side without a military detachment."[175] But recent developments had made it "close to being compared to Cairo." Unspoken in all of these pronouncements about provincial borders and agricultural improvement was the absence of locusts following their ravages in the 1860s.

THE FALL AND RISE OF ZOR

Yet even after a revolt over borders, and even after pronouncements praising these new borders, the status of the Jazira was not stable. And border-crossing nomads were not always to blame. In fact, sometimes at fault were the state officials whose administration was supposed to transform the region. For example, a number of disputes arose between the province of Aleppo and the special administrative district of Zor over where precisely the border between the entities existed, and where Zor could collect taxes on sheep in 1874. Because of the "border chaos" (*hudut karışıklığı*) – as it was termed – the treasury lost a significant amount of revenue on sheep, with people claiming the tax had been collected by those on the other side of the border as a means of tax evasion.[176] Such disputes pointed to the difficulty of dividing connected geographies and accounting for itinerant herds. Eventually, due to the malfeasance of local officials, the administration of Zor returned to Aleppo, with the special administrative district dissolved and the land relegated to the larger province's authority. Rather than functioning as an independent special administrative district, the region became part of the chain of command based in the large city far to its west.[177] Meanwhile, parts of the region that had once constituted Zor were returned to neighboring entities, with Nusaybin reattached to Diyarbekir, and Sinjar to Mosul.

The figure who had attempted to transform the Jazira – Midhat Pasha – continued to shape the empire as it entered a period of tumult. In 1875, the empire defaulted on loan repayments, which would eventually lead to European creditors having considerable power over the empire's finances.

[175] BOA, İ.MMS 46/1973, Administrative Council of Zor to Grand Vizier, 20 Mayıs 1289 (June 1, 1873).

[176] BOA, ŞD 2214/20, Mehmed Reşid to Grand Vizier, 28 Eylül 1290 (October 10, 1874).

[177] Al-Shahin, *Ahdath Khalida fi Tarikh Dayr al-Zur*, 23; BOA, A.}MKT.MHM 481/33, Meclis of Zor to Grand Vizier, 11 Teşrinevvel 1292 (October 23, 1876); ŞD 2219/3 Mehmed Reşid to Şura-yı Devlet, 11 Haziran 1300 (June 23, 1884).

Subsequently, two sultans were deposed, first Abdülaziz in favor of his son Murad (shortly after which Abdülaziz committed suicide), and then Murad on the claim of insanity in favor of his brother Abdülhamid II, the sultan who would define some of the empire's final decades. In the aftermath, Midhat Pasha served as grand vizier. In December of 1876, he managed to secure approval for an Ottoman constitution and parliament. The adoption of these measures was accompanied by shouts in the streets of "Long live the Sultan and Midhat!"[178] But the optimism did not last.

Midhat's power and the dream of a unified desert administration further receded in 1877 when the Ottoman Empire went to war with the Russian Empire as a result of simmering tensions in the Balkans. The squeeze on resources meant that the Ottomans could no longer maintain the military stations up and down the Euphrates that they had once hoped would guarantee the region's transformation.[179] Gone was the mule-mounted cavalry that had been "the terror of the Bedouins" and had vitiated their ability to "vanish into the desert at first sign" of soldiers.[180] What was more, there was no telegraph line between Deir ez-Zor and Aleppo. Given that the distance between the places was some eight days, the delays in receiving orders from Aleppo were consequently "interminable."[181]

While the Russo-Ottoman War (1877–1878) affected Zor, the conflict's results would also greatly shape the future of the empire. Abdülhamid used the conflict as justification for dissolving parliament and suspending the constitution that Midhat Pasha had helped to create. Meanwhile, the Treaty of Berlin, which concluded the war, transformed the empire's land and people. Having lost Bulgaria, Bosnia and Herzegovina, Romania, and Serbia and Montenegro in southeastern Europe as well as Kars, Batum, and Ardahan in eastern Anatolia, the empire not only had less land, but also fewer Christians. By one estimate, the losses amounted to "two-fifths of its entire territory and one-fifth of its population."[182] At the same time, the empire was obligated to offer special protections to Armenians in eastern Anatolia, which would come to be a frequent source of friction between outside powers, local people,

[178] Saliba, "The Achievements of Midhat Pasha," 317.
[179] Ababsa-al-Husseini, "Mise en valeur agricole et contrôle politique de la vallée de l'euphrate," 463.
[180] CADC, 4CCC/36, Consul in Aleppo to Saint Hilaire, September 12, 1881.
[181] TNA-UK, FO 424/123, Earl of Dufferin to Earl of Granville, July 29, 1881, Inclosure: Report by Captain Stewart on the Deir Sandjak and on Some of the Neighboring Districts, July 14, 1881.
[182] Shaw and Shaw, *Reform, Revolution, and Republic*, 191.

and the Ottoman state in the coming decades. In response to these and other measures, Abdülhamid oriented his rule increasingly toward the Muslim populace of the multiethnic empire while also aiming to maximize its resources.

In the Jazira, the Shammar maintained the opportunism they had practiced ever since 1871. The British traveler Lady Anne Blunt visited the tribe's branches in 1878. Like British officials in the future, she seemed to project her own anxieties about the demise of the English countryside onto the local landscape.[183] Regarding Farhan's settlement at Shirqat, she wrote, "Of all the wretched places, this, I think, is the wretchedest."[184] She decried the state of the environment, noting how "every blade of grass has been eaten down, and every inch of ground trampled and bemired for miles round."[185] Farhan's own quarters were "on the side of a bare heap of refuse, one of the mounds of Sherghat." In other words, Farhan's settlement was perched atop one of the region's distinctive *tall*, the archaeological sites that bespoke the region's potential for glory. By Blunt's accounting, however, Farhan's settlement could not be further away from the flourishing past. For all these reasons, Lady Anne Blunt declared that unlike Farhan, "I would not give up life in the desert" and "neither would I condescend to handle a spade, even in make-believe."[186] As a member of the English nobility – Lord Byron's granddaughter, in fact – Blunt could feel comfortable in such a pronouncement.[187]

But in her posturing Lady Anne Blunt may well have missed what was actually going on at Shirqat. It is telling that Farhan was not even around when she visited, suggesting that he had not entirely given up his life in the desert, as it were. Shirqat may have appeared grim, but it was also an opportunity for Farhan to obtain rent from the state. Government stipends and designated settlements did not bring about an existential change in virtue like reformers hoped. Rather, they constituted yet another strategy for gathering resources in the region. When locusts struck, one might move to another pasture. When drought struck and pastures disappeared, one might rely on a government subsidy. And in doing so, Farhan moreover managed to maintain connections much further afield. Part of Blunt's frustration at not meeting Farhan derived from the package that she had been entrusted to deliver to him: a basket of oranges and pomegranates from the Nawab of Awadh in Baghdad, who considered Farhan a "brother."[188]

[183] Dodge, *Inventing Iraq*, 69. [184] Blunt, *Bedouin Tribes of the Euphrates*, 188.
[185] Ibid., 187. [186] Ibid., 188. [187] Satia, *Spies in Arabia*, 63.
[188] Blunt, *Bedouin Tribes of the Euphrates*, 159–162.

When Blunt visited the other branch of the Shammar, she described matters in environmentally distinct ways. At the head of the other branch of the Shammar was ʿAbd al-Karim's successor and brother Faris, who had fled the Jazira with his and ʿAbd al-Karim's mother ʿAmsha in the wake of ʿAbd al-Karim's defeat, only to return in 1875. Blunt drew a stark contrast between the Shammar of Farhan and the Shammar of Faris, symbolized, it would seem, by the different environments that they inhabited. Farhan's camp had been a site of ruin. In contrast, Blunt described the area around Faris's camp as "white as snow with chamomile in full flower" and "new-born camels which every here and there peeped out of the herbage."[189] The verdant environment mapped onto Faris's virtue. In Blunt's terms, "a better-bred man would be difficult to find."[190] Faris moreover endeared himself to Blunt by insulting Farhan as "not a Bedouin at all" but rather "a mere fellah," or peasant.[191] Yet for all of Blunt's esteem for Faris, it was not he whom she deemed "the most important personage."[192] Rather it was ʿAmsha, "a sort of holy personage, and object of veneration with all the tribes of Northern Mesopotamia." In honor of her deceased son, she was still referred to among the tribe as "Mother of Abd ul Kerim."

But as pure as the encampment of Faris and ʿAmsha may have seemed to Blunt, the Shammar were not free of state influence. Indeed, the division between the Shammar fundamentally reflected the campaign of 1871, with Faris in Zor and Farhan in Mosul. In fact, Blunt herself described the border between the groups as being "the heart of Mesopotamia," which she meant specifically in cartographic terms, seeing as the dividing line was located "just at the top of the second O in our map."[193] In addition to both drawing salaries from the Ottoman state, Farhan and Faris were both involved in revenue collection. In what historian Talha Çiçek calls a "re-manifestation of the old tax farming system," the state compelled nomadic leaders – including the Shammar – to collect taxes in return for a share of the proceeds.[194] Among the Shammar, a racial division of labor took place, as enslaved and formerly enslaved Black men began to occupy new roles as tax assessors and collectors for the Shammar all across the Jazira.[195]

[189] Ibid., 227. [190] Ibid., 228. [191] Ibid., 231. [192] Ibid., 228. [193] Ibid., 218.
[194] Çiçek, *Negotiating Empire*, 222.
[195] Jwaideh and Cox, "The Black Slaves of Turkish Arabia during the 19th Century," 48.

Both Midhat and Blunt presented nomadism as more or less a question of virtue, a reflection of level of civilization. Midhat denigrated it, and Blunt praised it. Lost in these declarations was the pragmatism of these approaches. This value appeared in the winter of 1879 and 1880, when a horrible cold and devastating famine set in across the Jazira and beyond.[196] In these difficult circumstances, a German traveler named Sachau spoke of the benefits of the tent. Sachau pointed out how tents were actually quite "justified," since they were "much better suited than the house" for managing heat and avoiding vermin.[197] More immediately, in the midst of the cold winter, they allowed nomads to move, leaving, for example, the Khabur valley a "deserted, snow-covered desert."[198] The flexibility of the tent perhaps also helped to enable the kind of bravado Sachau encountered when he met Faris. The chief may have been more or less resident within the confines of the district of Zor, but upon welcoming Sachau, he gestured toward a broader space, telling the German that he was welcome in all of the Shammar domains, stretching from Mosul to Mardin to Urfa in the north and Raqqa to ʿAna to Hit in the south.[199] Sachau duly described Faris as "the desert king" (*Wüsten-König*).[200] The provincial borders pointed to change, but Faris's boast and environmentally specific title offered a reminder of how "the memory of ʿAbd al-Karim is very alive in the desert of Mesopotamia," Sachau wrote.[201]

The spring of 1880 witnessed reports of great suffering from the famine of the winter months. When the English artist Tristram J. Ellis passed through Shirqat on his way down the Tigris, he inspected the ancient ruins beside the erstwhile campground of Farhan. Inside, he found "dried locusts" that smelled "horribly."[202] The leftovers of invasions from nearly a decade before thus scented the spring in some places. Yet it was not the powerful insects that had shaped the region's suffering this time around, but rather unexpected cold. The historian ʿAbbas ʿAzzawi recalled how "the dead were witnessed in the streets, and girls and boys were sold" in Mosul and Baghdad.[203] But the bounds of suffering went much further. Bread riots emerged in Aleppo in March.[204] Outside of Aleppo, deaths among sheep were great, too. One traveler described how outside of the city he found "their skeletons at frequent intervals."[205]

196 Ertem, "Eating the Last Seed"; Ghazarian, "Ghost Rations."
197 Sachau, *In Syrien*, 265. 198 Ibid., 273. 199 Ibid., 266. 200 Ibid., 267.
201 Ibid., 303. 202 Ellis, *On a Raft and Through the Desert*, 117.
203 Al-ʿAzzawi, *Tarikh al-ʿIraq bayn al-Ihtilalayn*, vol. 8, 49.
204 TNA-UK, FO 195/1305, Aleppo Consulate to Henry Layard, March 10, 1880.
205 Ellis, *On a Raft and Through the Desert*, 40.

There were some 2,500 beggars in Diyarbekir out of a total population of 25,000 in the city.[206]

Reports from the Jazira's deserts underscored great suffering there, too, though it seems the flexibility of nomadism may have softened the edge to some extent. In May, one official suggested that "between Mosul and Mardin ... more than one half the sheep ... had perished during the past winter and spring."[207] The animals died in most cases from lack of pasture, but because they could ultimately be eaten, they also afforded their owners sustenance of last resort (rebuilding herds, of course, was a different matter). Across the Jazira, people sought what little food the land provided, with some searching for truffles in the desert and others "living on the wild thistle."[208] It seems people became nomads out of want. As one British official wrote, "From Aleppo to Diarbekir, from Diarbekir to Mosul ... it is the same everywhere ... ruined bridges, ruined barracks, ruined villages and towns, and a decreasing permanent population, a nomad one increasing at any rate in the proportion of its numbers." In response to famine, the geography of the Jazira and the significance of motion within it once again became apparent.

Ottoman officials in the region followed the famine by reverting to the idea of the special administrative district of Zor as a way of governing the region more effectively. In April of 1880, the local council of Zor called for their region to return to how it was "during the time of ... Arslan Pasha," the local official notorious for his tax collection practices.[209] The council resorted to a familiar argument. They called Zor the "natural" place for ruling what amounted to "the midpoint of the deserts of four provinces" (*dört vilayet çölleri mutevassıtı*). By bringing the edges together, the council argued, the Ottoman state could better protect the cultivated regions of those four provinces from nomadic depredations. In June of 1880, the grand vizier echoed these sentiments.[210] He explained that there had been no "benefit" from the attachment of Zor to Aleppo. Exacerbating the shift in borders was the lack of troops. He predicted that "returning" Zor's "independence" (*istiklaliyet*) and outfitting it with the proper number of troops would ensure "the gradual frightening of the

[206] CADC, 60CPC/6, Aleppo to Ministry of Foreign Affairs, March 12, 1880.
[207] TNA-UK, FO 195/1316, Trotter to Layard, May 3, 1880.
[208] TNA-UK, FO 195/1316, Trotter to Layard, March 22, 1880; TNA-UK, FO 195/1316, Trotter to Layard, May 3, 1880.
[209] BOA, ŞD 2434/69, Council of Zor to Grand Vizier, 3 Nisan 1296 (April 15, 1880).
[210] BOA, ŞD 2434/69, Mehmed Seyyid to Interior Ministry, 14 Haziran 1296 (June 26, 1880).

Arabs [*urbân*] into settlement." And so in 1880, Zor returned to its independent status, once again distinct from Aleppo province. European officials such as the British consul in Aleppo applauded the changes, declaring that "at present the greatest confusion is produced by the Arabs passing from the jurisdiction of one Governor General to that of another."[211]

The return to independent administration of Zor meant a return to borders that – crafted with the intention of containing nomadic migration and the desert – surprised some. A traveler by the name of Captain Stewart journeyed through the region and acknowledged that the reestablishment of Zor might be considered "absurd" given its "huge and unwieldy" proportions.[212] By design, it was close to almost everywhere: "within 21 miles of Mardin, 18 miles of Urfa, 30 miles of Aleppo, 36 miles of Hama, 72 miles of Damascus ... and to within 12 miles of the Sinjar Dagh." The strange proportions, though, were in deference to the region's unique political ecology. "The so-called desert is singularly rich," he wrote, and while one encountered plenty of "deserted villages," at certain places too, as "far as the eye can reach the country appears alive with herds of camel and sheep." In these dynamics, Captain Stewart understood the goal as so many others from Mustafa to Midhat to Arslan had explained it: "to pull all the districts through which the Bedouin wanders under one command."

Whatever high hopes were attached to the new borders, the Shammar continued to move in ways that contradicted the plans that people like Midhat Pasha and Cevdet Pasha had violently attempted to imprint on the map nearly a decade before. As the French consul in Aleppo observed, "the nomads who occupy the uncultivated plains of Mesopotamia are all in motion" and constituted "the plague [*le fléau*] of the country."[213] Later that same year he worried that if nothing were done, other nomads would follow "the contagious example" of the Shammar.[214] Altogether, he predicted that new borders would do little unless they included provisions for more troops in the region.

Left out of the pronouncements about the destructive nature of the nomads, however, was the devastating impact of the famine on them, and

[211] TNA-UK, FO 78/3128, Henderson to Granville, November 3, 1880.

[212] TNA-UK, FO 424/123, Earl of Dufferin to Granville, July 29, 1881, Inclosure: Report by Captain Stewart on the Deir Sandjak and on Some of the Neighboring Districts, July 14, 1881.

[213] CADC, 4CCC/36, Aleppo to Ministry of Foreign Affairs, June 6, 1881.

[214] CADC, 4CCC/36, Aleppo to Ministry of Foreign Affairs, August 1, 1881.

how the loss of so many animals in 1879 and 1880 had left them needing to recoup their losses in any way they could. Similar dynamics could be observed later in the year, when the British consul in Baghdad journeyed to the north. His visit was not only an occasion to comment on the "extraordinary perversity" of the Ottoman failure to develop the Tigris.[215] He also described Shirqat – long to be the centerpiece of Shammar settlement, complete with water pumps and a castle for Farhan – as merely "the principal summer encampment of the Shammar" and home to some "1,000 black tents." As he returned south, he claimed that the question of every cultivator for him was "Where are the Shammar?" He understood the comment as demonstrative of the "dread of them" that was "real and universal." Their unknown location may have frightened local residents (or, indeed, the British consul), but it was movement that likely allowed them to survive the famine. The harsh winter no doubt reminded them why Midhat's schemes on the Tigris were a threat to them perhaps even more than the Shammar were a threat to settled agriculture.

At the same time, events in the empire ensured that two of the key forces in the history of the Jazira would have their paths intersect once more. In 1881, Abdülhamid II ordered his minister of justice to arrest the governor of Aydın for the murder of Sultan Abdülaziz in 1876, even though the deposed sultan's death had been deemed a suicide; the minister charged with the arrest was Cevdet Pasha, and the governor to be arrested was Midhat Pasha.[216] The two men had seen the parts of the empire where their families hailed from lost in war, its populations torn apart by violence that brought new forms of sectarianism into being. They had served in numerous capacities all around the empire to prevent further losses through reform. It was not enough for Midhat, whom Abdülhamid deemed sufficiently a threat to deserve the trumped-up charges he faced. Cevdet voted for his execution.[217] Abdülhamid commuted the sentence and exiled Midhat and the other convicts to Ta'if in the Hijaz, where Midhat would be murdered three years later.

While Midhat found his life in question, the issues of commerce and governance that had consumed him were still very much up for debate in the Jazira. In 1882, a plan from Zor's local commander predicted that the establishment of "cordons" and the stationing of "regular troops mounted on mules" would restore security to the region that had so many ancient

[215] TNA-UK, FO 195/1409, British Resident in Baghdad to Grand Sec. of India, Calcutta, November 24, 1881.

[216] Finkel, *Osman's Dream*, 500. [217] Ibid., 501.

witnesses to its glory.[218] The plan echoed those of many over the years, but also spoke to the new urgency of collecting tax revenue in the wake of the Russo-Ottoman War, apparent in Libya as well.[219] The members of the Special Council (*Meclis-i Mahsus*) echoed these sentiments, calling for a return to the times of Arslan Pasha, whose accomplishments had turned to "ruins" thanks to the district's attachment to Aleppo again.[220]

Meanwhile, the Shammar continued to leverage their place on the edge into a kind of power. The French consul in Mosul wrote in the spring of 1883 how the Shammar found themselves in the "desert situated between the Mesopotamian part of the province of Mosul and its neighbors."[221] Because livestock traders in Mosul could only find pasture for their animals "in the desert" for part of the year, they had to negotiate with the Shammar. Farhan had been collecting taxes on them, and when government forces threatened to challenge him or other Shammar, the French consul noted that the nomads ably used the environment: "they flee into their deserts, where it is impossible to follow." An expedition against them later in the summer was judged to be "very mediocre."[222] The state apparently secured some restitution for sheep thefts, but it paled in comparison to the sheep possessed by the Shammar, estimated at some 150,000, which they paid very little tax on. The Shammar mastery of the environment on the edge of both provinces and cultivation enabled them to maintain some measure of power for themselves.

In the following years, the disjuncture between provincial borders, nomadic migration, and the environment persistently vexed Ottoman officials. The shari'a court judge of Zor wrote in 1884 of the beauty of the land.[223] Like so many others, he praised its promise and mourned the remnants of past civilizations strewn across the region. But he also added that the ruins were not necessarily of ancient vintage. He complained, instead, of how the Khabur had been settled and consisted of some thirty villages as recently as a decade before, but the motion of the Shammar had upended these settlements. Further north, officials of Diyarbekir protested that efforts to register people of the Kara Keçe nomadic group near

[218] BOA, İ.MMS 71/3295, District Governor of Zor, 15 Nisan 1298 (April 27, 1882).
[219] Çiçek, *Negotiating Empire*, 203; Ellis, *Desert Borderland*, 121.
[220] BOA, İ.MMS 71/3295, Meclis-i Mahsus, 6 Haziran 1298 (June 18, 1882).
[221] CADN, 166PO/D/54/7, Siouffi to Marquis de Noailles, May 24, 1883.
[222] CADN, 166PO/D/54/7, Siouffi to Marquis de Noailles, July 5, 1883.
[223] BOA, Y.EE 11/11, Zor shari'a court judge to Yıldız Palace, 21 Cumayziülevvel 1301 (March 19, 1884).

Siverek had prompted the people to flee into the "desert."[224] They were not simply fleeing into the desert; they were also fleeing across the provincial border into Viranşehir, a district of Zor. Viranşehir was some seventy hours from the district capital at Deir, while only ten from Diyarbekir. Yet it had been included in Zor because of the decades-long effort to encompass the desert and nomadic groups in a single provincial administration. As flight of people into other districts underscored, the dream of drawing borders precisely around the environment and thereby more effectively managing the region's population remained a fraught one, not least because deciding where the desert ended was challenging, if not impossible. Officials hoped borders would enable them to control the region, but as the Kara Keçe demonstrated, the same borders could be used for very different purposes, indeed, dividing and ruling the empire itself.

THE DESERT PROVINCE

Locust swarms were missing from the Jazira for most of the 1870s and early 1880s. But by the mid-1880s, the insects were back. And the dilemma of how to control them remained much the same as it had been in previous years, as locusts flew across provincial borders and emerged from places without settled populations to devour cultivation. In March of 1885, an archaeological expedition under the leadership of the American clergyman and scholar William Hayes Ward reached ʿAna on the Euphrates as part of the hunt for remnants of the ancient Assyrian Empire in the region. When the group tried to call on the Ottoman district governor, they could not find him. He was "gone with soldiers and people to kill locusts."[225] In March of 1886 when the locusts began to hatch, several thousand men together with the governor of Mosul journeyed an hour away from their city in order "to hunt down the larvae."[226] Yet however many locusts they destroyed, "it was nothing in comparison to how many remained." By mid-March, the locusts had wings, and there was little one could do to protect the fields. This did not mean humans could do nothing with regard to locusts. The French vice consul in Mosul, for example, seized the opportunity to study the insects. In the name of

[224] BOA, ŞD 1461/5, Council of Diyarbekir to Interior Ministry, 27 Teşrinisani 1301 (December 9, 1885).

[225] Peters, *Nippur*, vol. 1, appendix F: A Portion of the Diary of William Hayes Ward, Director of the Wolfe Expedition to Babylonia (1884–1885), 361.

[226] CADN, 166PO/D/54/7, Siouffi to Comte de Montebello, "Recueil d'observations sur les sauterelles, lors de leur passage à Mossoul, en 1886," October 2, 1886.

science, he even timed their periods of copulation (his conclusion: more than eighteen hours and forty-five minutes based on the nine couples and one "polyandrous threesome" he captured in bottles in his courtyard and observed).[227]

But for those with less time on their hands, the locusts presented a frightful spatial dilemma. Officials in Urfa, for example, sounded much like their colleagues from previous decades, as they blamed the locust invasion on the ineffectiveness of locust-control measures in other provinces.[228] They sought explanations from all points east, including the administrations of Baghdad, Mosul, and Zor. In Baghdad, officials candidly noted that despite "exceptional effort" against the locusts, control – especially when the locusts grew wings – was simply "beyond human ability." It was not just their wings that made them vaunted but their origins, which the Baghdad officials described as "places that people could not go to." By emerging from places that people could not normally go, the locusts proceeded to move in ways that people could not move. The entanglement of accessibility and locusts was even more clear in October of 1886. It was at this time that those charged with a road-building project in Diyarbekir province were sent back to their homes near Mardin. The reason for their departure was that they were ordered to collect locust eggs there.[229] The locusts thus catalyzed a circular dynamic. People did not work on the roads because of locusts. Yet people could also not reach the places where locusts laid eggs because they were difficult to reach.

The consequences of the insects' depredations were significant. In 1887, the insects crossed the Euphrates, consumed 35 percent of the grains of Urfa, and totally destroyed Aleppo's summer crops of cotton, sesame, and melon.[230] The invasion caused what the French consul in Aleppo called "touching scenes," as "men, women, and children roamed the fields under threat, some with a branch in hand, others waving kitchen utensils, all letting out cries ... foreshadowing the misery to come." He even suggested that this "scourge of Asia Minor" was perhaps as deleterious for Aleppo as the opening of the Suez Canal in 1869 had been, which rerouted trade via the sea that had once traveled overland. Aleppo wheat had in recent years developed a reputation

[227] Ibid.

[228] BOA, DH.MKT 1357/93, Interior Ministry to Grand Vizier, 17 Temmuz 1302 (July 29, 1886). On Mardin, see TNA-UK, FO 424/143, Devey to Thornton, June 22, 1886.

[229] BOA, DH.MKT 1373/108, 11 Teşrinievvel 1302 (October 23, 1886).

[230] CADC, 4CCC/37, Consul Gilbert in Aleppo, September 24, 1887.

with French and Italian merchant houses for use in pasta production. Though perhaps with less damaging consequences than in the case of cotton two decades before, locusts once again swooped in for a taste of products bound for export.

Amidst the devastation, even settled people had to use mobility to survive. In 1888, American missionaries in Mardin described how for three years in a row the insects had cut a swath of between fifty and a hundred miles all the way from Mosul to Urfa.[231] The French vice consul of Diyarbekir blamed the locusts in that district for causing "the depopulation of this country."[232] In other words, the apocalyptic prediction of the Diyarbekir notable in 1864 – that if locusts arrived again the governor would be the only person remaining in the province – appeared to have come true. The consul estimated that the population was barely a quarter of what it had been twenty years before.[233] Like nomads, settled populations had found mobility to be a practical solution, and as a result "one of the most fertile countries" of the empire stood "uncultivated."[234] Banning export of grains and facilitating its import from other districts helped to avert famine in some cases.[235] But still, in the early months of 1888, thanks to the "ravages of the locusts" people were selling "bedding, cooking utensils, and the rugs on which they sleep" in order to afford some millet and avoid starvation.[236] As of May 1888, 1,500 people were living on the streets of Diyarbekir, and Ottoman officials blamed the locusts of the previous year.[237] Missionaries in Mardin estimated that there were some 10,000 people starving across the region.[238] The band of fertile land stretching to the east and west of Mardin had been viewed as a bulwark against nomadism and the desert in the 1860s, most notably with the settlement of Ras al-Ayn. But the region had been overrun once again, this time by locusts.

[231] American Board of Commissioners for Foreign Missions (ABCFM) 76, Box 2, Alpheus Andrus, The Famine at Mardin, January 28, 1888.

[232] CADC, 5CCC/1, Consul in Diyarbekir, July 25, 1888.

[233] CADN, 166PO/D/22/1, Bertrand to Minister of Foreign Affairs, January 30, 1888.

[234] CADC, 5CCC/1, Consul in Diyarbekir, July 25, 1888.

[235] CADC, 214CCC/2, French consul in Mosul, July 2, 1886; June 2, 1887; July 5, 1888; BOA, A}MKT.MHM 490/21, Report of Meclis-i Mahsus, 5 Şaban 1303 (May 9, 1886); DH.MKT 1372/78, Interior Minister to Grand Vizier, 4 Teşrinievvel 1302 (October 16, 1886); DH.MKT 1379/116, Interior Ministry to Rusumat Emanet Aliye, 9 Teşrinisani 1302 (November 21, 1886).

[236] TNA-UK, FO 424/145, Boyajian to Wratislaw, February 6, 1888.

[237] BOA, A.}MKT.MHM 497/39, Diyarbekir Governor Sirri to Grand Vizier, 5 Mayıs 1304 (May 17, 1888).

[238] ABCFM 76, Box 2, Alpheus Andrus, The Famine at Mardin, January 28, 1888.

Even as Ottoman officials worked to fight locusts, they acknowledged that success in the matter required a good amount of luck. Officials charged with destroying locusts in Aleppo in 1888 were commended by the provincial governor because "the greater part of the crops" had been "saved."[239] However, their accomplishment derived not just from their own diligence, but also from, as the governor who commended them admitted, the absence of locusts coming from the "desert and the East." After all, the "inaccessible," "rocky," or "desert" regions that locusts came from remained difficult to control for even the most conscientious Ottoman officials.[240] As a result, it sometimes seemed like only a miracle could deliver people from their insect tormenters. In Mosul in 1889, officials wrote that they destroyed some 3,000,000 okka (8,500,000 pounds) of locust eggs in an effort to prevent another year of devastation.[241] Nevertheless, in the surrounding desert, "nothing was visible except locusts." People were saved, however, when stunningly the locusts began to move "as if driven by a spiritual leader or driver, attacking and storming the Tigris with strange, awesome movements ... and in a billion not one was able to be rescued, and all were destroyed." Thus, nearly thirty years after Veysi Pasha boasted of the Tigris flowing with locust carcasses thanks to the depredations of starlings, the mighty river once again formed a watery grave for the insects. In both cases, little that humans did seemed to matter, apart from prayers.

As with locusts, the management of nomads involved real limits to government control, in large part thanks to the spatial bounds of nomadic migrations across difficult environments and provincial borders. Nomads continued to move all across the locusts' geographic territory, including "Mosul, Baghdad, Basra, Mardin, Urfa, Deir, and the Euphrates," according to the French consul in Baghdad.[242] And though groups such as the Shammar continued to observe a rough boundary between Faris in Zor and Farhan around Shirqat, the groups' mastery of the desert more broadly remained an advantage.[243] When Ottoman troops attempted to pursue them in instances of sheep theft, they not only failed to retrieve the

[239] BOA, DH.MKT 1589/91, Aleppo Governor to Interior Ministry, 19 Kanunusani 1304 (January 31, 1889).

[240] BOA, DH.MKT 1699/57, Interior Ministry to Commerce and Public Utility Ministry, 3 Şubat 1305 (February 15, 1890); DH.MKT 1719/101, Aleppo Imperial Estates Administration to Interior Ministry, 27 Mart 1306 (April 8, 1890); DH.MKT 1721/67, Interior Ministry to Grand Vizier, 14 Nisan 1306 (April 26, 1890).

[241] BOA, Y.PRK.AZJ 15/27, Notables of Mosul to Grand Vizier, 21 Nisan 1305 (May 3, 1889).

[242] CADN, 166PO/D/54/7, 2 May 1889; Siouffi in Mosul to Montebello, May 16, 1889.

[243] CADN, 166PO/D/7/17, Pognon to Comte de Montebello, October 17, 1888.

stolen sheep, but – as in years past – sometimes even failed to get close to the nomads.[244] If what made control of locusts difficult was the way they appeared from "places that people could not go to," the same might be said of nomads.

Locusts and nomads were not simply taking advantage of similar spatial dynamics. Their movement was also reinforcing each other. In a dispatch to the grand vizier in April of 1890, the Ottoman interior ministry noted that the locusts in the "cultivated [*mamure*] and uncultivated places of the desert [*çöl*]" of Diyarbekir were beginning to emerge from eggs.[245] Though they remained "very small," it was difficult to manage the locust populations because of the way human populations continued to move in the region. "As for the people," the dispatch explained, it would be impossible to conscript them into locust-collection schemes because of how they were "scattered" (*dağınık*) "in this season." In response to the dilemma, officials called for a bounty system, in which twenty para were paid for the first okka (2.83 pounds) of locusts, and ten para for every subsequent one. The monetary rewards were an attempt to incentivize different responses to locusts and thereby produce a new relationship between people, insects, and the environment.

The summer of 1890 witnessed further reminders of these intersections. On June 23, 1890, the Shammar chief Farhan Pasha died of tuberculosis in Baghdad.[246] The British consul in Baghdad recalled him as "diplomatic," a quality evident in 1871 when Farhan stood down while ʿAbd al-Karim revolted. Yet the consul also described Farhan in terms similar to those applied to locusts, noting his penchant for "preying on settled people," an occupation all the more straightforward given his place in the "Jazira." Around this same time, another infusion of locusts appeared in the region. Just like the Shammar originally came from the Arabian Peninsula, these locusts did too.[247] Arriving from the south, the locusts were a species distinct from the variety of locust that always lived in the Jazira, appearing later in the year and having a different life cycle and appearance. The Ottoman governor of Baghdad complained in July of 1890 of the Najdi invaders as

[244] TNA-UK, FO 195/1647, Tweedie to White, July 20, 1889.

[245] BOA, DH.MKT 1721/67, Interior Ministry to Grand Vizier, 17 Nisan 1306 (April 29, 1890).

[246] TNA-UK, FO 195/1682, Tweedie to White, June 26, 1890.

[247] It would later be recognized as an instance of an invasion of *Schistocerca peregrina*, also known as *Schistocerca gregaria*. NHMA, Syria and Lebanon 6211 – 158, Subhi Hasibi, Report on the Anti-Locust Fight Effected in the Territories of States Adhering to the International Accord of 20 May 1926 during 1929.

"untimely" (*mevsimsiz olarak*) and "unprecedented."[248] Although they caused no damage to winter crops such as wheat and barley (which had already been harvested), the insects had a huge impact on summer crops and dates. And since they were already winged, there was little hope of fighting them. The governor noted that they were eventually "destroyed" not because of humans combating them but rather because "there was nothing to eat on the edges of water in the desert." In other words, it was the aridity of the Jazira that killed the locusts. Even in death, though, the locusts were believed to have had an impact. There was some suspicion that the insects played a role in the cholera epidemic in Iraq that had begun that year, since "thousands of them had drowned" in local well water.[249]

It may have seemed like there was nothing humans could do about these invaders, whom it appeared only the Jazira could kill (and even in death might communicate an often-lethal ailment). Yet in some cases, it does seem that pastoralists used their mobility and knowledge of the local environment to fight the insects. In 1890, one French diplomat making the journey between Damascus and Baghdad reported seeing "locusts as big as bees" within sight of Mount Sinjar.[250] That night he encountered "five bedouins from the Jabbur tribe" who had apparently "dedicated their day to killing the locusts." They did so – in a reminder of the fact that some pastoralists cultivated the land too – in order to save their harvests. They would continue "to make war against the pests" the following day, and so they preferred not to return all the way home and instead joined the camp of the consul and his party. It being Ramadan, they eagerly awaited the sunset, after which they ate mushrooms they had collected, alive to fight locusts another day.

But these instances were rare. Indeed, Ottoman administrators expressed a sense of futility when it came to locust control in the Jazira as opposed to other places in the empire. In December of 1890, the ministry of commerce and public works suggested that the province of Aydın – perched on the Aegean Sea – might serve as a model for locust destruction in Aleppo.[251] But Aleppo officials chafed at this effort at

[248] BOA, Y.PRK.A 5/80, Governor of Baghdad Sirri to Grand Vizier and Finance Ministry, 5 Haziran 1306 (July 17, 1890).

[249] CADN, 166PO/D/7/17, Pognon to Comte de Montebello, May 26, 1890. On cholera in Ottoman Iraq, see Bolaños, "The Ottomans during the Global Crises of Cholera and Plague."

[250] CADC, 79CPC/3, De Damas à Mossoul par le désert, Voyage effectué par Mr. Siouffe, Consul de France, August 8, 1890.

[251] BOA, DH.MKT 1789/2, Interior Ministry to Aleppo Province, 21 Teşrinisani 1306 (December 3, 1890).

imperial comparison. Aleppo, they argued, was "not comparable to Aydın province" (*orası Aydın vilayetine makis olmayıp*).[252] Aleppo, they said, was "surrounded by the desert and devoid of people," very different from Aydın's high population density and largely cultivated lands.[253] In other words, even if locusts afflicted different parts of the Ottoman Empire, divergent political ecologies ensured that locusts meant different things – and had to be killed in different ways – in different places.

Both locusts and nomads, then, presented particular spatial dilemmas for the Ottoman state in relation to regional ecology and in relation to the provincial division of the broader Jazira region. Both locusts and nomads emerged from places in between, arid areas where people moved but plows rarely ventured. Moreover, both locusts and nomads emerged from these ecologies on the edge to move across provincial borders, making their management difficult in an empire that had pinned hopes for centralization and reform on newly efficient provincial administrations. These spatial realities had vexed Ottoman authorities for decades and prompted them to imagine different borders for managing the region. Indeed, even in 1890, some observers saw fit to praise these boundaries as catalysts of development. The British consul in Aleppo in 1890 stated that "inroads made by the Shammars of Mesopotamia … practically came to an end some years ago."[254] He attributed the change in part to the creation of Zor and the way its borders encompassed a "vast extent of country practically uninhabited save during certain seasons by Bedouins." It was, he added, "the true and only centre of nomad politics."

The esteem of the British consul, however, did not stop Ottoman officials from continuing to imagine how to better draw borders around the Jazira and its moving people. There had been previous proposals, such as one in 1888 that aimed to resettle refugees and put nomads within the bounds of one province in the area "foreigners" called "Mesopotamia."[255] According to the plan, the province was to be called Hamidabad in honor of Sultan Abdülhamid II. A similar proposal emerged in 1890 when two officials named Ahmed Tevfik and Şevket called for consolidating the special

[252] BOA, DH.MKT 1797/37, Interior Ministry to Commerce and Public Utility Ministry, 18 Kanunuevvel 1306 (December 30, 1890).

[253] Locusts nevertheless had an impact on the Aegean region and elsewhere in Anatolia. On impact on Ottoman revenue generally, see TNA-UK, FO 424/109, Goschen to Earl Granville, November 23, 1880. On more general impact, see TNA-UK, FO 424/132, Wilson to Earl of Dufferin, February 21, 1882.

[254] TNA-UK, FO 195/1690, Report on the Vilayet of Aleppo by Consul Jago, June 1890.

[255] BOA, Y.PRK.AZJ 13/54, 29 Zilhicce 1305 (September 6, 1888).

administrative district of Zor and raising it to the full status of province (*vilayet*). They suggested returning the districts such as Sinjar, Nusaybin, Viranşehir, and Raqqa that had in previous years been removed from Zor.[256] They also proposed a name change. What had once been known as Zor would, in their vision, come to be known by its defining natural feature: it would be called Desert province (Figure 9). The Desert province was to be home to settlements of refugees and nomads coaxed toward sedentary life. Of course, the goal was to transform the region's environment in such a way as to make the province's name obsolete. They envisioned it eventually becoming home to some one million people, who would presumably find a way to make a life in the land that had for so long been described as "places that people could not go." The officials did so because the Desert province was, in their opinion, a "natural governing point" (*nokta-i hakime-i tabiiye*).[257] Many Ottoman officials had made a similar comment over the years. It was as if they had learned from the locusts and nomads of the Jazira, who for so long had used these realms in between to evade and afflict the Ottoman state's efforts at modernization. Yet Ottoman administration of the region would ultimately go in a different direction.

CONCLUSION

The continued challenges on the edges demonstrated how borders worked both as a tool of governance and a means of resistance. The environment functioned as an object around which borders were to be established, and a setting that incubated motion beyond the bounds of borders. In a variety of ways, the Ottomans had attempted to make borders into a reality as a means of controlling and transforming the Jazira in the wake of the reforms of the 1858 Land Code and the 1860s. They had envisioned cordons across which nomads might not migrate, and behind which seminomadic Kurdish groups might cultivate the land. They also attempted to use Chechen refugees as a bulwark of cultivation expanding into the desert. Then in 1871, they planned a district built specifically to encompass the desert, a project so controversial that it incited a revolt and

[256] BOA, Y.A.RES 55/38, Report of Commander of Military Reserves at Urfa Lieutenant General Şevket, 25 Teşrinievvel 1306 (November 6, 1890); Report of District Governor of Zor Ahmed Tevfik, undated.

[257] BOA, Y.A.RES 55/38, Report of Commander of Military Reserves at Urfa Lieutenant General Şevket, 25 Teşrinievvel 1306 (November 6, 1890).

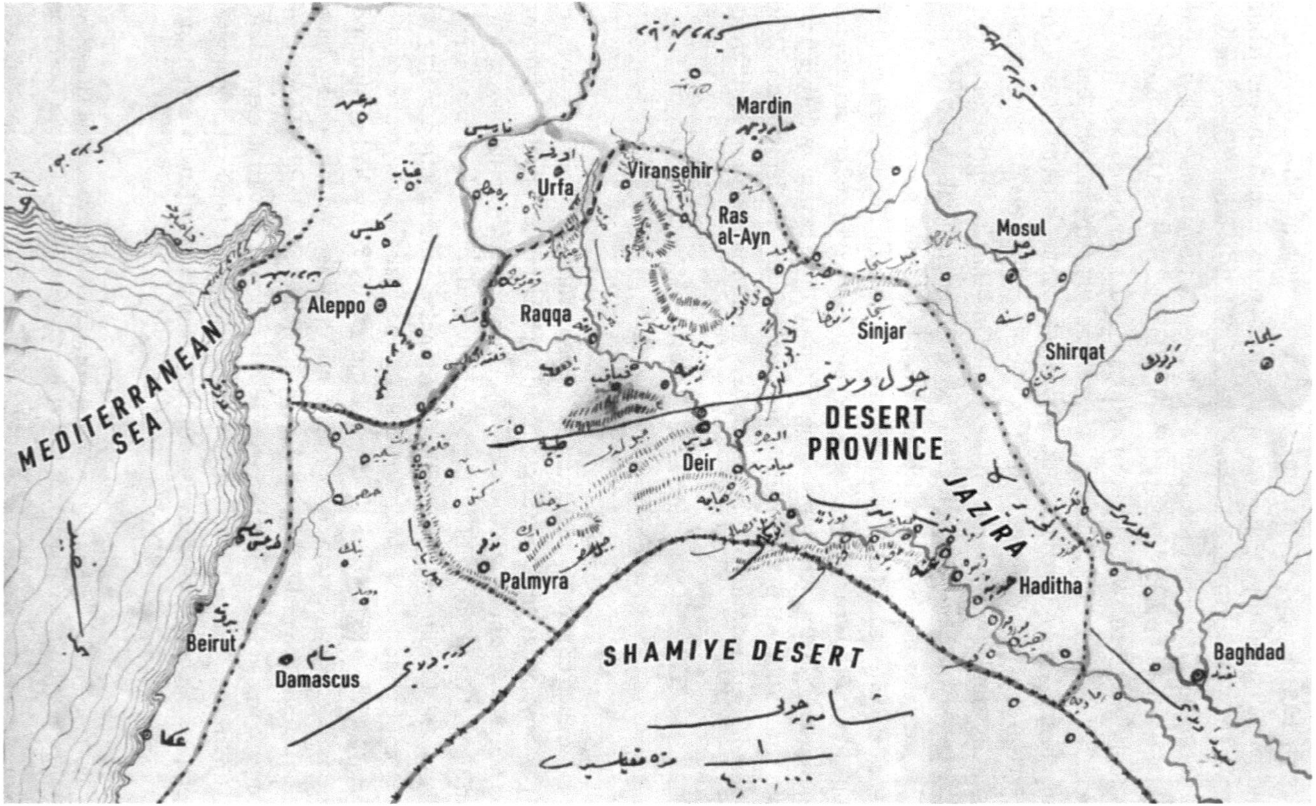

FIGURE 9 Proposed Desert province, 1890. BOA, Y.A.RES 55/38

a march on Baghdad, with the foremost Ottoman reformer of the late nineteenth century hanging the revolt's ringleader over a bridge in Mosul. Even after this apparent victory, control on the edges remained blurry, as tax collectors and census officials alike struggled with how to define where exactly their jurisdiction began and ended. Accordingly, officials pitched many alternative provincial borders, including both Hamidabad and Desert province.

Locusts animated nomadic motion. They prompted some like the Shammar to seek pasture away from the swarms of insects. They prompted others like the Kara Keçe to give up cultivation and practice pastoralism instead. But locusts were of course not the only reason. Motion was also in response to drought or cold, not to mention opportunism driven by the very same borders that were supposed to solve the dilemma of nomadism in the first place. Indeed, locusts seem to have been absent from the Jazira after the 1860s, when they had decimated cotton crops that many hoped to export in the midst of the global cotton shortage and ravaged provisions intended for Chechen refugees. Locusts' disappearance likely did not derive from effective control efforts. Ottoman officials were limited in their control of territory, so much so that the starling – believed to be attracted by a Sufi-blessed holy water – seems to have presented the most reliable deterrent to the insect swarms. By the mid-1880s, the insects were back, leaving famine and displacement in their wake by moving from the desert and across provincial borders.

The Shammar may have seemed like locusts in some ways, but they were treated quite differently. There was, for example, no plan for a cordon across which locusts would not pass. But there was such a plan for nomads. The failure to see locust motion as connected to nomadic motion bespoke a larger problem. Ottoman officials largely saw nomads as creatures lacking civilizational virtue. They could be redeemed through practices, such as tilling the land. But there was little attention to how it made sense to be a nomad both in relation to flows of capital and swarms of insects. These dynamics would persist even as Ottoman officials changed their policies in the Jazira in the coming years.

2

"Savage Swarms" (1890–1908)

The Ottoman official Ebubekir Hazim Tepeyran did not like nomads. Reflecting on his encounters with them while governor of Mosul between 1899 and 1901, he complained that they never took proper human names for themselves like Osman, Mustafa, or Mehmed. Instead, he claimed they only used names more appropriate for "dogs" or "oxen," such as Sahre (Rock) or Matar (Rain).[1] In saying so, Tepeyran seemed the epitome of a denigrating civilizational project that has alternately been called "Ottoman Orientalism" or "White Man's Burden wearing a fez."[2] He did not find much more attractive the Ottoman creation of paramilitaries known as the Hamidiye Light Cavalry Brigades (*Hamidiye Hafif Süvari Alayları*), formed largely from nomadic and seminomadic Kurdish groups beginning in 1891. He described them as "savage *sürüler*," with the latter word referring to groups of animals such as "herds" of livestock, "packs" of dogs, or, indeed, "swarms" of locusts.[3] As in the case of al-ʿUjayli's comparison of the Shammar to locusts, Tepeyran compared moving people and moving insects as an insult. But the language concealed significant material links and spatial overlaps between the groups in the Jazira.

As historian Janet Klein has argued, the history of violence and ethnicity in the late Ottoman period was largely a modern product of institutions such as the Hamidiye. As those associated with the Hamidiye usurped rural

[1] Tepeyran, *Hatıralar*, 460.
[2] Deringil, "'They Live in a State of Nomadism and Savagery,'" 312; Makdisi, "Ottoman Orientalism."
[3] Tepeyran, *Hatıralar*, 441.

landholdings – particularly but not only Armenian ones – they created antipathies that would eventually be conceptualized as "the agrarian question," deepening ethnic distinction and tension.[4] Klein makes these points while both arguing against a "causal or deterministic" approach to geography and emphasizing the importance of "spatial relationships" as Kurdistan changed in Adelman and Aron's terms from a "borderland to bordered land."[5]

But what did borders and land look like in this context? Tepeyran's comparison offers a clue. The meaning of borders and land derived from a geography that remained a patchwork of pasture and cultivation, crisscrossed by nomadic pastoralists, locusts, and government officials. Beginning in the 1890s, the Ottomans took a new approach in the Jazira, when they empowered the Millî confederation as a Hamidiye Brigade under the leadership of Ibrahim Pasha. Ibrahim and the Millî were charged with pushing south into the deserts of the Jazira. As locusts continued to fly and groups such as the Shammar continued to move, the Millî increasingly clashed with the Shammar. And the site where they did so – provincial borders – attested to the power of provincial borders dating back to 1871 and the creation of Zor. Meanwhile, Ottoman efforts at controlling locusts became more systematized, but they still remained limited by the unique geography of the Jazira and the lack of financial resources. The attempts at locust destruction also remained limited by an unwillingness to see humans and locusts as connected. Indeed, when Ibrahim Pasha and the Millî moved close to Diyarbekir because of locust swarms, his justifications were dismissed as merely an excuse. In other words, locusts and provincial borders profoundly shaped the nature of everyday life on the margins of empire. And they shaped, too, the kind of power that could emerge on the edge. But it was unclear how this power might remain alongside new visions of reform of the empire.

THE HAMIDIYE BRIGADES IN THE JAZIRA

The proposed Desert province of 1890 represented a variation on previous plans, which all had the aim of making provincial borders better match the environment and the humans who moved in it. Alongside these interventions, however, had been another option, one that had been practiced for many years before and even continued to be present in the years during

[4] Klein, *The Margins of Empire*, 128–169. See also Astourian, "The Silence of the Land."
[5] Adelman and Aron, "From Borderlands to Borders"; Klein, *The Margins of Empire*, 11.

which borders were discussed. This alternative was relying on people, rather than territorial units in which they moved. Examples of this approach appeared in the 1860s, when the Ottomans depended on the Shammar to protect the telegraph line, or when the Muntafiq captured ʿAbd al-Karim.

In the coming decades, the Ottoman state would take this approach in the Jazira through the Hamidiye Light Cavalry Brigades. In 1891, memos at Yıldız Palace spoke of putting the "Arab, Kurdish, and tribal" populations of the empire to use, perhaps in the form of an army of "several hundred thousand people."[6] Soon, such a policy was put in place. In the territory that was to be Desert province, there emerged a light cavalry brigade whose named derived from the sultan's name: the Hamidiye Light Cavalry Brigades. Modeled on the Cossacks, the Hamidiye Brigades are generally understood as a means of fending off potential Russian advances in eastern Anatolia, as well as keeping Armenian populations under control while co-opting Kurdish populations into the machinery of the state.[7] They were under the command of Zeki Pasha of the Ottoman Fourth Army, an officer of Circassian descent and the brother-in-law of the sultan.[8] Zeki was well acquainted with the edges of the empire, having previously served in both Epirus and Libya.[9] By the early 1890s, his authority extended over much of eastern Anatolia. The Jazira fell outside of this space, but brigades were still established on its northern edges. Most notably, the institution of the Hamidiye empowered the seminomadic Millî. Their forebears held governorships of Raqqa among other places, though the Millî seem to have thrown their lot in with the rebellious Ottoman governor of Egypt Mehmed Ali when his forces invaded greater Syria and fought imperial troops in the 1830s.[10] The Millî confederation boasted Muslims and Yazidis, and speakers of Arabic and Kurdish alike. But their power had dwindled by the 1860s, when they were among the groups targeted for settlement as part of the 1864 "line of cordon" plan.

It was at this point that their leadership was taken by a figure who had been called Berho as a child and Brahim as an adult – Ibrahim to the outside

[6] BOA, Y.PRK.AZN 5/23, 23 Mart 1307 (April 4, 1891).

[7] Klein, *The Margins of Empire*, 42; Deringil, "'The Armenian Question Is Finally Closed,'" 349.

[8] Klein, *The Margins of Empire*, 3, 27; Van Bruinessen, *Agha, Shaikh and State*, 187.

[9] TNA, FO 195/2274, Biography of Mehmed Zekki Pasha, January 17, 1893.

[10] Abu Bakr, *Akrad al-Milli Ibrahim Pasha*, 21; Winter, "The Other '*Nahdah*,'" 468–470; Husain, *Rivers of the Sultan*, 136.

world.[11] Fluent in Arabic, Kurdish, and Ottoman, the chief used his place on the edge – like so many before him – as a way of gaining power. The Millî base was at Viranşehir, a town halfway between Mardin and Urfa, and, more significantly for their purposes, halfway between their summer pastures at Mount Karaca and their winter pastures to the south.[12] Significantly too, Viranşehir had long been precisely on the border of Zor and Diyarbekir provinces and afforded its inhabitants all of the opportunities of that arrangement. Viranşehir literally meant Ruined City, perhaps appropriate given the links between ruins and dreams of glory in the Jazira. One of the period's definitive geographies warned that the Viranşehir of Ibrahim should "not be confused with the other rather numerous localities of the same name."[13] But the description of the city as ruined was also not entirely accurate. Armenian master craftsmen from Diyarbekir had built the headquarters of the Millî confederation there in the 1830s.[14] Ibrahim called the town something different: *qiza min*, "my dear daughter" in Kurdish. It was from this setting that he would come to command the Jazira in a new way. Ottoman officials hoped to use this person from the ruins. Almost all discussion of the formation of a brigade under the command of Ibrahim of the Millî mentioned the importance of his authority for controlling the deserts to the south of his base in Viranşehir. Or, as one memo put it, "the Jazira."[15] A report from the Fourth Army more specifically spelled out the plan for Ibrahim as control of the ʿAnaza and Shammar.[16]

But of course, controlling the Jazira and the Shammar meant controlling a space and group whose motion was enmeshed with those of locusts. The spring of 1891 had demonstrated once again how intricate these connections were. The "voracious creatures" moved from the desert of Zor into fields of winter wheat and barley as well as summer crops of cotton and sesame, devastating them all.[17] When they "crossed the Euphrates" into Aleppo province, the swarms looked like "compact clouds."[18] There were reports around Aleppo of starlings intervening.[19]

[11] Kıran, *Kürt Milan Aşiret Konfederasyonu*, 159. [12] Ibid., 156.

[13] Cuinet, *La Turquie d'Asie*, vol. 2, 311–312.

[14] Kıran, *Kürt Milan Aşiret Konfederasyonu*, 159.

[15] BOA, Y.PRK.ASK 71/79, May 26, 1891.

[16] BOA, Y.PRK.ASK 72/76, Fourth Army Command to Yıldız Palace, 25 Mayıs 1307 (June 6, 1891).

[17] National Archives and Records Administration (NARA), RG 84, Consular Records of Beirut, vol. 29, "Locusts in the Vilayet of Aleppo, Syria," July 14, 1891, 141.

[18] CADC, 4CCC/37, May 27, 1891.

[19] *No. 1017. Diplomatic and Consular Reports on Trade and Finance. Turkey. Report for the Year 1891 on the Trade of Aleppo* (London: Harrison and Sons, 1892), 5

But many insects survived. Near Aleppo, locusts devastated the sultan's own landholdings, which he had been amassing all across the empire since 1881 especially.[20] Officials smarted that thanks to the mobile nature of the insects, their own efforts on the properties were only as good as those efforts in the "desert" or "the other side of the Euphrates."[21] The insects reached as far as the outskirts of the western cities of Homs and Hama, where villages were on the verge of "becoming desert."[22] Moreover, the insects' depredations coincided with complaints about the presence of the Shammar. The governor of Diyarbekir pleaded for two or three divisions of cavalry in the desert to prevent the Shammar from "spreading out."[23] Because no troops arrived, the Shammar were reportedly "feeding their herds, and, to the great annoyance of the people, damaging cultivation."[24] Mobile groups like locusts and the Shammar both emerged from the desert of the Jazira and took advantage of their place on the edge.

Traveling across the Jazira at around this time, the German archaeologist Max von Oppenheim attested to the status of the Shammar since the revolt against borders in 1871. There remained, of course, the "border" from Mosul to Mayadin that separated the self-described southern and northern Shammar, Farhan and Faris, in the wake of the formation of Zor and ʿAbd al-Karim's revolt.[25] But there was also considerably less rebellious leadership. Oppenheim described Farhan as almost totally forgotten, despite only having died two years previously.[26] And as for Faris, Oppenheim estimated that he looked more like a "city Arab or Turk."[27] And yet they had not entirely bowed to authority. When Oppenheim visited Shirqat, he – like Lady Anne Blunt before him – noticed that no Shammar were to be found in the vicinity, although the land all around was "undoubtedly used for grazing animals and pitching tents."[28] Oppenheim declared the Shammar the "undisputed masters of the steppe of Mesopotamia from Urfa to … Baghdad."[29] He did so based not only on the Shammar herds, but also their collection of the *khawa* tax on other pastoralists in the Jazira (estimated at 150 percent of the tithe).[30] The

[20] Kiyotaki, *Ottoman Land Reform in the Province of Baghdad*, 20; Cole, "*Nafia* for the Tigris."
[21] BOA, İ.MMS 126/5406, 15 Eylül 1307 (September 27, 1891).
[22] BOA, Y.MTV 52/67, Temmuz 1307 (July 29, 1891).
[23] BOA, Y.PRK.UM 21/105, 16 Mayıs 1307 (May 28, 1891).
[24] TNA-UK, FO 424/169, Boyadjian to Acting Consul, June 8, 1891.
[25] Oppenheim, *Vom Mittelmeer zum Persischen Golf durch den Haurän*, 62–63.
[26] Ibid., 61. [27] Ibid., 41. [28] Ibid., 210–211. [29] Ibid., 64. [30] Ibid., 65.

practice seems to have been discontinued for some time but picked up again after Farhan's death.[31]

Hamidiye Brigade mastermind Şakir Pasha likely had these practices in mind in March of 1892 when he described how the light cavalry units might have a "good influence ... in the desert."[32] By April, the first step of plans for the Hamidiye Brigades to curb the Shammar took place in Diyarbekir in an elaborate ceremony that celebrated the transformed status of the seminomadic groups enrolled in the institution.[33] Some 500 cavalrymen of the Millî under Ibrahim Pasha's command paraded through the city of black basalt walls on the edge of the Tigris. The regular soldiers stationed in the city and local students crowded the streets by the "thousands," and they cheered on the seminomadic tribe turned state-supported militia. Ibrahim sacrificed camels and goats, and his followers chanted "Long live our Sultan Hamid." Zeki Pasha claimed that some among the Millî "had not seen a city before," underscoring the views of civilizational uplift associated with the brigades and echoing remarks made about the Shammar in previous decades. The attitude coincided with a broader project of incorporating pastoralists, including such institutions as the Imperial School for Tribes founded in Istanbul in 1892.[34]

If Ibrahim Pasha's troops were cause for celebration in the city of Diyarbekir, it was less clear how they would function outside of the city walls. In the summer of 1892, the Oxford-educated Anglican priest Oswald Parry observed Ibrahim Pasha's troops amidst the seasonal "carpet of flowers, mallows, anemones, cornflowers, and basalms" interspersed, of course, with "the old Assyrian mounds."[35] For Parry, the Hamidiye were "picturesque."[36] Yet they were after something mundane: grazing for their animals. In order to make enrollment in the brigades more attractive, the Ottoman state had exempted enrollees from the sheep tax.[37] In a foreshadowing of entanglements to come, later that summer Parry also observed what he described as "armies of locusts."[38] They were also after the grasses and crops of the Jazira.

[31] Çiçek, *Negotiating Empire*, 176.

[32] BOA, Y.MTV 60/62, Şakir Pasha, 8 Mart 1308 (March 20, 1892); Klein, "State, Tribe, Dynasty," 151.

[33] BOA, Y.PRK.ASK 81/34, Fourth Army Commander Zeki, 18 Nisan 1308 (April 30, 1892).

[34] Rogan, "Aşiret Mektebi."

[35] Parry, *Six Months in a Syrian Monastery*, 73.

[36] Ibid., 153.

[37] Klein, *The Margins of Empire*, 31; Ekinci, "Osmanlı Devleti Döneminde Milli Aşireti XVIII–XIX YY.," 200.

[38] Parry, *Six Months in a Syrian Monastery*, 116.

SHAMMAR, MILLÎ, AND THE HAMIDIAN MASSACRES

By 1894, tensions between the Hamidiye Brigades of the Millî and the target of their mission – the Shammar – had come to the surface. Provincial borders would play a key role in these disputes. As early as February 1894, Ibrahim complained that groups affiliated with the Shammar had raided the Millî in Diyarbekir province and taken animals.[39] The district governor of Zor dismissed the matter as an "old custom" among the "savage Arabs and tribes" (*bedevi olan urbân ve aşayır*). But it was actually more complicated. The Millî, after all, were a state-sanctioned irregular military unit. Meanwhile, the Shammar who had raided Ibrahim Pasha were from Mosul, meaning that their apparently timeless custom involved crossing state borders as a way of creating jurisdictional ambiguity.

As people complained about the Shammar, they also referred explicitly to the Jazira as the locus of disorder. After the Shammar stole animals in Mosul in April, residents of Mosul complained that the Shammar had used "the Jazira" as "a wandering place of banditry."[40] In response, Ottoman officials dispatched a military detachment.[41] But they also made clear that a broader campaign involving troops from Baghdad, Diyarbekir, and Zor would be necessary to have a lasting impact. The minister of the interior agreed. He blamed the Shammar for "scattering" settled populations but also knew that existing military forces at Zor were not up to the task of maintaining order.[42] The governor of Mosul called for cavalry brigades on the edge of the desert, including one in the ruins at Hatra.[43] An officer with the Sixth Army echoed this message, suggesting that only careful placement of troops would allow for the transformation of the "fertile and productive vast and empty lands" of "the Jazira located in Mesopotamia" (*Beyn ül-Nehryen'de vakia El-Cezire*).[44] Thus while Ottoman officials worked through the system of provinces, they also repeatedly referred by name to the broader geography within which the Shammar moved.

39 BOA, ŞD 2225/15, Zor District Governor Mehmed Salih to Interior Ministry, 13 Şubat 1309 (February 25, 1894).
40 BOA, BEO 382/28634, Abdullah and others to Grand Vizier, 29 Mart 1310 (April 10, 1894).
41 BOA, BEO 382/28635, 28 Mart 1310 (April 9, 1894).
42 BOA, BEO 547/41007, Minister of Interior to Grand Vizier, 30 Teşrinievvel 1310 (November 11, 1894).
43 CADN, 166PO/D/54/8, Alric to Boulinière, November 28, 1894.
44 BOA, ŞD 2166/3, Sixth Army in Mosul, 11 Nisan 1311 (April 23, 1895).

Disputes between the Millî and the Shammar increasingly involved Ottoman troops and the question of provincial borders. In the winter of 1894 to 1895, clashes between the Shammar and Millî over migration to Mount ʿAbd al-ʿAziz even witnessed Ibrahim Pasha seizing Shammar shaykh ʿAli ʿAbd al-Rizzaq – nephew of Shammar shaykh Faris – and taking him to Diyarbekir, a measure resulting in furious remonstrations on both sides, including a vow from Shammar shaykh Faris that "order among the Arabs will be completely destroyed."[45] By March of 1895, the grand vizier worried about further conflict between the groups "on the border of the district," especially during the sheep count.[46] By May, rumors of a Shammar attack on the Millî swirled, according to the acting governor of Diyarbekir, Enes.[47] A native of Salonica and a convert to Islam from Judaism, the acting governor's parents had been executed in connection with the killing of the French and German consuls in his home city in 1876.[48] Enes alleged that the Shammar planned to exploit the empire's administrative structures, being sure to attack Ibrahim only after "crossing the border of the province."[49] Accordingly, Enes sent troops to Viranşehir and moved troops from Mardin to the desert. When the Millî regiments of the Hamidiye Brigades were established, they had been envisioned as a means of pushing state control into the desert and over the Shammar. But it appears the deputization of the Millî had an impact in the other direction. Their clashes with the Shammar embroiled local administrations in the management of the desert and the nomads. Provincial boundaries functioned as a key part of this process.

There might have been tensions between the Millî and the Shammar, but the groups occupied a similar position geographically and economically. The ruins, empty lands, and lack of infrastructural networks decried by so many were precisely what allowed wool production to remain in their hands. As French officials wrote, "the country is made for the raising of animals."[50] Locusts helped. As in the 1860s, so too in the 1890s did

45 BOA, BEO 568/42574, Faris, 25 Kanunusani 1310 (February 6, 1895); Adamiak, "To the Edge of the Desert," 212–213. ʿAli ʿAbd al-Rizzaq remained in Diyarbekir prison until as late as 1896, when he complained that the prison food was going to kill him. BOA, ŞD 2672/461, ʿAli to Şura-yı Devlet, 3 Eylül 1312 (September 15, 1896).

46 BOA, Y.A.HUS 323/7, Grand Vizier to Seraskerlik, 15 Mart 1311 (March 27, 1895).

47 BOA, Y.A.HUS 328/82, Diyarbekir Acting Governor Enes to Grand Vizier, 9 Mayıs 1311 (May 21, 1895).

48 Verheij, "Diyarbekir and the Armenian Crisis of 1895," 102.

49 BOA, Y.A.HUS 328/82, Diyarbekir Acting Governor Enes to Grand Vizier, 9 Mayıs 1311 (May 21, 1895).

50 CADN, 166PO/D/54/8, Report on Commerce for 1891, February 12, 1892.

locusts destroy high-value export crops on a number of occasions, including cotton in Aleppo and experiments with poppies (for opium) in Mosul.[51] It was on these lands that emerged wools categorized by ethnicity, with "Arab wool" the "finest and most sought after," thanks, apparently, to desert pasture, while "Kurdish wool" referred to a production "of an inferior quality" that came from the mountains.[52] But outside of the Ottoman domains, the ethnic and environmental distinctions collapsed, as wool exported through Aleppo became rechristened "Smyrna Wools," as if their provenance were the Aegean port city of the same name.[53] Indeed, advertisements for Smyrna wool rugs appeared in newspapers everywhere from Louisville, Kentucky to Los Angeles ("we've more rugs than you think – unless you have recently visited our third floor rug department," read one ad).[54]

As wool found its way far from the plains stretching south from Anatolia, the Jazira – or at least its edges – was struck with violence that also had roots in much broader geographies. Dating back to the conclusion of the Russo-Ottoman War of 1877 to 1878 and the Treaty of Berlin, the Ottoman Empire had committed to offering special protection for its Armenian subjects, particularly in the six provinces of southeastern Anatolia in which they overwhelmingly lived. Yet still Armenians of these provinces struggled, often suffering under double taxation by both the state and local notables. In response, many emigrated, and in diaspora communities such as Tbilisi, Geneva, and Worcester, Massachusetts emerged groups such as the Armenian Revolutionary Federation and the Social Democrat Hunchakian Party, much to the concern of the Ottoman state. While transnational connections were crucial to these institutions, figures from closer to the Jazira also played a role and expressed their concerns in familiar terms. Diyarbekir native and poet Rupen Zartarian, for example, lamented how "our country is condemned to remain a desert."[55]

Violence of a new kind broke out in 1894 and threatened to engulf the Jazira. Northeast of Diyarbekir in the mountain district of Sasun,

[51] CADN, 166PO/D/54/8, Report on Commerce for 1891, February 12, 1892; CADC, 4CCC/38, Rapport commercial pour l'année 1895, April 8, 1896.

[52] CADN, 166PO/D/54/8, Report on Commerce for 1891, February 12, 1892

[53] CADC, 4CCC/38, Rapport commercial pour l'année 1895, April 8, 1896.

[54] *Courier-Journal*, April 19, 1901, 8; *Los Angeles Times*, August 21, 1902, 7; *The Sun*, September 14, 1908, 1.

[55] Berberian, *Armenians and the Connected Revolutions in the Russian, Iranian, and Ottoman Worlds*, 59.

Armenians refused to pay taxes to the state since they already paid customary tribute to local Kurdish notables. Local authorities declared that a revolt had broken out and carried out full-scale massacres of the local civilian populations, with casualties estimated at some 900 dead.[56] The Ottoman military commander of the Hamidiye Zeki Pasha received a medal of honor for his involvement in an early episode of the violence.[57] He subsequently authored a report that largely blamed it on Armenian agitation and also functioned as the blueprint for historiographical denial of the violence over the next century.[58] As early as summer of 1895, the head military official of the empire had warned that fighting between the Millî and Shammar might give "Armenian bandits" or foreign "consulates" pretexts to lodge complaints.[59] But when violence did in fact occur, it did not involve these groups. After the Ottoman state announced a renewed commitment to reforms, provincial power brokers took the opportunity to instigate attacks on Armenians beginning in November 1895, with pogroms throughout eastern Anatolia as part of what is commonly referred to as the Hamidian massacres. In the Jazira, violence occurred in Aintab, Urfa, and Diyarbekir, where the governor Enes Pasha was infamous for his support of forcible conversion and abduction of Armenian women.[60] The violence especially took place in regions where Diyarbekir notables such as the Pirinççizade family had landholdings, a premonition of the patterns of conflicts to come.[61]

There tended to be less violence, however, in the Jazira south of the Tigris, where Armenians usually lived only in towns rather than in both towns and the countryside as they did to the north. One missionary in Mardin described the approach of pastoralists in terms not unlike how one might describe locusts, recalling how from the city's hilltop one could see "the plains around black with Kurds day after day ... gathered for the purpose of massacre."[62] Armenians and Assyrians in nearby villages such as Tel Arman and Goliyê were massacred.[63] But in Mardin, Muslim notables as well as state troops prevented any attacks. In the nearby monastery Der Zafaran, defenders melted the printing press into bullets. In Nusaybin, the Hamidiye Brigades of the Arabic-speaking Tayy tribe

[56] Suny, *"They Can Live in the Desert but Nowhere Else"*, 109. [57] Ibid., 110.
[58] Miller, "Rethinking Violence in the Sasun Mountains," 98–99, 115.
[59] BOA, Y.MTV 121/114, Serasker Rıza, 29 Mayıs 1311 (June 10, 1895).
[60] Suny, *"They Can Live in the Desert but Nowhere Else"*, 118.
[61] Jongerden, "Elite Encounters of a Violent Kind," 74
[62] Harris and Harris, *Letters from the Scenes of the Recent Massacres in Armenia*, 105.
[63] Gaunt, "Two Documents on the 1895 Massacres in the Province of Diyarbekir."

reportedly protected local Christians from attack.[64] And so too was violence prevented in Viranşehir, the home of Ibrahim Pasha, located on the edge of the desert and the sown, Diyarbekir and Zor. In fact, Ibrahim's protection of Christians there in the midst of the massacres earned him a reputation that lasted for years as a protector of the oppressed and encouraged many Christians to relocate to his domains.[65] In villages further north in Anatolia, by contrast, Armenian villagers struggled under the threat of violence from extrastate and state sources alike, with one British official observing in Harput that tax collectors "swooped down on the vilayet like a swarm of locusts."[66]

The violence of 1895 and 1896 pointed to the way that political structures both within the Ottoman Empire and outside of it intensified ethnic distinction. But the nature of the violence also offered a reminder of how ethnic distinction remained complicated in the context of the Jazira. Violence in eastern Anatolia has sometimes been reduced to a phenomenon of the Hamidiye Brigades or Kurdish nomads in general against Armenians, a convenient explanation that positions the violence as a product of backward hordes and timeless conflict. Yet these simplistic explanations disguise the level of complicity at high levels of society and the role of property in the disputes. Indeed, one of the only figures who faced consequences for the violence was Diyarbekir notable Arif Pirinççizade.[67] He, like most of the city's large landholders, was Kurdophone and Muslim, but he likely would not have identified as Kurdish, a term more often used to refer to rural people or nomads.[68] Pirinççizade was not exiled – that would have been too much like a punishment. Instead, he was encouraged to depart the city for Mosul.[69] But he would return to Diyarbekir, where he and his family would be involved in further outbreaks of violence that would spread

[64] Verheij, "Diyarbekir and the Armenian Crisis of 1895," 136.

[65] Gaunt suggests that Ibrahim Pasha perhaps gained from the plundering of some Christian goods in the region. Gaunt, "Two Documents on the 1895 Massacres in the Province of Diyarbekir," 196; TNA-UK, FO 195/1930, Fitzmaurice to Currie, March 25, 1896; Jongerden, "Elite Encounters of a Violent Kind," 74.

[66] Sir P. Currie to the Marquess of Salisbury, October 27, 1896, Inclosure: Vice Consul Fontana to Sir P. Currie, October 14, 1896. *Turkey. No. 3 (1897). Further Correspondence Respecting the Asiatic Provinces of Turkey and Events in Constantinople.* (London: Harrison and Sons, 1897), 9. For another reference to provincial revenue streams and metaphorical "locusts which swarm round" them, see TNA-UK, FO 424/141, Report by Vice Consul Eyres on the Vilayet of Van, January 4, 1884.

[67] Verheij, "Diyarbekir and the Armenian Crisis of 1895," 130.

[68] Suny, *"They Can Live in the Desert but Nowhere Else"*, 116.

[69] Deringil, "'The Armenian Question Is Finally Closed,'" 363.

much more widely across the Jazira and employ the environment in punishing ways.

"THE DESTRUCTION AND DISAPPEARANCE OF A DISTRICT"

While locust invasions – if not metaphorical locust invasions – had died down by the mid-1890s, the Ottoman state still sought a modern means of fighting them. As with the Hamidiye, which was modeled on the Cossacks in the Russian Empire, the Ottomans attempted to borrow techniques for locust control from elsewhere. In the summer of 1895, one Ottoman official wrote to the French minister of agriculture expressing an "interest in knowing the measures adopted in Algeria for the destruction of locusts."[70] It seems that the French were not keen on sharing these details, as nearly a year later the Ottoman state reiterated its request for information.[71] The French were both an appropriate and a curious model for the Ottomans to follow. Like some Ottoman officials in the Jazira, the French in Algeria blamed nomads for the desert.[72] In expanding cultivation into these regions, however, they found their advances increasingly beset by locust invasions. But the French were also perhaps not the best example for the Ottomans, since as recently as 1888 a conference in the Algerian city of Oran had given way to the declaration by one expert that "I have been only talking to you of the Americans, Russians, and English and I have never cited the French."[73] It was because, he insisted – "alas!" – that the French had been "outdistanced by the foreigners," even as locusts were so bad in Algeria that they threatened "to stop colonization."

Emulating foreigners had limitations. To begin with, even countries more successful than the French relied on no small amount of luck. In the United States, tinkerers and inventors had come up with creative implements for fighting the insects that one scholar has likened to a "creative conspiracy between Rube Goldberg and the Marquis de Sade."[74] Yet it was not these devices of destruction that made the locust disappear. Instead, it was the plowing up of locust egg-laying grounds as cultivation expanded, a fact that scientists were not even sure of until 1990.[75] Moreover, the Ottoman state had to deal with both geographic variation and budget

[70] BOA, HR.SFR.4 535/2, Ziya Paşa to Gadaud, July 21, 1895.

[71] BOA, HR.SFR.4 535/2, Tevfik to Munir Bey, Ottoman Ambassador to France, April 20, 1896.

[72] Davis, *Resurrecting the Granary of Rome*.

[73] D'Herculais, *Les Sauterelles*, 38. For more on French Algeria and locusts, see Slobodkin, "Empire of Hunger," 35–45.

[74] Lockwood, *Locust*, 50. [75] McWilliams, *American Pests*, 79.

shortfalls. In the spring of 1896, the ministry of forests, mines, and agriculture constructed what it called a "statistical table" showing the entire domains of the empire and specifying whether locusts afflicted them or not.[76] Officials also emphasized the procedure through which the state should be notified of locust invasions and locust egg laying, with the message beginning at the lowest level of government and reaching to the provincial level. Meanwhile, money would flow to the lowest level of government to enable locust destruction. Even if this approach appeared more systematic, the minister of forests, minerals, and agriculture had to acknowledge some shortcomings. The Ottoman Empire simply did not have the resources that even the French wielded in Algeria, which amounted to some 3,000,000 francs annually. The Ottoman minister could call for locusts to be killed while they remained in an "ant-like state" (*karınca halı*) to limit damage, but this advice meant little when so many locusts – whether in "ant-like state" or not – hatched in places devoid of people.[77]

While the state struggled with locusts because of both geography and finance, they also faced similar challenges with Ibrahim Pasha. When he received a medal from the palace in honor of his service to the state, he responded in August of 1896 by expressing gratitude that they would bestow such an award on "a worthless individual in the wilderness [*beyaban*]."[78] His message was an expression of obsequious humility, but it also captured an aspect of the Ottoman plan for him. Because he roamed the "wilderness" and because he did not have too much power, he had value for the Ottoman state. Or so state officials wagered. In September and October of 1896, Ibrahim Pasha even journeyed to Istanbul.[79] Yet increasingly, the gamble on Ibrahim Pasha was not offering returns. In March of 1897, for example, the governor of Diyarbekir, Halid, complained that nearly half of the sheep belonging to the Millî in Mardin had not been counted for tax assessments.[80] He presumably referred to the sheep the Millî cared for on behalf of urban residents, since the Hamidiye guaranteed that its own members did not have to pay such taxes. When tax assessors arrived, associates of the local Hamidiye Brigade commander slipped away with sheep. When a soldier tried to pursue the sheep, he was threatened with guns. Here was a different meaning of Ibrahim's description of

[76] BOA, BEO 780/58474, Selim, 25 Mart 1312 (April 6, 1896).
[77] BOA, BEO 780/58474, Selim, 10 Nisan 1312 (April 30, 1896).
[78] BOA, Y.PRK.ASK 114/45, Ibrahim Pasha, 11 Ağustos 1312 (August 23, 1896).
[79] Ekinci, "Osmanlı Devleti Döneminde Milli Aşireti XVIII–XIX YY.," 267–268.
[80] BOA, DH.TMIK.M 31/37, Halid to Interior Ministry, 13 Mart 1313 (March 25, 1897).

a "worthless individual in the wilderness": people who used mobility and the environment to avoid taxes.

In the years that followed, the Millî and the Shammar would continue to clash, and both provincial officials and borders themselves would occupy a prominent place in these disputes. Despite coming to reconciliation on ending the imprisonment of Shammar shaykh ʿAli ʿAbd al-Rizzaq in May of 1897, disputes between the groups began again by the fall.[81] The district governor of Zor alleged that the Millî had stolen a number of animals from the Shammar.[82] The governor of Diyarbekir called the charge "manufactured."[83] Officials attempted to resolve the dispute by bringing the parties together. But after many delays they gave up because the Shammar were "far away."[84] The affair established a pattern in which governors advocated for their respective nomads as part of what historian Talha Çiçek has called "institutionalized cooperation" that entwined provincial governance and nomadic motion.[85] The case also foreshadowed how often distance would function as an excuse in future years.

The dynamic persisted later that year as cross-border motion took a new direction. In October of 1897, the Zor district governor complained that Ibrahim Pasha was recruiting for the Hamidiye Brigades among "the tribes near the border" of Zor and Diyarbekir.[86] In siphoning nomadic or seminomadic groups from Zor into Diyarbekir, Ibrahim threatened, according to the district governor, to cause the "destruction and disappearance of a district" (*bir sancağın mahv ve izmihlalı*). It would be ironic if Zor were to disappear because its nomadic groups were attracted to other districts by a state-supported paramilitary composed of seminomadic groups. The district had been formed to encompass nomadic motion within its boundaries. These questions of space and provincial administration were opaque to European observers. The British vice consul in Diyarbekir simply stated that "the great Mesopotamian plain" was "never so bad as now."[87]

Tensions over the border intensified in the midst of a historically cold winter in 1897 and 1898, described in Aleppo as "one of the worst ...

81 BOA, Y.PRK.UM 38/64, Halid to Baş Kitabet, 15 Mayıs 1313 (May 27, 1897).
82 BOA, DH.TMIK.M 39/21, Halid to Interior Ministry, 6 Eylül 1313 (September 18, 1897).
83 BOA, DH.TMIK.M 39/21, Halid to Interior Ministry, 27 Teşrinievvel 1313 (November 8, 1897).
84 BOA, DH.TMIK.M 39/21, İsmet Pasha to Diyarbekir Province, 29 Eylül 1313 (October 11, 1897).
85 Çiçek, *Negotiating Empire in the Middle East*, 127.
86 BOA, DH.TMIK.M 38/39, 14 Teşrinievvel 1313 (October 26, 1897).
87 TNA-UK, FO 195/1981, Waugh to Currie, October 10, 1897.

since 1880."[88] In December, Zor district governor Zuhdi renewed his claims about the Millî enrolling people in the Hamidiye and attracting them north across the border into Diyarbekir, this time near Ras al-Ayn, site of the Chechen settlement dating to 1866.[89] Here again was a reversal of sorts. Rather than acting as a bulwark against the desert and nomads to the south, Ras al-Ayn was seeing its people pulled north into Diyarbekir province. While bringing people north, the Millî were also pushing sheep south into Zor, presumably to avoid taxation. As Zuhdi noted, whenever tax assessors from Diyarbekir province arrived, the Millî moved toward Ras al-Ayn in the district of Zor.[90] "By crossing the border" from Diyarbekir into Zor "and mixing with the tribes," Zuhdi explained, the Millî and their sheep would be counted "neither here nor in Diyarbekir." To abet the effort, Ibrahim Pasha issued complaints that the Shammar had crossed the Khabur River.[91] The district governor of Zor found the complaint odd, seeing as the Khabur River was clearly in Zor, but Ibrahim used the specter of conflict between the groups and the Millî's long-standing pastures on the border of the province as justification. Borders had been and would continue to be flash points of conflict in the Jazira. But they could also be utilized differently, as a way to make people – or sheep – disappear.

The proximity of the tribes and the border figured into a number of disputes over the course of the spring. Yet again, the Shammar claimed camel theft, and yet again the governor of Diyarbekir dismissed it as a "fabrication."[92] As the disputes continued alongside losses of livestock due to the cold and cattle plague, Şakir Pasha, the mastermind behind the Hamidiye Brigades, weighed in.[93] He lamented that "two high officials like a district governor and a governor" were fighting over camels when the region was "in need of an increase of settlement and prosperity."[94] Of course, Şakir Pasha's remarks were disingenuous, not least because he had

[88] CADC, 4CCC/39, Agriculture, undated.
[89] BOA, DH.TMIK.M 46/31, Zuhdi to Interior Ministry, 3 Kanunuevvel 1313 (December 15, 1897).
[90] BOA, DH.TMIK.M 46/22, Zuhdi to Interior Ministry, 20 Kanunuevvel 1313 (January 1, 1898).
[91] BOA, DH.TMIK.M 46/22, Zuhdi to Interior Ministry, 4 Kanunuevvel 1313 (December 16, 1897).
[92] BOA, DH.TMIK.M 46/22, Zuhdi to Interior Ministry, 7 Kanunusani 1313 (January 19, 1898); DH.TMIK.M 46/22, Halid to Interior Ministry, 30 Kanunusani 1313 (February 11, 1898).
[93] TNA-UK, FO 195/2017, Barnham to Currie, March 31, 1898.
[94] BOA, DH.TMIK.M 50/32, Şakir to Interior Ministry, 3 Mart 1314 (March 15, 1898).

helped to empower groups such as the Millî in the first place. As discussions continued through the spring about which tribes and whose camels were where, Ibrahim Pasha took on a more aggressive tone. He alleged that the district governor of Zor was in cahoots with the Shammar to make fraudulent claims of animal theft and divide up any proceeds they secured through restitution.[95] Ibrahim moreover insisted that it was the Shammar who were a threat to him. He used provincial borders to make this case. Under the leadership of Faris, the Shammar had, Ibrahim insisted, engaged in "the violating of our borders" (*hududumuzu tecavüzle*) and set up camp only one-and-a-half hours from Ibrahim Pasha's tents outside of Viranşehir. It is unclear if Ibrahim referred to provincial borders or some other spatial unit, but in any case he called for the Shammar "to be expelled and distanced from our borders" and "returned to their customary pastures."

But borders continued to be transgressed. According to the district governor of Zor, "the borders of the provinces of both Diyarbekir and Zor are clear."[96] Yet many tribes "crossed the borders of the provinces with the excuse of grazing animals" and then used their motion to collect the *khawa* tax within those other provinces. In light of these tensions, the governor of Diyarbekir declared that neither group ought to be able to cross the provincial border between Diyarbekir and Zor.[97] This order was not respected. In late May, the governor of Diyarbekir described how Faris Pasha had come "within a distance of two hours of Nusaybin," which is to say near the border between Zor and Diyarbekir. Faris immediately began having disputes with the Tayy, resulting in both groups trampling upon nearby cultivated fields.[98] In other words, the Shammar remained very much on the edge, both in terms of provincial borders and in terms of cultivated and uncultivated areas.

While the Shammar continued to utilize the edge, Ibrahim Pasha seemed to have plans for provincial borders too. An official in Zor argued that through instigating conflict with the Shammar, Ibrahim Pasha aimed "to push out and distance the Shammar" from their customary realms of

[95] BOA, DH.TMIK.M 52/18, Ibrahim Pasha to Diyarbekir Province, 8 Şubat 1313 (February 20, 1898); DH.TMIK.M 51/53, Ibrahim Pasha to Interior Ministry, 28 Mart 1314 (April 9, 1898).

[96] BOA, DH.TMIK.M 51/53, Zuhdi to Interior Ministry, 13 Nisan 1314 (April 25, 1898).

[97] BOA, DH.TMIK.M 51/53, Diyarbekir Governor Halid to Interior Ministry, 2 Mayıs 1314 (May 14, 1898).

[98] BOA, BEO 568/42574, Diyarbekir Governor Halid to Grand Vizier, 15 Mayıs 1314 (May 27, 1898).

migration.[99] To convey this point, the official referred to the geography by the term that had for so long captured the region's connections but had not appeared on the map as a unit of Ottoman administration. The official said that he believed Ibrahim Pasha did not simply want the Shammar out of Zor but, indeed, out of the "Jazira" altogether. Thus even as the disputes between the Shammar and the Millî and various state officials took the form of arguments over where Zor and Diyarbekir ended and which people were where, the larger subject at stake was the amorphous regional moniker that had long haunted these provincial distinctions: the Jazira.

KHEDIVE OF THE DESERT

Amidst these disputes about provincial borders entered a figure who would come to be more associated with borders in southwest Asia than perhaps anyone else, so much so, in fact, that his name alone would eclipse the legacy of Ottoman deliberations and local negotiations over this space. Having often found refuge in his travels across the region during his youth as the only child of "a dysfunctional, landed Yorkshire family," he had taken leave from his studies at Cambridge to journey across the Syrian desert in the spring of 1899.[100] But he failed to receive a permit from Ottoman authorities, and so he took a more northerly route, crossing some of the provinces in which the Shammar and Millî were fighting. His path stretched from Aleppo to Baghdad, and then northward to Van and beyond. In his written account of the journey, the young man promised to avoid "the omniscience of the journalist or the globe-trotter."[101] Yet his travelogue revealed little reluctance to judge or capacity for self-doubt, very much in line with Kipling's "White Man's Burden," published the same year to encourage US colonization of the Philippines. The tendency to receive varying estimates of distance from passersby? "Really the most annoying form of stupidity I have ever met."[102] Uncultivated stretches beside the Euphrates? "Misery such as only the east can produce."[103] The preference of some to walk beside a road while it was under construction, rather than upon it? Further evidence that "if it is possible to put a thing to a use for which it was

99 BOA, DH.TMIK.M 51/53, Tahrirat Müdürü to Interior Ministry, 2 Haziran 1314 (June 14, 1898).

100 Barr, *A Line in the Sand*, 4; Sykes, *Through Five Turkish Provinces*, vii.

101 Sykes, *Through Five Turkish Provinces*, viii. 102 Ibid., 35. 103 Ibid., 37.

never intended an oriental will do it."[104] And regarding the region's various religious and ethnic groups? "Even Jews have their good points but Armenians have none."[105] When he began his journey, this child of the British Empire was nineteen years old. His name was Mark Sykes.

As Sykes passed through a region whose borders he would be indelibly linked with, tensions mounted over the role of the Hamidiye Brigades. On the one hand, figures such as the governor of Mosul, Ebubekir Hazim Tepeyran, described the Hamidiye Brigades as "a long nightmare."[106] On the other hand, people such as the German archaeologist Max von Oppenheim – who first encountered Tel Halaf, future site of his excavations, in 1899 – described Ibrahim Pasha as a sage operator who turned his place on the edge into political power. Oppenheim allowed that Ibrahim Pasha's continuous migration prompted city dwellers to "live in dread of him," but the German also noted that the Millî chief had transformed Viranşehir into a market center, appealing enough to convince Muslim and Christian merchants alike to have "built permanent houses amidst the ruins."[107] What the dueling images of Ibrahim Pasha revealed was the ambiguous perception of the chief.

Increasingly, however, the vision of Ibrahim Pasha as a miscreant and the Jazira as a spoiled land prevailed, even as it seems clear that it was far from Ibrahim Pasha's fault. Upon receiving news that a group of nomads had set up between Mardin and Siverek, the French vice consul in Diyarbekir captured the fatalistic mood attached to the place: "voila, a new ingredient of trouble in a region already in complete anarchy."[108] Other reports described widespread flight from villages in the countryside to places such as Urfa and Suruç, where people begged for bread.[109] While most accounts focused on the deleterious impact of Ibrahim Pasha, they largely left out the drought that ravaged harvests in eastern Aleppo, Mosul, and Diyarbekir, a dynamic that no doubt contributed to the suffering across the region.[110] The environment could not be held accountable, but Ibrahim Pasha could, or so it seemed, and in deference to the many complaints about him, the Ottoman government acquiesced to sending a commission to investigate.[111] Few had confidence in the step,

[104] Ibid., 39. [105] Ibid., 80. [106] Tepeyran, *Hatiralar*, 436.

[107] Oppenheim, *Tel Halaf*, 2.

[108] CADN, 166PO/D/22/1, Diyarbekir to Constantinople, July 28, 1899.

[109] TNA-UK, FO 424/199, O'Conor to the Marquess of Salisbury, August 30, 1899, Inclosure: Falanga to O'Conor, 18 August 1899.

[110] CADC, 4CCC39, Rapport commercial pour l'année 1899, October 20, 1900.

[111] CADN, 166PO/D/22/1, Diyarbekir to Constantinople, October 11, 1899.

believing that Ibrahim had local officials in his pocket. These perceptions seem to have been warranted. The most immediate result of the investigation of Ibrahim Pasha was actually the removal of one of the local officials who had most irked the Millî chief: Zuhdi Bey, the district governor of Zor, who had often advocated on behalf of the Shammar.

Zuhdi Bey's replacement in Zor Ahmed Şükrü continued to complain about the Millî chief and border crossing, and the Millî continued to find other borders to cross and take advantage of. In January 1900, the new district governor of Zor claimed that Ibrahim Pasha had come within two hours of Ras al-Ayn in January with the aim of collecting taxes on the sheep of local residents.[112] For fear of conflict between the Millî and "the tribes here," the district governor called for Ibrahim Pasha "to be removed from the borders of the district." But as had been evident in the past, Ibrahim Pasha derived his power from the borders – whether they were between provinces or between cultivated and uncultivated lands – so any measure aimed at removing him from them was not likely to succeed. In May, complaints surfaced in Derik of Ibrahim Pasha moving with a group of over 1,000 tents onto cultivated lands, where they allowed their animals to feed on crops.[113] European consuls struggled to intensify their already dire predictions. The British consul in Aleppo opted for "the same anarchy but more pronounced."[114] In the 1860s, the Ottomans had hoped to stop the motion of the Shammar through cordons, and then to use groups like the Millî or Chechen refugees as bulwarks against nomadism. By 1900, as Ibrahim Pasha's tax-free animals fed on cultivated lands, the reversal of Ottoman policies in the region was readily apparent.

If state support of Ibrahim Pasha represented a reversal of sorts, it also represented a return, as his motion across borders invited comparisons to the winged marauders who also arrived in the Jazira at the same time. After an absence of several years, locusts struck in Urfa in 1900. And in 1901, they appeared again, as the land along the Euphrates all the way from Rumkale down to Deir ez-Zor crawled with newly hatched locusts.[115] Tepeyran, the governor of Mosul, recalled how "black clouds"

[112] BOA, DH.TMIK.M 80/47, Zor District Governor Ahmed Şukrü to Interior Ministry, 26 Kanunuevvel 1315 (January 7, 1900).

[113] BOA, DH.TMIK.M 87/36, Derikli Ahmed and Derikli Haci Osman to Grand Vizier, 12 Mayıs 1316 (May 25, 1900).

[114] TNA-UK, FO 195/2073, General Report upon the Vilayet of Aleppo for the Quarter Ended 30 June 1900.

[115] TNA-UK, FO 195/2095, Barnham to O'Conor, Enclosure: General Report on the Vilayet of Aleppo, May 13, 1901.

of locusts left the city in "darkness."[116] He was a man with little patience for superstition, and so while he acknowledged the role of the starling in fighting locusts, he also complained that Sufi shaykhs made money off of the blessed water believed to attract the locust-eating birds.[117] No money was necessary, the dyspeptic bureaucrat wryly noted, for the storks that did the same duty for free. While Tepeyran's observations about locusts attested to the way they continued to afflict the region, Tepeyran himself also participated in the legacy of likening nomadic groups to the insects. When he referred to the Hamidiye Brigades as "savage swarms" (*vahşi sürüler*), he played on a durable metaphor used by many at the time. One British consul, for example, referred to the Hamidiye near Cizre as "like locusts, pillaging the settled population, burning the forests, and generally leaving a track of desolation."[118] Tepeyran also used choice words to convey how the Hamidiye committed "endless injustices" (*hudutsuz zulümler*).[119] The word that he used to mean "endless" literally meant "borderless," which was in fact an accurate designation given the way that Ibrahim Pasha – like the locusts and like so many nomads before him – had used his place on the borders to thwart the very structures through which Ottoman officials envisioned reforming the region. He alternately brought people north from Zor into Diyarbekir for enrollment in Hamidiye Brigades and sent sheep south to avoid tax assessors in his home province.

But Ibrahim Pasha's emergence in the Jazira would not go unchallenged, and in 1901, an unprecedented conflict broke out between the ascendant Ibrahim and the Shammar. The spark was a Millî raid in January of 1901 against the Kara Keçe, another seminomadic Kurdish tribe that had been tagged for settlement in the 1864 plan.[120] Tensions persisted between the groups throughout the early months of 1901 and took the familiar form of provincial governors arguing about the blurry overlaps between nomads, borders, and the desert. In February, the district governor of Zor once again suggested that the people of Ras al-Ayn were "being ... taken to the border of Diyarbekir."[121] Reports further suggested that Ibrahim Pasha and the Millî had staked tents some nine hours north of Zor and had plans to attempt to recruit local tribes to their

[116] Tepeyran, *Hatıralar*, 404. [117] Ibid., 405.
[118] Tepeyran, *Hatıralar*, 441; TNA-UK, FO 424/203, Tyrrel to O'Conor, October 7, 1902.
[119] Tepeyran, *Hatıralar*, 441.
[120] TNA-UK, FO 195/2104, Francis Jones to E. de Bemsen, January 22, 1901.
[121] BOA, DH.TMIK.M 100/37, Ahmed Şükrü to Interior Ministry, 27 Kanunusani 1316 (February 9, 1901).

side.[122] The district governor of Zor wrote that there were real fears that Ibrahim's actions could spoil "the security of the desert entirely." Added to the familiar concerns of Ibrahim using pastoral ambiguity to attract Zor's population were the ominous signs of conflict between the Shammar and Ibrahim Pasha. While the district governor of Zor complained about Ibrahim Pasha's proximity to Zor, the governor of Diyarbekir complained about the proximity of the Shammar to Diyarbekir.[123] What was more, the governor of Diyarbekir alleged that the Shammar had sent a courier to encourage the Kara Keçe to unite in a fight against Ibrahim Pasha, leaving the Millî in a state of "great anxiety and fear." To prevent "any great unfortunate event" (*büyük bir fenalığı*), he called for the Shammar "to be returned" to the Sinjar region, safely distant from Diyarbekir on the border of Mosul and Zor.

The Shammar were not forced to return, and officials scrambled to prevent any fighting between the groups. By early April, the Millî had struck a Shammar encampment of some 300 tents near Ras al-Ayn on the Zarkan River.[124] Subsequent reports from Zor officials alleged that the Millî horsemen had killed some fifteen men, eleven women, and twenty-three children and made away with 1,200 camels and 20,000 sheep.[125] There were rumors, too, that the Millî had attacked and violated the women of the Shammar camp, which Ottoman officials deemed "very contrary according to tribal customs."[126] Officials feared Shammar reprisals against the Millî, with word of some 1,600 horsemen being summoned for this purpose.[127] In an echo of 1868, the Shammar had painted a camel black and dispatched young women clad in black to help summon other branches of the Shammar – indeed, the branches whose division had been accentuated by the formation of Zor in 1871.[128] Efforts to mediate between the groups in late April failed, as they could barely even agree upon a place to meet to discuss

[122] BOA, DH.TMIK.M 100/11, Zor District Governor Şükrü to Interior Ministry, 28 Kanunusani 1316 (February 10, 1901).

[123] BOA, Y.PRK.UM 53/126, Diyarbekir Governor Halid, 8 Mart 1317 (March 21, 1901).

[124] BOA, DH.TMIK.M 102/10, Ahmed Şükrü to Interior Ministry, 19 Mart 1317 (April 1, 1901).

[125] BOA, DH.TMIK.M 102/10, Ahmed Şükrü to Interior Ministry, 21 Mart 1317 (April 3, 1901).

[126] CADC, 166PO/D/1/87, Pognon to Constantinople, May 27, 1901; BOA, Y.PRK.UM 53/93, Ahmed Şükrü, 3 Nisan 1317 (April 16, 1901).

[127] BOA, DH.TMIK.M 102/10, Ahmed Şükrü to Interior Ministry, 21 Mart 1317 (April 3, 1901).

[128] Tepeyran, *Hatıralar*, 461.

the matters.[129] When they did meet, Shammar chief Faris declared that if the animals and property taken from the Shammar by Ibrahim Pasha were not requisitioned, Faris would consider obtaining "English or Russian citizenship" (*İngiliz ve yahut Rusya tabiiyeti*).[130] The opportunistic use of foreign citizenship was a growing concern of the Ottoman state at the time, whether with respect to Central Asians in Istanbul, South Asians in the Hijaz, or Armenians in Anatolia.[131] Faris's vow concerned local officials so much that they scrambled to ascertain whether there were any English or Russian citizens among the Shammar instigating conflict (there were not).

Finally, the great battle that officials had feared took place. According to the account of officials in both Mardin and Zor, on May 5, 1901 the Shammar came within three hours' distance of the Millî. Believing that the Shammar planned to attack, Ibrahim Pasha struck first, and his forces – totaling 10,000 of his own fighters along with irregular forces from Mazıdağı and Diyarbekir – did so while the Shammar were in their most vulnerable state: unloading the contents of their 500 tents.[132] Ibrahim Pasha's forces fired their famous breechloading Martini rifles.[133] The bullets were "like rain," and they not only struck members of the Shammar but also spooked the Shammar camels. The ensuing stampede crushed "no fewer than 100 women and children ... under the hooves of the camels."[134] The creatures so crucial to the mobile life of the Shammar in the Jazira had become a tool of their own destruction.

In the wake of the battle, camels would continue to play a role as kingmakers, confirming Ibrahim Pasha's resounding power in the region. In response to the attack, many of the Shammar fled into the desert. The Millî made off with some twenty human prisoners as well as, by one probably exaggerated account, some 200,000 animals, mostly sheep but

[129] TNA-UK, FO 195/2104, Jones to O'Conor, April 16, 1901, April 30, 1901.

[130] BOA, Y.PRK.UM 53/126, Ahmed Şükrü to Yıldız, 9 Nisan 1317 (April 22, 1901).

[131] Can, *Spiritual Subjects*, 94–124; Gutman, *The Politics of Armenian Migration to North America*, 123–153; Low, *Imperial Mecca*, 88–90, 104–111; Yılmaz, "Governing the Armenian Question through Passports in the Late Ottoman Empire."

[132] BOA, Y.PRK.UM 54/8, Mardin District Governor and Administrative Council of District of Mardin, 5 Mayıs 1317 (May 18, 1901).

[133] For more on these rifles, see Bennett, "The 'Aynalı Martini'"; Ghazarian, "Ghost Rations," 146–157; Öztan, "Tools of Revolution."

[134] BOA, Y.PRK.UM 54/8, Interrogation Reports, 27 Nisan 1317 (May 10, 1901); Mardin District Governor and Administrative Council of District of Mardin, 5 Mayıs 1317 (May 18, 1901).

also camels.[135] At first, the Millî had only been able to capture baby camels, while the mother camels remained with the Shammar. But a few days after the battle, the mother camels fled the Shammar in search of their offspring and thus fell into Ibrahim Pasha's hands too. It very much seemed like the Jazira and its creatures were on Ibrahim Pasha's side. Accentuating this fact was the way Ibrahim Pasha's troops celebrated their victory. "God grant victory to the Khedive of the Desert ... God grant victory to the Sultan of the Open Lands Ibrahim!" The environmental specificity of the epithets attested to the nature of the Millî chief's power, supplanting the Shammar's long-standing claims to this title. But the appellation of "sultan" concerned many who believed the title was one of the "holy rights" reserved for the Ottoman head of state, and not appropriate for a provincial power broker, perhaps especially given the fact that foreign observers increasingly referred to Ibrahim's domains as "a little empire of his own."[136]

As Ottoman officials investigated the clash between the Millî and the Shammar, testimonies revealed the role played in the conflict by ambiguities of distance and space enabled by nomadic pastoralism. In the many pages of interrogations on the incident, numerous observers raised questions about whether the Shammar actually intended to attack the Millî. To be sure, the Shammar had many reasons to do so, given the confrontations between the groups over the previous few months. But one observer suggested that the Shammar sought "a place that had grass since there was no grass remaining for the animals" where they had previously been.[137] No fewer than four other people made the same point, all applying the adverb "merely" to emphasize the innocence of the Shammar movement toward Ibrahim Pasha.[138] Such claims were plausible given the presence of locusts and a drought that was the worst in "living memory" in northern Syria, leaving "the Euphrates ... fordable everywhere."[139] But plans to attack the Millî and efforts to find pasture were not mutually exclusive. One observer made this point exactly,

[135] BOA, Y.PRK.UM 54/8, Mardin District Governor and Administrative Council of District of Mardin, 5 Mayıs 1317 (May 18, 1901).

[136] TNA-UK, FO 424/202, N. O'Conor to the Marquess of Lansdowne, September 5, 1901, Inclosure: Vice Consul Freeman to O'Conor, August 22, 1901; TNA-UK, FO 195/2095, Barnham to N. O'Conor, May 13, 1901; BOA, Y.PRK.UM 54/8, District Governors of Mardin and Zor to Dar al-Saadet, 7 Mayıs 1317 (May 20, 1901).

[137] BOA, Y.PRK.UM 54/8, Testimony of Hanna, 26 Nisan 1317 (May 10, 1901).

[138] BOA, Y.PRK.UM 54/8, Testimony of Tuma, Husayn, Yakub, and others, 28 Nisan 1317 (May 12, 1901).

[139] TNA-UK, FO 195/2095, Barnham to O'Conor, October 21, 1901.

suggesting that the Shammar had decided to "threaten and pressure Ibrahim Pasha by grazing their animals in the pastures of Viranşehir."[140] The lack of clarity pointed to the kind of flexibility embedded in pastoralism, as motion could be construed as innocent or threatening depending on the circumstances.

But also remarkable was the source of so much of the testimony regarding Shammar movement, which came almost entirely not from the Shammar, presumably far away in the desert. Instead, the testimonies emanated from around eighty largely Armenian Catholic and Assyrian merchants from Mardin traveling with the Shammar at the time of the Millî attack.[141] The presence of these blacksmiths and butchers not only suggested how the Shammar perhaps did not plan to attack the Millî (or at least did not plan to attack them at that moment), but also the degree of interconnection between settled and nomadic populations, Christians and Muslims. The natives of Mardin among the Shammar had moved with the nomads' camps for some two weeks, but others had been with the group for considerably longer. Their presence with nomads moreover foreshadowed how, in a very different context years later in the same deserts around Ras al-Ayn, some would manage to survive the genocide of World War I by living – and moving – with groups like the Shammar.

"I WONDERED IF AT THE BACK OF ALL THAT UPHEAVAL ... THERE WAS A LOCUST"

Locusts did not destroy the Jazira in 1901 as they had in years past. But many suggested that the nomadic groups to whom locusts were compared did. The French vice consul in Diyarbekir called the situation "the most complete anarchy," describing how the region had become largely "deserted" as a result of being "infested ... with brigands."[142] The roads between Aleppo and Mosul were "essentially closed," and the district governor of Zor estimated that tax was assessed on only one in forty sheep.[143] There were even rumors of intervention from Ibn Rashid and the relatives of the Shammar in the Najd who had remained when their ancestor had journeyed to the Jazira nearly one hundred years before.[144]

140 BOA, Y.PRK.UM 54/8, Testimony of Yusuf Terzibaşı, 28 Nisan 1317 (May 11, 1901).
141 BOA, Y.PRK.UM 54/8, Testimony of Hanna, 26 Nisan 1317 (May 10, 1901).
142 CADC, 206CPCOM/75, Diyarbekir to Ministry of Foreign Affairs, June 28, 1901.
143 BOA, DH.TMIK.M 105/42, Zor District Governor to Interior Ministry, 6 Haziran 1317 (June 19, 1901).
144 CADN, 166PO/D/1/87, Pognon to Constantinople, May 27, 1901.

Ibrahim Pasha insisted that the problem was in fact the Shammar.[145] He showed his keen sense of politics when he dispatched a delegation of Christian leaders of Viranşehir to Diyarbekir to convey concerns about the Shammar "constantly attacking them."[146] As in years past, whether in response to nomads or locusts, the Millî moved to the elevated pastures of Mount Karaca, the same region that had been the limit of locust invasions and a setting of conflict with the Shammar in 1861.[147] The crops that the Millî left behind on the edge of the desert were burned.[148]

Meanwhile, restitution by the local administration led to government buildings being overrun by animals. Diyarbekir officials displayed "an extraordinary zeal" as they seized animals that the Millî had stolen from the Shammar.[149] The efforts gave way to what the French vice consul called "a fine spectacle," as the courtyard of the government building of Diyarbekir filled with sheep and camels "who bleat and cry." Government support for groups such as the seminomadic Millî represented a departure from Midhat Pasha's dreams of the Jazira as a sedentary agricultural heartland, and few images conveyed this reality more clearly than the hooves and calls of animals virtually overtaking the seat of government in Diyarbekir.

Many again raised the question of how to manage nomads more effectively, given the unique challenges of the environment and provincial borders in the Jazira. Familiar assessments and suggestions arose, including the difficulty of stationing troops in the desert during the summer, the "extremely vast" extent of provincial borders, and the possibility of establishing an agile mule-mounted regiment at the provincial borders.[150] Others proposed using other nomadic groups – rather than Ottoman soldiers – as a means of resolving the fighting between the Shammar and the Millî. The foremost advocate of the policy was the

145 BOA, DH.TMIK.M 106/9, Delegation of Suryani and Armenian Christians to Grand Vizier, 17 Haziran 1317 (June 30, 1901). In another show of the mutually beneficial arrangement between Ibrahim and local Christians, American missionaries would smart that the Christians of Viranşehir were "more fanatical than the Moslems," as they had prevailed upon Ibrahim Pasha to expel missionaries from the place: CADC, 206CPCOM/75, Vice Consulate of Diyarbekir to Ministry of Foreign Affairs, July 18, 1901.

146 ABCFM 16.5, vol. 42, Annual Report of Mardin Station, 1903.

147 CADC, 206CPCOM/75, Diyarbekir to Ministry of Foreign Affairs, July 18, 1901.

148 CADN, 166PO/D/22/2, July 18, 1901.

149 CADC, 206CPCOM/75, Diyarbekir to Ministry of Foreign Affairs, August 16, 1901.

150 BOA, Y.PRK.ASK 172/71, Fourth Army Commander Mehmed Zeki, 28 Temmuz 1317 (August 10, 1901); Y.PRK.UM 55/98, Diyarbekir Council, 29 Ağustos 1317 (September 11, 1901); Y.PRK.UM 55/98, Diyarbekir Governor Halid, 31 Ağustos 1317 (September 13, 1901).

military commander of Urfa, Mehmed Lütfü. He blamed the Shammar for "engaging in the widening of the border" (*tevsiye-i hudud ederek*) through their motion in Zor, and he believed they intended to expand their realms into places such as Diyarbekir and Urfa.[151] To prevent this fate, he called for them to be removed from Zor and returned to Mosul. Into Zor, meanwhile, would come the ʿAnaza, longtime enemies of the Shammar. The district governor of Zor reasoned that given the "breadth and size of the district" and the "lack of soldiers," something had to be done if the "security of the desert" were to be preserved.[152] By December, he acquiesced to the proposal for the ʿAnaza to be attached to Zor, and the Shammar to Mosul.[153] His province was Zor, but he acknowledged the broader region it was a part of when he articulated the goal of these measures: to "prevent the Jazira from being an axis of motion."

If Ottoman forces sought to control motion, other forces conspired to maintain it. In January of 1902, Ibrahim Pasha returned to Viranşehir from the pastures to the south. He typically stayed "in the desert with his flocks and herds until the middle of April," but he feared being vulnerable to the Shammar there.[154] What was more, locusts had arrived, and while officials had been appointed to oversee the destruction of locust eggs, there remained the specter, as always, of swarms emerging from the desert.[155] The tensions between Ibrahim Pasha and the Shammar thus had to do with grazing, locust swarms, and the violence of the previous year.

These tensions also intersected with ethnicity. It was not explicitly invoked by those in the heat of the conflict. But British observers presented the tensions in these terms, noting that the Ottoman state had charged groups such as the Millî to mark the "frontier line along the southern border of Kurdistan … and the Arab country."[156] Nor did agents of the British empire have a problem with this approach; on the contrary, one official wrote, "there is much to recommend the system chosen." Ottoman officials did not necessarily see policy in these terms. After all, the call for the

[151] BOA, Y.PRK.ASK 175/26, Urfa Military Commander Mehmed Lütfü to Grand Vizier, 14 Eylül 1317 (September 28, 1901)

[152] BOA, DH.TMIK.M 116/8, Ahmed Şükrü to Interior Ministry, 21 Teşrinisani 1317 (December 4, 1901); DH.TMIK.M 111/36, Şükrü to Interior Ministry, 24 Eylül 1317 (October 7, 1901).

[153] BOA, DH.TMIK.M 146/42, Zor District Governor to Interior Ministry, 12 Kanunuevvel 1317 (December 25, 1901).

[154] TNA-UK, FO 195/2125, Anderson to O'Conor, January 28, 1902.

[155] BOA, BEO 1789/134110, Council of State, 19 Kanunusani 1317 (February 1, 1902).

[156] TNA-UK, FO 195/2125, Anderson, Report on Diyarbekir Vilayet, March Quarter, 1902.

ʿAnaza to be attached to Zor rested on conflict between Arab tribes. But the Hamidiye did represent a concerted effort on the imperial level to cultivate ties with the empire's Kurdish populations, a dynamic so pronounced that Abdülhamid enjoyed the title of Bavé Kurdan, or Father of Kurds.[157]

Ultimately, the interior ministry quashed the plan of replacing the Shammar with the ʿAnaza and instead offered a starkly different vision of managing the Jazira. The interior ministry argued that few gains would be made from moving the ʿAnaza to Zor and the Shammar to Mosul without more troops to enforce settlement.[158] Moreover, no progress could be achieved as different tactics were adopted all across the region, with some administrations attempting to settle nomads and others attempting to attach them to new administrations. The ministry of the interior called, then, for the Shammar to be limited to the region between the Tigris and the Euphrates, and the ʿAnaza to the region known as the Shamiyya to the southwest of the Euphrates, after which they called for forts to be built along the river and mule-cavalry units to maintain order. The officials thus echoed the plans of many previous officials, and they also echoed the certainty with which officials such as Midhat Pasha had viewed nomads with disdain: "the moving inhabitants [*sekene-i seyyare*] as well as the country of the Jazira," concluded the ministry, were not "a benefit for the state."

If this premise shaped Ottoman policy, it offered little guidance for how to practically manage concerns that the Shammar would strike back against the Millî in revenge for their losses under camel hooves in 1901. In this regard, the policy seems, once again, to have been to occupy the same interstitial places that the nomads had long utilized to carve out their own spaces of autonomy. A report for Yıldız Palace called for mule cavalry – renowned for their "famous power among the Arabs [*urbân*]" – to be stationed "in the desert" for the fifteen to twenty days of seasonal migration to prevent conflict.[159] The report also suggested using provincial borders as a way to maintain separation. It called for the stationing of troops near the Zarkan River, since it formed "the border of Diyarbekir province with Zor district" and also was "a natural transit point for the Shammar to cross the border."

While Ottoman officials argued about policies and struggled with the challenges of the Jazira, Ibrahim Pasha wasted little time in enlarging his

157 Ateş, *The Ottoman-Iranian Borderlands*, 231.

158 BOA, DH.TMIK.M 113/74, Interior Ministry to Grand Vizier, 5 Mart 1318 (March 18, 1902).

159 BOA, Y.MTV 230/47, 16 Nisan 1318 (April 29, 1902).

power. In 1902, motion remained the rule in the Jazira, with Christian leaders of Mardin decrying the convergence of "drought, cholera, fire, and locusts."[160] And with these disasters came the problems of years past, including rumors of imminent clashes between the Millî and Shammar, together with the simultaneous need and impossibility of stationing troops in the desert thanks to the "heaviness of the air."[161] Meanwhile, Ibrahim Pasha expanded. He sent his subordinates to villages "within a distance of eight miles from Diarbekir," where they threatened violence if they did not receive annual payments.[162] Such measures would increasingly embroil Ibrahim Pasha in another set of conflicts over the coming years, as many of these villages were owned by Diyarbekir notables. Ibrahim's power came not just from guns, but also butter. To ensure support from high levels of the Ottoman state, Ibrahim also funneled the dairy product of his flocks to the imperial kitchens as well as horses to the sultan.[163]

Provincial borders remained a space of conflict, and the disputes had fiscal consequences. In early January of 1903, some 20,000 tents of the Shammar and Millî were pitched in the vicinity of Ras al-Ayn, as they once again faced off in the borderlands of Diyarbekir and Zor provinces.[164] As in years past, Ibrahim crossed into Zor to take advantage of administrators unsure of how to handle him, while also bringing "the settled people of Zor" into his ranks.[165] Eventually, tensions were defused, as Ibrahim Pasha returned to Viranşehir – once again, earlier than usual – and the Shammar decamped south to al-Shaddadah.[166] Officials in Aleppo province offered an ominous assessment, warning that conflict "increased day by day" and "the total destruction [*bütün bütün hirabiyetini*] of this district" loomed.[167] As Ibrahim Pasha returned to the region near Urfa, so many troops were devoted to maintaining order between the nomads

160 BOA, DH.MKT 492/65, Zor District Governor to Interior Ministry, 20 Mart 1318 (April 2, 1902); DH.TMIK.M 126/29, Christian Leaders of Mardin to Interior Ministry, 25 Nisan 1318 (May 8, 1902).

161 BOA, DH.TMIK.M 127/40, Diyarbekir Governor Faik to Interior Ministry, 23 Haziran 1318 (July 6, 1902); DH.TMIK.M 127/2, Military Commander, 17 Ağustos 1318 (August 30, 1902).

162 TNA-UK, FO 195/2125, Anderson to O'Conor, Reports for Vilayets of Diyarbekir and Kharput, July 2, 1902.

163 CADN, 166PO/D/22/2, August 13, 1902.

164 TNA-UK, FO 195/2137, Barnham to O'Conor, January 13, 1903.

165 BOA, Y.MTV 241/6, District Governor of Zor to Interior Ministry, 9 Kanunusani 1318 (January 22, 1903).

166 TNA-UK, FO 195/2137, Barnham to O'Conor, February 24, 1903.

167 BOA, Y.MTV 241/6, Aleppo Province to Interior Ministry, 4 Şubat 1318 (February 17, 1903).

that none remained to assess the tax on sheep.[168] Meanwhile, much of the plains south of Mardin appeared empty due to the "depredations of Ibrahim Pasha's men," as many had left their villages for the safety of the hills to the north.[169]

Locusts appeared in the spaces in between. At the same time as officials further west worried about nomads, the governor of Mosul sounded the alarm about the insects. Locusts had been appearing for the previous four years, he explained, and had laid eggs in "places far and wide."[170] Though they could be managed in settled areas, there was little hope for the many areas "empty of improvement or residents." In these places, of course, "it was not known whether eggs had been laid" and "even if it were known" the regions were often "desert or mountains," and a challenge for the labor-intensive work of locust-egg collection and destruction. Nor were these invasions limited to the eastern portions of Mosul. The governor of Aleppo warned in December of 1903 that the locust invasions of the year had resulted in eggs all across the provinces' northern and eastern portions, among them the districts of Urfa, Aintab, Kilis, Bab, Maara, Raqqa, and Manbij.[171] Reports from Birecik on the Euphrates warned that more than three-quarters of the district's territory contained locust eggs.[172]

Mark Sykes – back in the Jazira again after a stint in South Africa during the Boer War – was one person who took note of the insects. While traveling from Mosul to Shirqat, he saw a swarm of locusts that had not yet developed wings, stuck between the Tigris and a cliff too steep for them to climb.[173] "Some black ants were using them as food," Sykes observed, and the ants were killing the locusts "by sawing off their heads, which with the springing legs were carried away, the soft bodies being left behind and covering the ground thickly." Yet in most places in the Jazira, the locusts avoided such gruesome dismemberment. When Sykes saw tensions emerging between people shortly after, he wrote, "I wondered if at the back of all that upheaval and explosion, that smashing and incursion, that slaughter and wreck there was a locust nibbling and

[168] BOA, Y.MTV 241/48, Military Commander, 23 Şubat 1318 (March 8, 1903).
[169] TNA-UK, FO 195/2147, Anderson to O'Conor, April 27, 1903.
[170] BOA, DH.MKT 705/14, Mosul Governor Nuri to the Interior Ministry, 21 Nisan 1319 (May 4, 1903).
[171] BOA, DH.MKT 804/66, Aleppo Governor Mecid to Interior Ministry, 15 Kanunuevvel 1319 (December 28, 1903).
[172] BOA, DH.MKT 804/66, Memo of Aleppo Governor Mecid, 25 Kanunuevvel 1319 (January 7, 1904).
[173] Sykes, *Dar-ul-Islam*, 181.

gnawing and hopping in the rear."[174] Sykes had gotten and would continue to get many things wrong about these places. But in that image of a minuscule locust as the premise of all that followed, he gestured to the ways that the landscape was coproduced by humans and locusts alike.

Indeed, locusts and nomads both took advantage of mobility to carve out a measure of power on the edge. One German traveler who visited the region in the spring of 1903 placed these spatial and environmental dynamics in focus. Ibrahim Pasha may have been "the uncrowned king of Kurdistan," he intoned, but his seat of power did not appear conventionally opulent.[175] True to its name, Viranşehir retained ruins of walls and paving stones from Roman times and left "a dismal impression" with "decay and dirt wherever you looked."[176] But tellingly – and like so many travelers' visits to Shirqat in the east – the seat of Ibrahim Pasha's power did not even contain Ibrahim Pasha. He had decamped for "the grazing areas" to the south, where travelers such as the German "could not follow him."[177] Like locusts, Ibrahim Pasha was comfortable at some remove from cities. There, he could make something out of a place that seemed worthless to outsiders.

This dynamic was further underscored when, in lieu of meeting Ibrahim Pasha, the German traveler met one of his deputies, a Yazidi convert to Islam named Hüseyin Kanco (Kurdish: Hasanê Kenco) in his "castle" of "limestone block walls" east of Viranşehir.[178] The structure sat atop a hill, likely an outgrowth of ruined cities of the past, indeed, the same ruins that had convinced Ottomans and others that the Jazira could be something else. But Kanco and Ibrahim Pasha used this environmental feature to create a kind of power not immediately amenable to the environmental transformation that so many envisioned. After all, Hüseyin Kanco's castle was surrounded only by the "spring green landscape."[179] With mobility, locusts and nomads alike took advantage of an environment that looked like a wasteland.

While nomads and locusts further entrenched themselves on the edges of provincial borders and cultivated lands, Ottoman officials struggled to respond. In 1904, the minister of forests, minerals, and agriculture

[174] Ibid., 190. [175] Wiedemann, "Ibrahim Pascha Glück und Ende" (1909), 54.

[176] Wiedemann, "Ibrahim Pascha Glück und Ende" (1908), 37. [177] Ibid.

[178] The structure is located in present-day Atlı near Derik and known as Kasrı Kanco. Güldoğan, *Diyarbekir Tarihi*, 179; Wiedemann, "Ibrahim Pascha Glück und Ende" (1908), 37.

[179] Wiedemann, "Ibrahim Pascha Glück und Ende" (1909), 52.

expressed concerns about the amount of locust eggs in the eastern portions of Aleppo near "the desert," while no less a figure than the war minister declared anxiety at the prospect of fighting between the Shammar and Millî once again interrupting the sheep count.[180] Disputes among the Shammar over leadership led to clashes around Mosul.[181] Ibrahim Pasha even threw his support behind one of the parties involved.[182] With the distraction of the Shammar by these disputes and the death of Faris in 1904, Ibrahim Pasha took advantage so as to expand his power by collecting tribute from smaller confederations.[183] Fear of his depredations was so great that cultivators and nomads alike of the Harran pitched tents in the vicinity of Urfa, leaving the countryside "totally empty of people," with deleterious consequences for "both the imperial treasury and the people."[184] At the same time, locusts swooped in and, according to the British consul in Aleppo, did "considerable damage to the green crops in this neighbourhood."[185] As mobility across the Jazira once again confounded Ottoman officials, they attempted to reinforce the provincial borders that divided the region. To respond to Ibrahim Pasha's "crossing of the border" from Diyarbekir into Aleppo province, officials arranged to have 150 cavalry stationed "at the border."[186] Within a region cut up by borders, clinging to them was the only remedy some Ottoman officials could imagine to control these broader circuits of motion.

Ottoman officials eventually ended up succeeding in their efforts to quell the conflict, but Ibrahim Pasha's place with respect to the state remained ambiguous. In October, hostilities ended between Ibrahim Pasha and the smaller confederations as officials from Diyarbekir and Urfa successfully brokered a deal between the groups in Viranşehir.[187] Meanwhile, Ibrahim Pasha pushed east, with reports that his men had struck various villages in Mardin and Diyarbekir, again challenging the

180 BOA, BEO 2265/169810, Selim, 8 Muharrem 1322 (25 March 1904); Y.PRK 216/65, Serasker, 4 Nisan 1320 (April 17, 1904).
181 Çiçek, *Negotiating Empire*, 192–195. 182 Ibid., 194.
183 Ibid., 183, 184 n. 25; CADN, Vice Consul in Mosul to MAE, May 12, 1904, June 16, 1904; 166PO/D/22/2, Vice Consul in Diyarbekir to MAE, June 3, 1904.
184 BOA, Y.A.HUS 474/71, Governor of Aleppo Mecid to Yıldız, 14 Haziran 1320 (June 27, 1904); Y.A.HUS 474/82, Aleppo Governor Mecid, 15 Haziran 1320 (June 28, 1904).
185 TNA-UK, FO 424/206, Sir N. O'Conor to the Marquess of Lansdowne, July 19, 1904, Inclosure: Report by Consul Barnham upon the Administration of the Aleppo Vilayet.
186 BOA, Y.PRK.ASK 221/19, Aleppo Governor Kazim and Representative of Aleppo Extraordinary Command Lieutenant General Bekir Sıdkı, 4 Recep 1322 (September 14, 1904).
187 BOA, Y.MTV 266/67, Aleppo and Adana Extraordinary Command, 28 Eylül 1320 (October 11, 1904).

power of the notables of the province.[188] The notables felt doubly threatened because they suspected that their telegrams of complaint did not make it to the central government thanks to Ibrahim Pasha's allies in the bureaucracy. Rumors circulated about how the state might force him to move by, for example, taking his Hamidiye Brigades to fight in Yemen or the Arabian Peninsula.[189] But it seems Ibrahim Pasha was reluctant. Not only did he refuse to go, he also refused to even provide camels for the effort. The majority of his camels were females, he protested, and thus could not be employed "to transport from Ankara to Baghdad loads as heavy as war munitions."[190] Perhaps some of them were the mother camels that found their way into his hands three years before, when they searched so feverishly for their offspring that they escaped from the Shammar. Or perhaps it was just an excuse. The state wished to mobilize his human and nonhuman forces across an imperial geography, but Ibrahim Pasha himself evidently wished to remain in the landscape that had played such a significant role in his political power. And he would do so, even as his power increasingly threatened politically connected forces on the edge of that same space.

"THE MOST INTERESTING PERSON IN THE JAZIRAH"

In previous years, the motion of locusts and nomads had intersected, as had the language treating them and the methods the state used to deal with them. But in 1905, they would come to be linked in a new way, as Ibrahim Pasha's power led to a new challenge to his rule. As in so many years before, Ottoman officials all around the Jazira knew that locusts remained a risk. As the governor of Aleppo explained, no matter how many eggs had been collected in districts such as Kilis, Aintab, and Raqqa in the previous year, "the districts neighboring the desert" such as Bab and Manbij could still be struck by locusts appearing "in a winged state" from the arid interior.[191] Accordingly, he called for extensive funding to support local locust commissions, and their purchase of tools including shovels (for digging up the eggs), petroleum (for setting the eggs ablaze),

188 CADN, 166PO/D/22/2, October 14, 1904.
189 TNA-UK, FO 195/2173, Wilkie Young to Townley, December 15, 1904.
190 CADN, 166PO/D/22/2, December 30, 1904.
191 BOA, BEO 2505/187809, Governor of Aleppo Province to Grand Vizier, 25 Kanunusani 1320 (January 7, 1905).

and quicklime (for deodorizing the destroyed locust eggs, the stench of which many believed caused disease).[192]

Locusts in the desert threatened the Ottoman state in a way like the presence of Ibrahim Pasha in the desert threatened the Ottoman state. In the western portions of the Jazira, authorities worried about springtime fighting between the ʿAnaza, Millî, and Shammar so much that the war minister called for troops to keep peace between them "in the pastures in the desert."[193] Meanwhile, in an echo of the efforts to transfer Ibrahim Pasha's power elsewhere in the empire the previous year, Zeki Pasha attempted to enact orders to send Ibrahim to Yemen with two companies of his Hamidiye Brigades.[194] The effort was likely disingenuous, seeing as Zeki was one of Ibrahim's allies. Zeki noted, perhaps knowingly, that "on account of Ibrahim Pasha being in the desert and at a great distance from a telegraph center, a response had not been received as of yet." The allure of Ibrahim Pasha was that he might project Ottoman power into environments in which they struggled to wield power. But the flip side of this dynamic was that when the Ottomans wanted to send Ibrahim Pasha elsewhere, he could easily slip away into the desert with the comfortable excuse that he was too far away from the telegraph to receive orders.

But Ibrahim Pasha shared power in the desert with locusts, and they would cause him – or at least give him an excuse – to clash with the notables of Diyarbekir, a group he had run afoul of more and more over the previous few years. By late July of 1905, the governor of Diyarbekir reported that "on account of an abundant number of locusts striking in the direction of the desert this year, and the pasture in these parts being totally wiped out, the Millî tribe has been forced to approach the vicinity of the villages on the edge of Diyarbekir."[195] Subsequent reports confirmed the conditions, attesting to "no grass" and a "disaster" in the desert thanks to the locusts.[196] Others speculated about more nefarious reasons for Ibrahim Pasha's movement. The grand vizier believed that

[192] On "bad air" and locust carcasses, see BOA, DH.MKT 1910/32, Interior Ministry to Province of Aleppo, 31 Kanunuevvel 1307 (January 12, 1892).

[193] BOA, DH.TMIK.M 197/16, Serasker, 2 Mayıs 1321 (May 15, 1905).

[194] BOA, Y.PRK.ASK 230/53, Mehmed Zeki, Fourth Army Command, 2 Haziran 1321 (June 15, 1905).

[195] BOA, DH.ŞFR 349/38, Diyarbekir Governor Nazim to Interior Ministry, 7 Temmuz 1321 (July 20, 1905).

[196] BOA, BEO 2635/197608, Minister of Interior Affairs, 18 Temmuz 1321 (July 31, 1905); Y.PRK.ASK 231/87, 22 Temmuz 1321 (August 4, 1905).

Ibrahim Pasha had not returned to his typical summer pasture of Mount Karaca and instead entered the cultivated regions of Diyarbekir because he wished to display power in an effort to avoid dispatch to Yemen.[197] The notables of Diyarbekir, for their part, cast an especially suspicious eye on the claim that locusts prompted Ibrahim Pasha's motion. In their view, locusts could not be spoken of as a real thing, but rather only as "the excuse of locusts" (*çekirge bahanesi*).[198] They were an environmental explanation that offered Ibrahim Pasha justification for pursuing his own interests. For years, the notables of Diyarbekir had watched as Ibrahim Pasha's power expanded such that it seemed "no one remained" for him to recruit, and he "would essentially rival the power of the government." He did so, they charged, by gathering "Kurds, savage tribes, and criminals" in Zor, Mosul, and Aleppo. What they left unsaid was that Ibrahim Pasha's power expanded in the desert through his manipulation of the border between Diyarbekir and Zor. Moreover, he had come closer to the black basalt walls of Diyarbekir in the previous years, where his men often collected tribute from villages owned by Diyarbekir notables. From the notables' view, then, locusts concealed the nomadic chief's real motivations of amassing wealth.

To make this point clear, they leveraged one of the few ways their humble city could exert power over the rest of the empire: the telegraph network. Because Diyarbekir was the only station through which all of the empire's messages from Iraq flowed, Diyarbekir could also be the place where all of this information stopped (a line between Baghdad and Aleppo had not yet been completed).[199] And so in early August, the notables dispatched criers carrying orders that all shops be closed and all the residents of Diyarbekir gather at the telegraph house (Christians, fearing a repeat of the massacres of 1895, fled to their homes).[200] The notables may have been the same ones involved in the violence of the 1890s, and many of them would be involved in the violence of the genocide a decade later. In 1905, however, they were concerned not with Diyarbekir's Christians but rather with Ibrahim Pasha. With the empire's telegraphic flow cut off, they grabbed the state's attention. They attempted to

[197] BOA, Y.A.HUS 490/13, Grand Vizier, 10 Temmuz 1321 (July 23, 1905).

[198] BOA, Y.MTV 277/38, Letter from Notables of Diyarbekir, 22 Temmuz 1321 (August 4, 1905).

[199] TNA-UK, FO 195/2187, Barnham to O'Conor, March 27, 1905; Beysanoğlu, "Ziya Gökalp'ın Yaşam Öyküsü," 330.

[200] TNA-UK, FO 424/208, N. O'Conor to the Marquess of Lansdowne, August 29, 1905, Inclosure: Vice Consul Shipley to Sir N. O'Conor, August 12, 1905.

discredit Ibrahim Pasha – in a sleight of hand of their own – by denigrating him as uncivilized.

Penning most of the risibly outrageous communiqués was a young man by the name of Mehmed Ziya. His opposition to the Millî chief was both personal and political. Hailing from a Kurdophone family in Çermik, Ziya's father Tevfik had been the editor of *Diyarbekir Gazetesi*, the same Tanzimat-era publication that had hailed the monster watermelons of the Khabur as evidence of progress in the desert.[201] Tevfik also served as a census official.[202] In this position, he had allegedly rejected bribes offered to him by Ibrahim Pasha and had in fact died shortly after spending a considerable amount of time between Nusaybin and Viranşehir attempting to assess taxes on the Millî and other pastoralists. "Worn down in the desert," explained Ziya's brother Nihat.[203] Tevfik's death left the family in debt. Ziya buried himself in books, voraciously reading Descartes, al-Ghazali, and crime novels.[204] He also cultivated political connections, including with Abdullah Cevdet, the doctor and founder of the Committee of Union and Progress (CUP), who was stationed in Diyarbekir in response to the 1895 cholera epidemic.[205] At around the same time, Ziya attempted suicide, apparently in despair over what he perceived as a gulf between materialist philosophy and religious practice, the fact that the family's economic problems would not allow him to go to Istanbul for education, and his maternal uncle – local notable Arif Pirinççizade – banning him from seeing Abdullah Cevdet.[206] Cevdet saved Ziya, and the bullet the young man had fired in the attempted suicide remained in his head the rest of his life.[207] Subsequently, Ziya briefly attended veterinary school in Istanbul – tuition was free for boarding students – before being expelled for his political activities and returning to Diyarbekir.[208] In the coming decades, he would be renamed Mehmed Ziya Gökalp, under which he would become "the most important figure in the creation of Turkish nationalism."[209]

201 Asena, "Arkadaşım Ziya Gökalp" (1975), 323.
202 Gökalp, "Ağabeyim Ziya Gökalp'ın Hayatı" (1975–1976), 434, 436. 203 Ibid., 438.
204 Asena, "Arkadaşım Ziya Gökalp" (1977), 158.
205 Gökalp, "Ağabeyim Ziya Gökalp'ın Hayatı" (1977), 172.
206 Gökalp, "Ağabeyim Ziya Gökalp'ın Hayatı" (1977) 172.
207 Asena, "Arkadaşım Ziya Gökalp" (1977), 159; Heyd, *Foundations of Turkish Nationalism*, 26.
208 Beysanoğlu, "Ziya Gökalp'ın Yaşam Öyküsü," 327.
209 Jongerden, *The Settlement Issue in Turkey and the Kurds*, 22.

But in 1905, Ziya's main concern was calling Ibrahim Pasha names. He and his allies called Ibrahim Pasha an "enemy of civilization" (*duşman-ı medeniyet*).[210] They called him "an ignorant savage" (*cahil bir bedevi*).[211] They said that he left no "hope for the future of Kurdistan and the Jazira," as if his power could only be conceived of in these toponyms that did not exist on the Ottoman provincial map. They even mentioned a rumor that Ibrahim Pasha wore on his pinky finger a seal bearing his name that had been in possession of the eponymous son of Mehmed Ali Pasha of Egypt, so as to emphasize separatist ambitions alongside lack of civilization.[212] The insult, of course, did not point to some insurmountable dichotomy between urban notables and nomadic pastoralists but rather corresponds to what historian Mostafa Minawi has described as the denigration of nomads as "a rhetorical tool."[213] After all, the conflict between the groups stemmed not over some civilizational difference between them, but rather from their clash over rent. Ibrahim Pasha and the Diyarbekir notables were both local power brokers fighting over who could extract revenue and protection payments from villages.[214] Ibrahim relied on the volcanic soils of Mount Karaca to nourish his flocks; for protection, the Diyarbekir notables relied on walls built of basalt taken from the very same Mount Karaca region.[215] The distance drawn in their rhetoric was great precisely because the distance between their structural situations as collectors of surplus value was small. As the British vice consul in Diyarbekir wrote, "it is beyond doubt ... that the town population, were it not for the Notables, would remain indifferent to the Ibrahim Pasha question."[216]

[210] BOA, Y.A.HUS 491/26, Mufti of Diyarbekir and 25 others to Grand Vizier, 24 Temmuz 1321 (August 6, 1905).

[211] BOA, Y.A.HUS 492/75, Diyarbekir Notables to Grand Vizier, 18 Ağustos 1321 (August 31, 1905).

[212] BOA, Y.A.HUS 491/97, Mina Efendi to Grand Vizier, 25 Temmuz 1321 (August 7, 1905); Y.A.HUS 491/53, Nazim, Diyarbekir Governor, 28 Temmuz 1321 (August 10, 1905).

[213] Minawi, "Beyond Rhetoric," 78.

[214] Elsewhere in the empire, locusts were even implicated in the ways that local notables gained control over villages. In Mosul, the combination of "drought, locusts, and blight" allowed notables or tax farmers to swoop in and pay taxes in return for title deeds, such that "it is quite an exception to come across a village in the neighbourhood of Mosul which is thus not owned by one of the notables of Mosul." TNA-UK, FO 424/239, Hony to Lowther, June 18, 1913.

[215] Diken, *Ahım Var Diyarbakır*, 18.

[216] TNA-UK, FO 424/208, N. O'Conor to the Marquess of Lansdowne, August 29, 1905, Inclosure: Vice Consul Shipley to Sir N. O'Conor, August 12, 1905.

Ibrahim Pasha and his allies responded to the charges by exploiting the ambiguity afforded by locusts and the Jazira environment. Ibrahim's staunchest protector in the Ottoman infrastructure, the military commander Zeki Pasha, insisted that the claims against Ibrahim were baseless. Zeki Pasha emphasized that the Millî chief had come to the vicinity of Diyarbekir because of the "locust catastrophe" and could not be reasonably forced to return to "places of dry soil without water or pasture."[217] All of this was not to mention the Shammar, whom Zeki described as waiting for Ibrahim Pasha there. In his own defense, Ibrahim Pasha wrote that the Millî were in a parlous state, with the "time to migrate in the direction of the desert" having arrived but many people on the verge of starvation.[218]

Conditions did seem difficult. In late September of 1905, a massive dust storm struck near Mardin, with clouds of the Jazira's distinctive "reddish brown" dirt some 2,000 feet in height and ten miles in length.[219] The phenomenon was so terrifyingly magnificent to one American missionary in Mardin that he advised using a magnifying glass when inspecting the photograph he took of the event so as to get some sense of the true scale. It was in that way only that the viewer might appreciate how the cloud of fertile soil dwarfed trees. The cataclysm not only pointed to the arid environmental conditions that perhaps prompted Ibrahim Pasha to approach Diyarbekir in the first place. It also pointed to the ambiguous nature of the environment, which Ibrahim Pasha knew well how to use. When an investigative commission arrived in October of 1905 to adjudicate the conflict between him and the notables, the Millî chief went "straight to the desert," as if the dust of the region could protect him from punishment at the behest of the Diyarbekir notables.[220] The dust may have in part saved him, but so, too, did his connections with Ottoman officials, as general sentiment held that the proceedings were a "farce."[221]

With yet another of Ibrahim Pasha's challengers vanquished, the chief seemed untouchable. The notables of Diyarbekir, for their part, continued their tactic of rhetorical insult. They caustically remarked that they were not asking for "commercial factories or agricultural machines like other

[217] BOA, Y.PRK.ASK 232/2, Zeki Pasha, 30 Temmuz 1321 (August 12, 1905).

[218] BOA, Y.PRK.AZJ 51/12, Ibrahim Pasha, 6 Eylül 1321 (September 19, 1905).

[219] ABCFM 76, Andrus to the Platts, January 23, 1906.

[220] BOA, Y.MTV 283/23, Investigative Commission to the Grand Vizier, 13 Teşrinievvel 1321 (October 26, 1905).

[221] TNA-UK, FO 424/208, N. O'Conor to the Marquess of Lansdowne, November 28, 1905, Shipley to O'Conor, November 8, 1905.

provinces" but rather merely requested "security and safety."[222] Still, they would get no satisfaction, at least in the short run. By February 1906, Ibrahim was rumored to be expanding his domain of power "all the way to the banks of the Euphrates," prompting many in the countryside of Urfa to leave behind their fields for the city.[223] Officials in Aleppo claimed they could not do anything about these problems because all of their soldiers were tied up in the desert assessing the sheep taxes. The war minister, for his part, complained that because of "nothing being done" by the government, "the tribe's complete impertinence and violations were increasing day by day."[224] Meanwhile in the east, clashes between the Shammar and other confederations along the border between Zor and Diyarbekir left their chief Hadi – grandson of Farhan – suddenly dead.[225] In a further stroke of luck for Ibrahim Pasha, three regiments of the Shammar made common cause with the Millî and struck against the Shammar in Zor.[226] Having survived the campaign against him by the notables of Diyarbekir the previous year, and watching his erstwhile foes the Shammar fall in battle and even join his side, Ibrahim Pasha could be confident from his place on the edges of Urfa in 1906, where he was trying to collect tribute from yet another tribe.[227]

From this vantage point, his power seemed so great as to rival the power of locusts. Indeed, an agent of the Jewish Colonization Association named Niege was scouting territory along the proposed path of the Baghdad Railway during the summer of 1906. To be built with German capital, construction had begun on the southern Anatolian portion of the line at Ereğli in 1904, although it would not successfully make it through the Taurus Mountains and into the Jazira until World War I.[228] There, the reign of two powers loomed, Niege explained. First, there were the "wretched insects" that had been there for twenty-five years, and because they could not be gotten rid of, "more and more

[222] BOA, Y.MTV 280/128, Diyarbekir Notables to Military Commander, 8 Teşrinisani 1321 (November 21, 1905).
[223] BOA, Y.MTV 284/2, Aleppo Province, 8 Şubat 1321 (February 21, 1906).
[224] BOA, Y.MTV 284/2, Military Commander, 12 Şubat 1321 (February 21, 1906).
[225] BOA, DH.TMIK.M 219/7, Governor of Diyarbekir Encoded to Interior Ministry, 11 Mart 1322 (March 24, 1906); DH.TMIK.M 221/30, Acting Governor Lieutenant General Bahri, 7 Nisan 1322 (April 20, 1906).
[226] BOA, DH.TMIK.M 219/7, Governor of Mosul Mustafa to Interior Ministry, 8 Nisan 1322 (April 21, 1906).
[227] BOA, Y.MTV 288/73, From the Fifth Army Command, 26 Haziran 1322 (July 9, 1906).
[228] *Servet-i Funün*, 18 Teşrinisani 1320 (December 1, 1904); Fraser, *The Short Cut to India*, 51; McMeekin, *The Berlin–Baghdad Express*, 40, 44.

villages and property owners had renounced cultivation" and chosen animal husbandry as a way of "utilizing the exuberance of the natural grasses."[229] Second, there was Ibrahim Pasha, who had emerged as "the king of upper Mesopotamia in a region that extends to the border of Kurdistan in the north and that of Arabia in the south," encompassing "all of the country that is found between the Tigris and the Euphrates." The agent had no respect for Ibrahim, declaring his men "only live by theft and pillage," but he also gestured to how locusts had made Ibrahim Pasha's power too, by helping to create a landscape where mobility mattered.

Around the same time, Ibrahim Pasha left a better impression on Mark Sykes. The young British traveler was not necessarily wiser and certainly not humbler than he was during his earliest jaunt in the Jazira, but he had a different approach all the same. The trip began, he wrote, by seeking the "portions of the map which were the whitest."[230] It was in these unmapped areas that Sykes hoped to find refuge from modernity, as would many British agents in the region in the coming decades.[231] Sykes concurred with many in his estimation that the Jazira had fallen since "the time of Caesar, Julian, or the Caliph Mansur," as its wealth had given way to nothing more than "a few broken mounds . . . abandoned to the nomads, their flocks and pastoral warfare."[232] But Sykes viewed this not as a waste so much as an antidote. He called "the Jazira the merriest and most entertaining little kingdom of disorder."[233] The nomads raided one another "not because they are savage; but because it is such fun." It was here that Sykes hoped to find deliverance from such modern ills as "old age pensions, Noncomformist consciences, suffragettes, maffickings, professional politicians, trusts, excursions, halfpenny papers, hysteria, and appendicitis."[234] His opinions had changed in young adulthood. The places he once regarded with disdain had transformed into the places in which he sought redemption.

Sykes found what he sought in Ibrahim Pasha, whom he deemed "without a doubt, the most interesting person in the Jazirah."[235] The young man encountered the Millî chief in his enormous tent, which amounted to "over 100 poles and measured 1,500 square yards of

[229] TNA-UK, FO 371/155, Board of Trade to Undersecretary of Foreign Affairs, September 24, 1906, Inclosure: Report by an agent of the Jewish Colonization Association, describing a journey in the neighborhood of the projected track of the Bagdad Railway, in which an interesting account is given of the country between Aleppo and Ourfa.

[230] Sykes, *The Caliph's Last Heritage*, 297. [231] Satia, *Spies in Arabia*, 65.

[232] Sykes, *The Caliph's Last Heritage*, 301–302. [233] Ibid., 302. [234] Ibid., 303.

[235] Sykes, "Journeys in North Mesopotamia," 385.

cloth."[236] Sykes said that Ibrahim Pasha asked if it was true that "Sarah Bernhardt's traveling tent was larger than his," the longtime symbol of tribal backwardness transformed into a playful contest with a French actress. Ibrahim Pasha also asked Sykes how the Algeciras conference had gone, his awareness gleaned from a steady supply of newspapers.[237] Sykes described the chief as "a tall, slimly-built man ... with a large lumpy nose, rather piercing eyes, a fairly high forehead, and a large mouth," as well as a "sword-cut under his eye" (Figures 10a and b).[238] Ibrahim Pasha "in repose" seemed "sinister" thanks to "that settled frown common to all ... who live in a sunlit desert," yet "in conversation" he became "almost genial" thanks to his "continual smile." Sykes was enamored. In Ibrahim, Sykes saw shades of Timur and Attila as well as the early Ottomans and Mithridates.[239] Sykes gushed, too, of Ibrahim Pasha's "sufficiently picturesque" habit of moving with his caravan astride a white camel beside his private guards and fifty dogs that specialized in hunting gazelles. To be certain, Sykes had not done away with his tendency to denigrate the people of the region. Once he reached an area of Zor that was, in his words, "as little known as the inmost recesses of the Sahara," he channeled his racism toward people in Europe to convey his racism for people outside of it.[240] He described the Shammar as "a sort of cross between a gypsy mountebank and a Jew money-lender."

Ibrahim displayed his awareness of the importance of European friends in the summer of 1907, when a visit from a British official prompted the singing of the Millî chief's praises. When the British vice consul of Diyarbekir arrived, he found Ibrahim Pasha encamped about a day's distance from Viranşehir, surrounded by "immense droves of camels and numerous herds of cattle and sheep."[241] Yet the consul's discussion with Ibrahim Pasha, as in the case of Sykes, did not seem to fit with the image of the chief painted by the cascade of telegraphs from Diyarbekir. Ibrahim Pasha spoke of how he had "registered in his name large tracts of country on the future line of the Baghdad Railway," and established "a flour-mill, worked by petroleum." The figure whom Diyarbekir notables decried as the sheep-owning scourge of civilization looked quite different in the eyes of the British consul: a large landholder and property speculator grinding wheat with advanced technology.[242]

[236] Sykes, *The Caliph's Last Heritage*, 317. [237] Ibid., 318. [238] Ibid., 317.
[239] Ibid., 326. [240] Ibid., 442.
[241] TNA-UK, FO 424/213, Sir N. O'Conor to Sir Edward Grey, September 20, 1907, Inclosure: Vice Consul Heard to Sir N. O'Conor, July 17, 1907.
[242] On complaints about Ibrahim Pasha's property acquisitions after his death, see Ozok-Gündoğan, "A Peripheral Approach to the 1908 Revolution in the Ottoman Empire," 199.

FIGURE 10A AND B Ibrahim Pasha (center) and other Millî leaders. Used with the permission of Istanbul University's Nadir Eserler Kütüphanesi

THE CONSTITUTION AND THE DEATH OF IBRAHIM PASHA

Aware of these growing ambitions, the notables of Diyarbekir lashed out at Ibrahim Pasha once again in November of 1907. Just as in 1905, they did so by occupying the telegraph house of the city. And also as in 1905, they seemed to be in a contest to outdo each other in their elaborate descriptions of Ibrahim Pasha's wickedness. One compared him to the "Germanic tribes" that caused "misery" to the "Roman Empire," while another brought up Genghis Khan.[243] Still another likened him to Timurlane, adding, in a citational flourish, that more details could be found in the sixth volume of Cevdet Pasha's history.[244] Others were even more blunt. One called Ibrahim Pasha worse than "plague and pestilence."[245] A week later, another message upped the stakes, calling him worse than "plague, pestilence, and cholera."[246]

Not all complaints were so grandiosely grim, with one telegram in fact relating the dilemma of Diyarbekir notables to recent rather than ancient history. The letter admitted that the region had benefited from efforts to cut down on the power of the notables of Kurdistan such as the Bedirkhans as well as the power of nomadic leaders such as ʿAbd al-Karim of the Shammar.[247] They recalled fondly how the government had established districts like Ras al-Ayn and Yenişehir (the government's preferred anodyne toponym for Viranşehir, meaning "New City"). Policies such as these had ensured the "spreading of cultivated lands into the desert and not ... the desert spreading into the cultivated lands."

To cure this pestilential force, petitioners called for Ibrahim Pasha to be removed from the space in which he enjoyed power. Where he should be removed to was open to question. Petitioners from one village had lost their sheep to the man, and had subsequently been forced to move to Diyarbekir, where they survived "by sleeping in the courtyards of

[243] BOA, DH.TMIK.M 257/33, Diyarbekir Notables to Interior Ministry, 4 Teşrinisani (November 17, 1907); DH.TMIK.M 257/33, People of Sinan Village to Interior Ministry, 11 Teşrinisani 1323 (November 24, 1907).

[244] BOA, DH.TMIK.M 257/33, Diyarbekir Notables to Interior Ministry, 5 Teşrinisani 1323 (November 18, 1907).

[245] BOA, Y.MTV 303/123, Diyarbekir Notables, 4 Teşrinisani 1323 (November 17, 1907).

[246] BOA, İ.DH 1460/31, Diyarbekir Notables to Meclis-i Vükela, 10 Teşrinisani 1323 (November 23, 1907).

[247] BOA, Y.PRK.ASK 252/8, Chaldean Archbishop, Chaldean Priest, Armenian Catholic Bishop, Members of the Ulema, Rabbi, and One Hundred Others of Diyarbekir to the Sultan, 4 Teşrinisani 1323 (November 17, 1907).

mosques" in the "mud and the rain."[248] Their one request to the government was that they be "sheltered in a place that Ibrahim Pasha cannot reach." Others maintained that Ibrahim Pasha ought to be exiled.[249] Another specified that Ibrahim's punishment ought to be "migration to Ottoman Europe" (*Rumeli'ye muhaceret*).[250] Their demand represented an inversion of the vision that had for so long prevailed across the empire and in the Jazira, in which migrants from the Balkans and the Caucasus would migrate there and help restore the region. Instead, Ibrahim Pasha was to go to the places that the agents of the Jazira's transformation were coming from. Of course, exile as punishment was common in the Ottoman Empire. But movement into a faraway place had a special resonance with respect to Ibrahim Pasha, who had for so long used his mastery of the Jazira's space to be both far and near at will.

In fact, these powers would serve Ibrahim Pasha well in yet another repeat of 1905. As in years past, he disappeared upon the arrival of the investigative commission sent in response to the complaints. By early December of 1907, Ibrahim Pasha was rumored to be at a distance of fifty to sixty hours from Diyarbekir, near Mount ʿAbd al-ʿAziz. Because it was unclear what district he was in, telegrams summoning him were sent to both Urfa and Zor.[251] Neither message reached him. He still retained some support from people like Zeki Pasha, the commander of the Fourth Army, who dismissed complaints as lies concealing "the personal interests" of the Diyarbekir notables.[252]

Yet there were also signs that Ibrahim Pasha had become too big to succeed, as it were. The commander of the imperial armory, apparently affected by the stream of telegrams from Diyarbekir, worried that Ibrahim Pasha was like Mehmed Ali in Egypt in using his bribes of "tins full of butter by the thousands" to buy support in Istanbul.[253] And Ibrahim

[248] BOA, Y.A.HUS 516/136, Jibare Muhtarı Hasan and others, 6 Teşrinisani 1323 (November 19, 1907).

[249] BOA, DH.TMIK.M 257/33, Diyarbekir Notables to Interior Ministry, 3 Teşrinisani 1323 (November 16, 1907); BOA, DH.TMIK.M 257/33, Diyarbekir Notables to Interior Ministry, 4 Teşrinisani 1323 (November 17, 1907).

[250] BOA, Y.PRK.ASK 252/8, Chaldean Archbishop, Chaldean Priest, Armenian Catholic Bishop, Members of the Ulema, Rabbi, and One Hundred Others of Diyarbekir to the Sultan, 4 Teşrinisani 1323 (November 17, 1907).

[251] BOA, Y.PRK.ASK 252/8, Encoded Telegram from Talat, 20 Teşrinisani 1323 (December 3, 1907).

[252] BOA, Y.PRK.ASK 252/8, Zeki to Edhem Pasha, 6 Teşrinisani 1323 (November 19, 1907).

[253] BOA, Y.PRK.ASK 252/8, 9 Teşrinisani 1323 (November 22, 1907).

Pasha even seemed to have noticed a change in the air. His personal secretary expressed concern in correspondence that his typical gift of butter and animals might be held up at İskenderun as it had been the previous year.[254] It is unclear what happened with the butter, but by mid-December, the war minister conveyed a plan to dispatch Ibrahim Pasha to the Hijaz.[255] The stand-off continued into March of 1908. Reports suggested that he was "found on the west side of Mount ʿAbd al-ʿAziz in the Jazira" and trying to bring more tribes to his side "by all means of tricks."[256] The war minister warned that Ibrahim would be kept under watch until he reached Aleppo for his assignment to the Hijaz. Zeki Pasha sang an old tune. He once again explained away Ibrahim Pasha's defiance as being an innocent result of him being far away from telegraph stations.[257]

In this shroud of ambiguity, Ibrahim managed to maintain his autonomy well into the spring. For many of the previous years, he had deftly operated on the edge, using Ottoman provincial borders as a way of both expanding his own power and protecting himself. The spring of 1908 represented how far he had come. He had long pushed into the Khabur Valley south of Ras al-Ayn, and he did so with renewed vigor in March of 1908.[258] He was accompanied by unexpected allies, given his previous battles in the region: at his side were branches of the Shammar. His arms allegedly came thanks to the Chechens at Ras al-Ayn. With allies and arms, Ibrahim Pasha apparently pressured "tribes in the Jazira" such as the Baggara, Jabbur, and Shirabi to join his side. At the same time, he claimed to be so sick that he could not ride a horse and therefore could not come to Aleppo as ordered by the state.[259] When a judgment of the investigative committee finally appeared later in the spring, Ibrahim Pasha was exonerated.[260] The British consul in Diyarbekir reasoned that

254 BOA, Y.PRK.ASK 252/83, Secretary of Millî Ibrahim, 23 Teşrinisani 1323 (December 6, 1907).

255 BOA, Y.MTV 304/53, Military Commander Rıza, 1 Kanunuevvel 1323 (December 14, 1907).

256 BOA, Y.MTV 305/159, Military Commander Rıza, 30 Kanunuevvel 1323 (January 12, 1908).

257 BOA, Y.PRK.ASK 253/38, Zeki of the Fourth Army, 4 Kanunusani 1323 (January 17, 1908).

258 BOA, Y.PRK.ASK 255/26, Bekir Sıdkı, Aleppo and Adana Extraordinary Command to Head Secretary, 27 Şubat 1323 (March 11, 1908).

259 BOA, Y.MTV 307/90, Military Commander, 4 Mart 1324 (March 17, 1908).

260 TNA-UK, FO 424/215, Barclay to Grey, May 2, 1908, Inclosure: Heard to Barclay, April 8, 1908.

rather than him being a cause of turmoil as so many had charged over the years, it was actually the removal of Ibrahim Pasha that "would no doubt . . . throw the southern portion of the vilayet into a state of anarchy, lasting perhaps for years."

With the challenge of the Diyarbekir notables defeated and locusts arriving once again in the spring of 1908, it appeared as if much remained the same in the Jazira. The insects flew across provincial boundaries, with locusts from Baghdad, Mosul, and Zor laying waste to cultivation.[261] The campaigns in the spring were so extensive that they used up the empire's entire budget for the destruction of locusts and crows – some 85,000 kuruş – by June, forcing officials to apportion another 50,000 kuruş for Mosul and Aleppo specifically to respond to the locusts.[262] But even these efforts could not protect cultivated lands, including the imperial estates around Aleppo, where in early June the locusts appeared just as in 1891 "flying from Mosul, Zor, and the deserts around these areas."[263] The locusts were devouring the sultan's holdings that had come to be seen as emblematic of the excesses of his rule. And at the same time, the mobile humans such as Ibrahim Pasha who – because of their support from the palace – were seen too as emblematic of the sultan's excesses also had reached an apogee of power.

Ibrahim's confidence seems to have been so great that he even agreed to leave the place from which he had derived so much power. He was to accompany a regiment of the Hamidiye directly commanded by his son Ismail, charged with protecting the Hijaz Railway.[264] As historian M. C. Low has suggested, the "massively ambitious infrastructure project" was aimed at facilitating pilgrimage, but also centralizing Ottoman control of a space long managed by local power brokers.[265] In other words, Ibrahim Pasha's forces were being dispatched to a space not unlike the Jazira, a region where local notables, mobile populations, and a difficult climate invited governmental exception.[266] The chief, having fended off challenges from notables and rival nomads alike, had obtained

[261] BOA, DH.MKT 2687/58, Commerce and Public Works Minister to Interior Ministry, 16 Teşrinisani 1324 (November 29, 1908).
[262] BOA, BEO 3344/250785, Forests, Mines, and Agriculture Minister Selim, 9 Haziran 1324 (June 22, 1908). Tural, "Osmanlı Orman ve Maadin ve Ziraat Nezareti'nde Büroktraik Reform (1908–1914)."
[263] BOA, ML.EEM 685/22, 11 Haziran 1324 (June 24, 1908).
[264] TNA-UK, FO 195/2272, Longworth to Barclay, July 20, 1908.
[265] Low, *Imperial Mecca*, 254.
[266] The governor Osman Nuri described Hijaz as among the "hot provinces" of the empire and suggested that they required special mechanisms of governance. Ibid., 177.

worldly power in the Jazira. His power could be counted not only in camels and sheep and butter and Martini rifles but also in political connections to Yıldız Palace and investments in land along the expected path of the Baghdad Railway. Denigrated as a backward nomad much as the Shammar had been decades before, he was also, like the Shammar, quite agile and aware of opportunity, a businessman and political operator. All that was left, it seemed, was his spiritual commitment to pilgrimage, which, according to one British report, he planned to complete alongside his son before returning to Viranşehir.[267] By late June of 1908, he left his hometown bound for Medina, accompanied by sacrifices of animals, much the same way his journey from Diyarbekir upon induction into the Hamidiye Brigades had begun nearly two decades before.[268] From Urfa, they decamped to Aleppo, and then they traveled by train to Damascus where they arrived in late July.[269]

Then an event occurred that would forever change the empire, and the place of Ibrahim Pasha within it. In late July of 1908, the CUP – whose member Doctor Abdullah Cevdet had saved Mehmed Ziya's life in Diyarbekir over a decade before – instituted a coup through which they forced the sultan to reinstate the constitution and parliament that he had suspended in 1878. Their slogan was "liberty, equality, fraternity." They promised to clean up the excessive secret police of the sultan and to deliver on the promises of modernity that Ottomans saw all around them. It was initially unclear how the new developments would affect Ibrahim. Indeed, his son held a feast in Damascus in honor of the constitution shortly after its announcement.[270] But a feast would not salve the unease among many about the connection between the chieftain and the old regime. Both Ibrahim Pasha and his long-time protector Zeki were recalled to Istanbul. As one Istanbul newspaper remarked, no "desert chief" relished receiving that message, but Ibrahim Pasha initially made it seem like he might comply. He headed for Aleppo, ostensibly on the way to the port of Alexandretta.[271]

Then on August 30, he bolted east.[272] A journey that had begun as a pilgrimage to Mecca had turned into an attempt to return to the

[267] TNA-UK, FO 195/2272, Longworth to Barclay, July 20, 1908.
[268] BOA, Y.A.HUS 523/13, Telegram from Diyarbekir, 17 Haziran 1324 (June 30, 1908); Y. MTV 312/18, Serasker Rıza to Yıldız, 19 Haziran 1324 (July 2, 1908).
[269] TNA-UK, FO 195/2272, Longworth to Barclay, July 20, 1908.
[270] TNA-UK, FO 424/217, Lowther to Grey, November 10, 1908.
[271] *Tanin*, 27 Ağustos 1324 (September 9, 1908); 5 Eylül 1324 (September 18, 1908).
[272] TNA-UK, FO 424/216, Surtees to Lowther, September 9, 1908.

landscape in which he had for so long wielded power. Imperial troops tried to head off his flight, but he managed to cross the Euphrates on September 2 near Tel Ahmar, at which point his actions instigated a massive and unified military operation of the sort long called for but rarely witnessed in the Jazira.[273] The troops pursuing him hailed from Raqqa, Aintab, and Deir ez-Zor, while others were sent from Diyarbekir, Mardin, Harput, Suruç, and Midyat.[274] Yet still Ibrahim Pasha made it to Viranşehir, where some 3,000 fighters armed with government weapons gathered.[275] Rumors suggested Ibrahim Pasha and his men buried some of his great wealth in iron boxes underground.[276] The district governor of Zor observed that many of his nomadic allies abandoned him and had even begun to "loot a portion of his livestock and possessions."[277] The order to capture Ibrahim "alive or dead" arrived on September 19.[278] Observers alleged that Ibrahim had attempted to surrender in Viranşehir, only to be forced into fighting when Ottoman soldiers fired on his unarmed men.[279]

Ibrahim eventually managed to make a break from Viranşehir, hoping to find freedom in the Jazira's pastures. He sought refuge with one of his few steadfast allies, Hüseyin Kanco and the Sharkiyat.[280] When they, too, were forced to surrender, Ibrahim Pasha and his family continued moving east. By some accounts, he even encountered his long-time enemies and sometime allies the Shammar on the path to Sinjar near the Zarkan River, also long a setting of conflict between the groups.[281] Ibrahim did not get much further. Like ʿAbd al-Karim thirty-seven years before, Ibrahim Pasha had sought refuge in the Jazira. Unlike ʿAbd al-Karim, Ibrahim Pasha met his end before being captured by Ottoman troops. It was not his enemies or his allies that got him, but rather dysentery, a disease of water felling the figure who had so long used movement to take advantage of a region noteworthy for its lack of water.[282] He succumbed somewhere between Tel Aswad and

[273] Ibid. [274] TNA-UK, FO 424/217, Lowther to Grey, September 28, 1908.
[275] BOA, DH.MKT 2683/20, Memo on Ibrahim Pasha, 14 Eylül 1324 (September 27, 1908).
[276] Ibrahim, *Amir Umaraʾ Kurdistan*, 69.
[277] BOA, BEO 3401/255053, Zor District Governor to Interior Minister, 7 Eylül 1324 (September 20, 1908).
[278] TNA-UK, FO 424/217, Lowther to Grey, November 10, 1908.
[279] TNA-UK, FO 195/2284, Heard to Lowther, October 13, 1908.
[280] TNA-UK, FO 424/217, R. Harris to Grey, October 23, 1908, Inclosure: Andrus to Harris, October 4, 1908; Abu Bakr, *Akrad al-Milli Ibrahim Pasha*, 54.
[281] Abu Bakr, *Akrad al-Milli Ibrahim Pasha*, 55.
[282] TNA-UK, FO 424/217, Lowther to Grey, November 10, 1908.

al-Hasakah beside the Jaghjagh River.[283] Before he died, he allegedly said, "If a person dies, let him die on his soil. There are no feelings of loneliness" (*Mirov dimre jî bira li ser axa xwe bimire xerîbî nayê kişandin*).[284] His body remained there, while his wife Khansa and sons headed to Nusaybin, where they surrendered. Yet even in death, mystery prevailed. Some suggested that "the rumor of his demise was spread by the crafty chieftain himself, in order to slip away safely with a good portion of his treasures."[285] Even the Ottoman authorities gave credence to this view, as they disinterred the grave to confirm that Ibrahim Pasha had not escaped yet again.[286]

Gone was a pasha whose power had intersected with the range of locusts and whose movements had even been arguably determined by locusts. In his absence, his erstwhile allies and enemies, "Kurdish and Arab" alike "gathered" – in the words of the American missionary Andrus – "like locusts."[287] The British consul in Diyarbekir utilized a similar comparison: groups like the Shammar "swarmed over the frontiers of Ibrahim Pasha's country" and then "swept the country like … locusts."[288] The land once again turned into "a desert" as they looted, killed, and raped. According to Hüseyin Kanco, 150 Yazidis of the Sharkiyat tribe were taken prisoner and held in a cave.[289] Upon instruction from Arif Pirinççizade of Diyarbekir – architect of massacres of Armenians in the 1890s and longtime rival of Ibrahim Pasha – they were all slaughtered "like sheep," with pregnant women allegedly even having their bellies cut open, their fetuses thrown out. The corpses were left to "the dogs and wolves." Somehow thirteen of the group managed to escape, and they carried the story with them. Thus, while the missionary Andrus was happy to blame the violence on locust-like nomads, it was in fact one of Diyarbekir's most elite urban power brokers who gave the orders for the gruesome scene.[290] In a pattern that would repeat itself

283 British sources suggest Ibrahim intended to seek refuge in Kuwait, where he hoped to receive British protection. TNA-UK, FO 424/217, Lowther to Grey, November 10, 1908.

284 Kıran, *Kürt Milan Aşiret Konfederasyonu*, 198.

285 TNA-UK, FO 195/2272, Longworth to Lowther, October 3, 1908.

286 Asena, "Arkadaşım Ziya Gökalp" (1979), 235.

287 TNA-UK, FO 424/217, R. Harris to Grey, October 23, 1908, Inclosure: Andrus to Harris, October 4, 1908.

288 TNA-UK, FO 424/217, Lowther to Grey, November 10, 1908.

289 BOA, DH.MKT 2670/55, Hüseyin Kanco and Beşar of the Sharkiyat Tribe, 10 Teşrinisani 1324 (November 23, 1908).

290 TNA-UK, FO 424/217, R. Harris to Grey, October 23, 1908, Inclosure: Andrus to Harris, October 4, 1908.

within less than a decade, the notables of Diyarbekir arranged for massacres in the deserts to the south of the city and then proceeded to seize the property of those killed.[291]

The same Mehmed Ziya who had helped write many of derogatory telegraphs protesting Ibrahim Pasha had become a CUP activist, and he took the opportunity of Ibrahim Pasha's death to write more insults. His "Epic of the Bandit Ibrahim" carried much of the message of his previous writings. He charged that the Hamidiye Brigades amounted to licensed thievery.[292] He rendered the Hamidiye as "rabid wolves" who transformed "Diyarbekir, Urfa, and Mardin" into "desolate pasture."[293] Ibrahim Pasha would have likely disagreed with the characterization of him as a lupine figure or the pasture as "desolate." Indeed, in a petition to Ottoman authorities, Ibrahim's son Abdülhamid pointedly described the "desert" as "wonderfully fertile," and his family's flocks as "a mine of wealth" that only inspired the jealousy of Diyarbekir notables.[294] But Ziya did accurately gesture to the scope of Ibrahim Pasha's power, which stretched across provincial borders, beyond which the leader deftly recruited people and moved sheep. But clearly part of Ziya's project – like that of his telegraphs in Diyarbekir in 1905 and 1907 – was to position Ibrahim Pasha as an unreasonable enemy of cultivation and civilization. In closing, Ziya conveyed Ibrahim Pasha not only as a figure who transformed the environment, but one who even made the environment fear him: "Unplanted became all the fields / The green plain turned to thorns / And even these miserable ones shake with fear." Ziya envisioned progress in the wake of Ibrahim Pasha, and many who similarly saw Ibrahim Pasha as "a pest," as one British colonial official put it, agreed.[295]

CONCLUSION

In previous decades, the Ottoman administration had managed the locust-stricken Jazira by attempting to encourage the Shammar to settle and cultivate the land. When that endeavor did not work out as planned, they attempted at least to have them respect provincial borders. In the 1890s, the Ottoman state took a different approach with Ibrahim Pasha and the

[291] Kaiser, *The Extermination of Armenians in the Diarbekir Region*, 58.
[292] Gökalp, "Şaki Ibrahim Destanı," 12. [293] Ibid., 15.
[294] TNA-UK, FO 424/217, Lowther to Grey, November 10, 1908.
[295] Lukach, *The Fringe of the East*, 241.

Hamidiye Brigades. They charged him with moving south into the desert and the Jazira in order to control the Shammar. Ibrahim Pasha did so, moving his herds in response to locust swarms and provincial borders alike. In the process, the provincial border between Zor and Diyarbekir emerged as a flash point of conflict between Ibrahim Pasha and the Shammar, as the parties opportunistically crossed the border, whether in search of pasture, in order to evade taxes, or to raid their rivals with relative impunity. Through deft manipulation of the Jazira's political ecology and with support from buttered-up officials at Yıldız Palace, Ibrahim Pasha became hailed as the Sultan of the Open Lands. He amassed enough power to transform himself from a simple shepherd in "the City of Ruins" to an astute operator who charmed foreigners such as Mark Sykes, enraged Diyarbekir notables such as the Pirinççizades, and disappeared into the desert whenever it suited him.

His death signaled a new political moment, but did not change the structural possibilities afforded by the lines drawn by locust swarms and provincial administrators alike. In December of 1908, the interior minister of the Ottoman Empire wrote to the grand vizier and, with a message carried by many others before him, gushed about the possibilities of "the fertility of the vast lands on either side of the Khabur River."[296] Emptied because of nomads, the lands could once again be settled, he argued, and offer "incalculable benefits to the state treasury." The German traveler Wiedemann saw redemption from another path: the Baghdad Railway. The legends of Ibrahim Pasha would soon be drowned out as "locomotives roar between Aleppo and Mosul," connecting the region's soil to the rest of the world.[297] In hailing the end of Ibrahim Pasha as a transformative event for the Jazira, Ziya, the interior minister, and Wiedemann all left out another force beneath the ground in the Jazira. Even as Ibrahim Pasha was being pursued to his death, this force had been laying the groundwork for a restoration. In August of 1908, the Council of State had written of how locusts arriving "from the direction of Zor, Baghdad, and the desert" had destroyed cotton crops and summer vegetables in Aleppo.[298] The newly installed Minister of Forests, Minerals, and Agriculture, Dmitri Mavro Kordato, took special care to ensure Aleppo was protected, but

[296] BOA, BEO 2044/153256, Interior Minister to Grand Vizier, 3 Kanunuevvel 1324 (December 16, 1908).
[297] Wiedemann, "Ibrahim Pascha Glück und Ende" (1909), 54.
[298] BOA, BEO 3419/256418, Council of State, 3 Ağustos 1324 (August 16, 1908).

he came up against that dilemma that had for so long ensured the locusts remained in power even as sultans – both of the desert and of the empire as a whole – came and went.[299] Regions remained beyond settlement because that was where the locusts laid their eggs, and locust eggs could not be destroyed because they were located in regions beyond settlement.

[299] BOA, BEO 3419/256418, Minister of Forests, Mines, and Agriculture, 1 Eylül 1324 (September 14, 1908).

3

"Weren't We A Lot Like Those Creatures?" (1908–1918)

In August of 1915, Vahram Dadrian trudged along the road of deportation. He was – like hundreds of thousands of other Ottoman Armenians – displaced from his home in Anatolia. Also like them, he was bound for the deserts to the south, the largely Arabic-speaking regions of what is now Jordan, Syria, and Iraq. He described the scene in his diary. "Today, on the road, we ... saw locusts, which covered everything with their green bodies."[1] He added, "Thousands of them were crushed under our feet and wheels." The locusts loomed for Dadrian as an analogy for himself and other Armenian deportees. "The sight of them made me think," he wrote, "Weren't we a lot like those creatures, scattered everywhere, abandoned on every road, while bloody heels crush us without mercy and kill us?" Dadrian articulated a vision of locusts different than those of decades past. He did not describe others lurking along the outside as locusts, omnipotent in their destructive capacity. Instead, Dadrian described himself and his family as locusts for the way they were being pushed to the outside and crushed, powerless. While Dadrian's vision of the locust analogy was unique, the dynamics of the Armenian genocide intersected with long-standing patterns in the Jazira that centered borders and the mobility of locusts, nomads, and refugees who crossed them.

The environment that Dadrian wrote of has occupied a curious place in scholarship on the genocide. Denialist accounts have suggested that death was an unintentional byproduct of the environment and the "savage" people who lived there, or left these details out

[1] Dadrian, *To the Desert*, 36.

altogether.[2] In response to these arguments as well as the continued state denial of the violence of the genocide, scholars have been forced to prove – often at great cost – that the violence happened. Resulting accounts move with convoys of deportees to places like the Jazira and conduct the grim and important work of counting the dead.[3] But they less often attend to what the Jazira was before or after the violence of the extermination campaigns. With titles such as Ronald Grigor Suny's 2015 *"They Can Live in the Desert but Nowhere Else"*, the environment is ubiquitous in these accounts, yet also largely unexamined. Building on these pathbreaking works, a number of scholars have begun to highlight the complicated place that the Jazira environment occupied in the violence of the genocide, and also how Armenians themselves used this environment to survive.[4]

Indeed, the violence of the genocide connected with longer-term discussions about the best way to manage the arid landscape of the Jazira, its mobile people, and the locusts in whom Vahram Dadrian saw a symbol for his own suffering. The reinstatement of constitutional rule in 1908 led to new hope for the management of both moving people and locusts. But optimism was quickly dashed. Narrowing visions of ethnicity and politics gave way to unprecedented violence in the Jazira. Ultimately, the district whose borders inspired revolt in 1871 and whose borders were adeptly used by Ibrahim Pasha in the 1890s and early 1900s became transformed into the bounds of the killing fields of the Armenian genocide. Before, amidst, and during the violence, people continued to offer counterproposals for spatial units that might allow better management of the region. At the same time, locusts continued to swarm across the region. In fact, German officials dispatched to fight the insects even believed the Armenian genocide had exacerbated the locust catastrophe. Yet amidst the effort both to send Armenians to the Jazira to die and to kill the locusts that incubated in the Jazira, some managed to survive. The desert landscape that was meant to kill in fact allowed some to live, including not only Armenian orphans with nomadic pastoralists such as the Shammar,

[2] Gürün, *The Armenian File*, 215; Halaçoğlu, *Facts on the Relocation of Armenians*, 83; McCarthy, *The Ottoman Peoples and the End of Empire*, 110; Shaw and Shaw, *Reform, Revolution, and Republic*, 315.

[3] Akçam, *The Young Turks' Crimes against Humanity*; Kévorkian, *The Armenian Genocide*; Adanır and Özel, *1915*; Suny, Göçek, and Naimark, *A Question of Genocide*; Kurt, *The Armenians of Aintab*.

[4] Dündar, *Modern Türkiye'nin Şifresi*, 250–257; Kévorkian, "Earth, Fire, or Water"; Mouradian, *The Resistance Network*; Semerdjian, "Bone Memory."

but also an Armenian agronomist who concealed his identity as he himself became charged with killing the locusts of the Jazira.

LOCUSTS AND NOMADS IN THE CONSTITUTIONAL ERA

The 1908 reinstatement of the constitution and the election of an empire-wide parliament marked a period of excitement in the Ottoman domains, and the Jazira was included. Abdülhamid II remained the sultan, but a considerable amount of power devolved to the CUP, the party that led the revolution and emerged triumphant. The CUP's origins were in the Imperial Medical School and the various European capitals to which its adherents were exiled under Abdülhamid II.[5] The party was on some level committed to the positivist orientation evinced by the banner carried by students of the Imperial Medical School to hail the political change of 1908, which read "Science is the salvation of the nation" in French.[6] For many, the triumph of 1908 was a cause for celebration following the repression of the Hamidian period. In Cairo, Ottoman exiles were so thrilled with the multiethnic commitment to renewing the empire that they carried the Muslim modernist thinker Rashid Rida on their shoulders and had him embrace an Armenian bishop.[7] Rida later declared that the Ottoman state surpassed the revolutionary beacon of France, seeing as the Ottoman revolution uniquely brought together so many different groups of people under the banner of equality.[8] The hopes for the empire as a whole extended to the Jazira, which many believed would be renewed following the death of Ibrahim Pasha and the construction of the Baghdad Railway. Rumors spread of Germans or Russian Jews being settled along the path, and they were credible enough for one Istanbul newspaper to even run an interview with the empire's chief rabbi on the topic.[9]

Yet whatever the real changes and new hopes in the empire, the dilemma facing Ottoman administrators in the Jazira remained much the same as it had been for decades before. Few things demonstrated these continuities more clearly than a letter from Aleppo's Governor Reşid in early 1909. While the empire was celebrating, Reşid grappled with an old foe, directing efforts to collect and destroy some 1,000,000

[5] Hanioğlu, *The Young Turks in Opposition*, 18. [6] Ibid., 20.
[7] Der Matossian, *Shattered Dreams of Revolution*, 2. [8] Campos, *Ottoman Brothers*, 59.
[9] Fraser, *The Short Cut to India*, 91–92; "Müsevilerin El-Cezire'de İskânı ve Hahambaşı Mülâkatı," *Tanin*, June 12, 1909, 2–3.

okka (2,830,000 pounds) of locust eggs and plow fields so as to dislodge eggs from the ground.[10] Yet according to Reşid, the districts to his east – Zor, Diyarbekir, and Mosul – had failed to do their part. As a result, all of the work in Aleppo was "in vain." In blaming other districts for the locusts, Reşid was not simply searching for an excuse. Nearly a month before, officials in Zor had complained of how locals "with their ignorant mentality" had refused to show officials where locust eggs could be found.[11] Of course, even if local people had complied with Ottoman officials' requests, it might not have helped, since so many locust eggs were likely located in places that humans had no awareness of.

As the locusts of the Jazira dashed hopes, there were other reasons across the empire for pessimism. Armenians in particular had high hopes for the constitutional regime. But these efforts to improve the place of Armenians in the empire would be undercut by the Counterrevolution. Following the assassination of CUP critic Hasan Fehmi on the Galata Bridge in Istanbul in April of 1909, Ottoman troops loyal to the old regime mutinied in Istanbul. They began to demonstrate in front of Ayasofya, where they were joined by lower-ranking soldiers as well as religious clerics and students.[12] The offices of CUP-affiliated newspapers – including *Tanin* – were destroyed.[13] The violence spooked CUP leaders like Talat, the Edirne-native, erstwhile telegraph operator and teacher of Turkish language at the Alliance Israelite school.[14] For fear of his life, Talat hid in the home of his comrade, the Armenian Revolutionary Federation leader Aknuni.[15] A military force dubbed the "Action Army" (*Hareket Ordusu*), however, came to Istanbul from Salonica and snuffed out the hopes of a return to the old order. By late April, Sultan Abdülhamid II had been deposed, and his brother Mehmed Reşad V took over.

Further complicating matters, at around the same time, violence broke out in the southern Anatolian city of Adana. As historian Bedros Der Matossian has detailed, disputes between Armenians and others in the

[10] BOA, DH.MKT 2727/75, Aleppo Governor Reşid to Interior Ministry, 19 Kanunusani 1324 (February 1, 1909).

[11] BOA, DH.MKT 2716/20, Mehmed in the name of the District Governor of Zor to Interior Ministry, 6 Kanunusani 1324 (January 19, 1909).

[12] Suny, *"They Can Live in the Desert but Nowhere Else"*, 165; Der Matossian, *Shattered Dreams of Revolution*, 152; Finkel, *Osman's Dream*, 516.

[13] Der Matossian, *Shattered Dreams of Revolution*, 153.

[14] Kieser, *Talaat Pasha*, 41–42.

[15] Suny, *"They Can Live in the Desert but Nowhere Else"*, 165.

cotton boomtown – inflamed by word of the Counterrevolution – turned to riots.[16] CUP officials fanned the flames, distributing copies of the local party's organ *İtidal* (meaning, ironically enough, Moderation). The periodical included an article from Burhan Nuri in which he declared, "If the Armenians intend to form a state . . . they should look for it . . . in the desert lands of Africa."[17] The situation devolved further when the Action Army arrived in Adana from Istanbul and there set off a new round of violence when it massacred Armenians. Local people joined, leaving thousands of people dead, and much of the city's Armenian quarter torched.[18] By late 1909, the Ottoman state had hanged the alleged culprits in the city's cotton market.[19] But the event suggested to many that the constitutional regime might not bring about the change so many had envisioned.

With nomads, new hopes met the same old challenges of the past, too. In April of 1909, "lack of rainfall in the desert" and the consequent "lack of vegetation" had pushed many, among them the Shammar, to the north toward Viranşehir, the erstwhile perch of their deceased enemy, Ibrahim Pasha.[20] The Shammar movement in turn set in motion members of the Millî, who fled to their pastures at Mount Karaca. In light of this seasonal motion, the commander of the Ottoman army reserves in Zor seized upon the enthusiasm of the new regime. In June of 1909, he echoed many when he lamented how the region might be developed were it not for the nomads who came from surrounding provinces to "the borders" of Zor so as to avoid government officials.[21] But the nomadic migration was also, according to the officer, a dynamic not fit for "the age of the constitution." So he took action. He suggested that a commission ought to be formed of people of the provinces for which Zor was a "midway point" – Aleppo, Baghdad, Basra, Damascus, Diyarbekir, and Mosul – and especially those who were well versed in "nomadic conditions and customs."[22] He even invited the Shammar leader Mashʿal – son of Faris – to Deir ez-Zor to the intended meeting. There, Mashʿal waited "for a long time." But not a single official from the neighboring provinces arrived. Officials in Zor received a telegram from Basra and a letter from Diyarbekir explaining their absences. Yet as the army officer explained to the interior ministry,

[16] Der Matossian, *The Horrors of Adana*, 97–133. [17] Ibid., 141. [18] Ibid., 134–152.
[19] TNA-UK, FO 195/2307, British Vice Consulate in Adana, December 15, 1909.
[20] BOA, DH.MKT 2789/23, Diyarbekir Governor Süleyman to Interior Ministry, 24 Mart 1325 (April 6, 1909).
[21] BOA, DH.MKT 2838/79, Commander of the Reserve Brigades in Deir to the Commander of the Fifth Army, 25 Mayıs 1325 (June 7, 1909).
[22] BOA, 2838/79, Zor District Governor to Interior Ministry, 26 Mayıs 1325 (June 8, 1909).

"until now no response has been received from the other provinces." In other words, the Jazira – a coherent arid space divided by provinces but connected by nomadic motion – continued to perplex Ottoman officials.

Such developments did not seem to augur well for the long-discussed transformation of the Jazira, where the issues of locusts and nomads came together in the writing of Karen Jeppe. The Danish missionary had devoted herself to the uplift of the Armenians of the Ottoman Empire. Her vision of transformation included the land, and in 1909, she purchased a farm outside of Urfa for this purpose.[23] A regular writer in the Danish journal *Armeniervennen* (Friend of the Armenians), she adopted an Armenian son and wrote his biography in a series of columns. It was in this context that she wrote, "As soon as the first crops appeared in the gardens, the Arabs came calling and behaved not unlike locusts as they descended on the fields and stripped them bare. And if this was not enough, they allowed their animals to graze in the wheat fields."[24] Her words brought together the enduring perceptions of how ethnicity, mobility, and cultivation intersected. The British traveler and writer Gertrude Bell similarly observed how the motion of locusts and people were connected at this time. In doing so, she exhibited her sense of "the timelessness of the East," which would suffuse her later stints as intelligence officer and political administrator in interwar Iraq.[25] Long before this posting, though, as Bell reveled in former environs of ancient empires, the Jazira's reigning insects interrupted her fantasy. She noticed that nomads "were all ferrying across" the Tigris "to escape the locusts."[26]

The constitution did not only offer a new impetus for reform; it also offered new institutions through which officials could speak out on behalf of the Jazira's exceptional status. One of these was an elected parliament. In this institution in February of 1910, Saffet Bey of Urfa complained that "our just constitutional government" offered "aid ... in the name of ... humanity" when disasters occurred elsewhere.[27] But the state could shamefully offer little help within the Ottoman domains with respect to the locusts that afflicted his region. The Diyarbekir deputy Feyzi Efendi – son of Arif Pirinççizade, and cousin and close friend of Ziya Gökalp – echoed the tone of dissatisfaction, complaining that "there are many measures for destroying locusts in Europe, in France,

[23] Kauffeldt, "Introduction," in Jeppe, *Misak*, xvii, lxxi. [24] Jeppe, *Misak*, 144.
[25] Satia, *Spies in Arabia*, 79.
[26] Gertrude Bell Archive (GBA), Gertrude Bell to unknown, April 27, 1909.
[27] MMZC, İ:41, C:1, 2 Şubat 1325 (February 15, 1910), 325–326.

and in Algeria."[28] For these deputies, the Ottoman Empire – and the Jazira in particular – seemed left behind.

Other deputies from the region made clear why policies that worked for other places in the empire did not work in the Jazira. When an official from the ministry of agriculture proposed offering a monetary reward to the person who first spotted a locust, Nafi Pasha of Aleppo mocked him, asking if he thought "the locusts are like a few bandits coming and attacking a place."[29] Nafi Pasha – described by one newspaper as "the best Arab orator in Parliament" – condescendingly continued, describing how "a cloud five hours long and two hours wide" was far from a "secret issue" that required a paid informant to spot.[30] In part because of these types of assertions from the representatives of the Jazira, other deputies began to complain that the law in discussion "does not vary according to place."[31] To control locusts in densely populated and cultivated Izmir was very different than controlling locusts in the Jazira, where low population density and patchy cultivation meant that locusts could almost always find a refuge inaccessible to human hands or plows.[32] Nafi Pasha seized on the point and reiterated the importance of being attuned to geographic variation. "Our province of Aleppo is the point of the extreme limit of Anatolia. In Anatolia it is one thing, but our places on the banks of the Euphrates are something else."[33] In other words, the new institutions of constitutional governance allowed for newly public complaints about pests, part of a growing sense that humans ought to be free from the deprivations associated with insect invasions. The same institutions also opened a space for new articulations of the sense that the Jazira was different. But discussion of the Jazira's exceptionalism did little to change the core issue. In June of 1910, near Mardin "the villagers were making frantic efforts to frighten away" the locusts "by beating drums and shouting."[34] By July, the governor of Aleppo described how "an unimaginable amount of insects" had begun to arrive "in winged form everyday."[35] As usual, they came "from the direction of the desert that surrounds the province."

[28] Ibid., 327; Jongerden, "Elite Encounters of a Violent Kind," 66.
[29] MMZC, İ:47, C:2, 15 Şubat 1325 (February 28, 1910), 538.
[30] Der Matossian, *Shattered Dreams of Revolution*, 135.
[31] MMZC, İ:41, C:1, 2 Şubat 1325 (February 15, 1910), 538.
[32] This is of course not to say that locusts did not have an impact on Izmir. In some seasons, locusts even threatened to make the railway through the region unprofitable. See TNA-UK, FO 424/240, Rathmore to Sanderson, November 14, 1905.
[33] MMZC, İ:41, C:1, 2 Şubat 1325 (February 15, 1910), 538.
[34] TNA-UK, FO 424/224, Matthews to Lowther, June 18, 1910.
[35] BOA, İ.OM 14/14, Aleppo Governor Fahri to Interior Ministry, 22 Haziran 1326 (July 5, 1910).

LOCUSTS AND THE FALL OF HÜSEYIN KAZIM

While these dilemmas were familiar, they would also take on new meaning in relation to the promises of the constitutional regime. In August of 1910, a new governor arrived in Aleppo, and he embodied many of the hopes of the new era.[36] His name was Hüseyin Kazim. A graduate of the École d'Agronomie in Paris, he had helped found the CUP-affiliated newspaper *Tanin* in Istanbul (one of his first articles in the paper, fittingly, dealt with plant seeds).[37] Yet Kazim was not content to occupy himself with intellectual matters. He was an experienced governor, having served previously in Serres. And when he arrived in Aleppo, he vowed to make changes. In a front-page newspaper article, he declared that large landholders would no longer be able to exploit peasants. On his watch, the downtrodden would also "taste the pleasure of life thanks to the constitution."[38] The notables heard Kazim's message, and they did not like it at all.[39] Among them was the Jabri family, and its foremost member, Nafi Pasha. The same figure who railed about the difficulties of controlling the Jazira's locusts in the Ottoman parliament was also a large landholder.[40]

Nafi and his allies would not directly challenge the new governor's stated position to empower peasants, however. Instead, they used locusts. Their scheme emerged from a cold winter. The cold was not limited to the Jazira, as even around Istanbul wolves ventured out of surrounding areas toward the cities in search of food (one pack grew so bold as to attack a gendarme on guard duty; a French-language opposition newspaper remarked "quel appétit!").[41] But matters were especially dire in the Jazira. The Tigris froze so solidly between Diyarbekir and Mosul that caravans of camels could cross on the ice.[42] The British traveler Gertrude Bell wrote to her mother of riding for hours "through heaps of dead sheep and donkeys."[43] Some estimated that 80 percent of Aleppo's sheep died.[44] And humans fared little better, with the British consul in Aleppo writing of

[36] "Halep Valiliği," *Tanin*, 15 Ağustos 1325 (August 28, 1910).
[37] Hüseyin Kazim, "Tohumluk meselesi," *Tanin*, 24 Ağustos 1325 (September 6,1909), 1–2.
[38] *Al-Taqaddum*, October 1, 1910, cited in Watenpaugh, *Being Modern in the Middle East*, 100.
[39] Lewis, *Nomads and Settlers in Syria and Jordan*, 52.
[40] TNA-UK, FO 195/2272, Longworth to O'Conor, January 10, 1908.
[41] *Mècheroutiette*, March 1911, 64.
[42] *Sada Babil*, February 5, 1911; Wigram and Wigram, *The Cradle of Mankind*, 27.
[43] GBA, Gertrude Bell to her mother, April 14, 1911.
[44] Widell, "Historical Evidence for Climate Instability," 56.

"shepherds ... found standing, huddled in their sheepskins, frozen to death."[45] "The oldest living has seen nothing to compare to it," wrote an American missionary in Mardin.[46] For a long time afterward, it was known among Kurds as the year of the snow.[47]

In addition to causing suffering, the winter also affected the government's locust destruction efforts. Kazim claimed that officials still managed to destroy almost a million okka (2,830,000 pounds) of eggs.[48] But in some places, the snow was too deep and the ground too frozen for the typical measures of plowing or egg collection to occur (to say nothing of the vaunted locust water, which does not seem to have been employed).[49] Combined with the usual challenges of reaching distant places, the snow ensured that, in the words of Kazim, comprehensive locust control was simply "beyond human ability" – much as it always had been.[50]

Kazim's enemies pounced, using the locusts as pretext for fighting their reformist governor. They complained to the ministry of the interior of "imminent disaster" thanks to their governor being "busy with journalism" – a reference, of course, to the caustic editorial with which Kazim began his term – rather than the practical matters of managing the Jazira's environment.[51] "The locust does not destroy itself," the notables sarcastically remarked. Of course, the issue of locusts stretched far beyond the lands under the purview of Hüseyin Kazim. Around the same time, a missionary in Mardin warned that there too the "everlasting grasshoppers are at work."[52] But despite the fact that the scale of the problem went beyond the province of Aleppo, the notables' opposition still had an effect. Kazim was recalled to Istanbul to explain his conduct. In the meantime, the interim governor of Aleppo made nice with the

[45] TNA-UK, FO 195/2366, British Consul in Aleppo Fontana to Lowther in Constantinople, March 20, 1911.

[46] ABCFM 16.9.7, v. 28, Mosul Microfilm A467, Reel 717, Thomas to Barton, February 6, 1911.

[47] Hay, *Two Years in Kurdistan*, 26.

[48] BOA, DH.MKT 104–1/6, Aleppo Governor Hüseyin Kazim to Interior Ministry, 14 Şubat 1326 (February 27, 1911).

[49] BOA, DH.MKT 104–1/6, Urfa District Governor Karim to Interior Ministry, 22 Kanunuevvel 1326 (January 4, 1911).

[50] BOA, DH.MKT 104–1/6, Aleppo Governor Hüseyin Kazim to Interior Ministry, 14 Şubat 1326 (February 27, 1911).

[51] BOA, DH.MKT 104–1/6, Notables of Maraş to Interior Ministry, 11 Mart 1327 (March 24, 1911).

[52] ABCFM, 16.9.7, vol. 28, Microfilm A467, Reel 717, Thomas in Mardin to Barton, May 22, 1911.

notables and confided to a friend in Istanbul that Kazim's return would be "most inexpedient."[53] The friend in Istanbul allegedly passed the message along to the interior minister, who promptly passed the message to his friend, who was none other than Hüseyin Kazim. In response, Kazim flatly ordered his subordinate to "continue to wage war against the locusts."

Kazim's injunction attested to the way locusts mattered in both old and new ways in the constitutional period. The power of locusts meant that they had served as pretexts for other actions before, such as when the notables of Diyarbekir had charged that Ibrahim Pasha used locusts as a justification for challenging their control over the city's hinterland. But the hopes attached to the constitution and the relatively free press added new layers to these struggles. And with Kazim, that complaint about locusts accomplished what the notables desired. The reformist governor was defeated, despite the esteem he had earned in the province. Upon returning to Aleppo in late May, he was met by a crowd of 50,000, so rapturous that they reportedly removed the horses from his carriage and began to pull it themselves; Kazim opted to walk.[54] By July of 1911, he had been removed from his position, reformist zeal defused by disingenuous invocation of the locust.[55] The notables were correct that "the locust does not destroy itself." But the locust could evidently destroy others, including threats to the power of the notables.

Elsewhere in the empire at this time, humans were making locusts speak in other ways. In 1911, a satirical newspaper was published in Istanbul under the name of *Locust* (*Çekirge*) (Figure 11). The journal's masthead listed the editor as "locust" and illustrator as "starling," alongside anthropomorphic images of each of them. Each issue was referred to as a "leap" (*sıçrayış*) and featured the locust and the starling – unlikely friends, perhaps – commenting on the day's news in sardonic tones. In April of 1911, in perhaps the most absurd of many absurd discussions, the locust and the starling talked about a news item both of the moment and of existential import for their multispecies friendship: locust destruction.[56] The locust warned the starling that the bird might lose its insect friend. But the starling remained unaffected, suggesting that the locust deserved destruction because of

[53] TNA-UK, FO 195/2366, Fontana to Lowther, June 3, 1911. [54] Ibid.
[55] TNA-UK, FO 195/2366, Fontana to Lowther, October 24, 1911.
[56] *Çekirge Dergisi*, 31 Mart 1327 (April 13, 1911), 2.

FIGURE 11 *Locust* journal, 1911

the way it disrespected property. Pilfering grain here and there could hardly be considered theft, the locust argued. It had been tantamount to an act of liberation the previous year, the locust claimed, when his fellow insects consumed grain that had been waiting on a train platform. The locust insisted that they were saving the foodstuffs from spoiling. The conversation underscored on the one hand how the relationship between the locust and the starling was sufficiently well known that it could be a premise of humor. But the nature of the conversations between the locust and the starling revealed critical views of Ottoman society. Humor was one part of the conversation between winged friends, but it also opened up a broader critique of Ottoman politics.

While locusts became a vehicle of criticizing the reformist governor Hüseyin Kazim and inefficient Ottoman infrastructure, there was also a growing awareness of the nature of insect invasion and the human ability to control it on a global scale. In Britain, the idea of economic entomology had begun to take off, and a specialist by the name of H. Maxwell-Lefroy, fresh off of work in the Caribbean and India, gave an inaugural lecture on the topic at the Imperial College of Science and Technology. He was both bold and cautious. So great was the power of science that Maxwell-Lefroy declared that control of insects might soon allow humans to eradicate disease altogether.[57] At the same time, he warned that "there is no magic we can use, no one perfect simple remedy that fits all cases, no universal insecticide" (nevertheless, he would die in search of such a silver bullet in 1925, accidentally ingesting chemical insecticides he was experimenting with).[58] Ottoman scientists could not help but carry some of the same confidence, plugged in, as they were, to global networks of entomology. They had long discussed cutting-edge research, such as the locust-proof wheat being experimented with in Mardin in the 1890s.[59] But around the time of Maxwell-Lefroy, they seemed to carry the newfound confidence of the British entomologist and his like around the world. Ottoman scientific journals not only covered stories as wondrous as the experimental use of the coccobacillus bacteria against locusts by a Pasteur Institute-affiliated French-Canadian entomologist in Guatemala.[60] They also experimented with the tactic of locust death by bacteria-induced diarrhea at the empire's own Halkalı Agricultural School in 1913.[61] Though they ultimately deemed the method ineffective, it nevertheless pointed to a new horizon for imagining human control of the non human world.

"IS IT NOW THAT WE ARE INVENTING SUCH THINGS?"

While locusts became deployed in political clashes, so too did nomads. With the season's terrible cold and loss of sheep by – in the words of one observer – the "millions on millions," some saw an opportunity to discipline the Shammar.[62] To this end, an expedition of 1,500 horsemen, two batteries of artillery, and two machine guns set out from Mosul in January under the command of Rıza Bey.[63] But the cold hit the troops, too, and by February they were "immobilized by the snow" in Shirqat, the erstwhile home of Farhan Pasha's settlement.[64] There, the troops survived on "dates and sheep" for weeks, before eventually surprising the Shammar and forcing them to acquiesce and pay their tax arrears.[65] Rıza also seems to

[57] TNA-UK, CO 885/20, Inaugural Lecture in Applied Entomology by H. Maxwell-Lefroy, March 2, 1911.

[58] Fleming, *The Entokil Man*, 1–2.

[59] *Orman ve Maaden ve Ziraat Mecmuası*, 1 Temmuz 1310 (July 13, 1894), 5–6.

[60] *Ticaret ve Ziraat Nezareti Mecmuası*, 30 Nisan 1329 (May 13, 1913): 82–87.

[61] Süreyya, *Haşerat-ı Ziraiye Mecmuası*, 220–226.

[62] ABCFM, Thomas to Barton, February 27, 1911.

[63] CADN, 166PO/D/54/10, Vice Consul of France in Mosul to French Ambassador to Ottoman Empire, January 20, 1911.

[64] Ibid., February 3, 1911. [65] GBA, Diary, April 3, 1911.

have made ʿAsi, son of Farhan, the paramount shaykh of the tribe and assessed taxes on Shammar sheep.[66] The British traveler Gertrude Bell related the novelty of the events to long-standing dilemmas of motion and borders in the Jazira. She wrote that "none of this has ever been possible before because the Shammar were divided up into 3 parts, responsible to Baghdad, Mosul, and Deir and they always slipped out of one by saying that they had satisfied the other." She called the expedition a "complete success" and, in a letter to her mother, vowed that "the whole desert will shortly be as safe as any city." She went so far as to say, "I shall write a long article for some leading journal when I get home and call it The Pacification of the Desert."[67] If Bell did in fact write such an article, it would have been premature. British officials later wrote that the expedition was recalled in mid-April and caused surprise, since some had expected permanent military bases would be established in the desert.[68] Nor did the balance sheet of the expedition appear to justify it, as expenses greatly outweighed the amount of tax revenue collected. In other words, the Jazira remained an environment where people – and Ottoman expenditures – could disappear. Revenue related to wool was at the center of these continued disputes. Shammar power likely also related to wool exports in Mosul, the value of which had nearly doubled between 1874 and 1912.[69] But the government also benefited from these practices, as revenues from the sheep tax had increased ten times between the 1860s and 1895.[70]

As Ottoman officials attempted to control the movement of the nomads for whom Zor had been established, the administrative nature of Zor was changing in relation to its challenging environment. Though Chechens had made a life in Ras al-Ayn after their settlement in 1866, they also suffered from numerous epidemics, so much so that one observer described how the growth of the graveyard outpaced that of the settlement.[71] Most blamed the diseases on the difference between the environment of the desert and the mountain homes to which the Chechens were accustomed.[72] Ottoman

[66] For more on ʿAsi and his interactions with the Ottoman state, see Çiçek, *Negotiating Empire*, 192–196.
[67] GBA, Diary, April 3, 1911; Gertrude Bell to her mother, April 14, 1911.
[68] TNA-UK, FO 424/228, Marling to Sir Edward Grey, July 4, 1911, Enclosure: Quarterly Report on the Affairs of Mesopotamia for the Quarter Ended June 30, 1911.
[69] Shields, *Mosul Before Iraq*, 170. [70] Ibid., 177.
[71] Cameron, *Our Future Highway*, 163.
[72] Fraser, *The Short Cut to India*, 91–92; Lewis, *Nomads and Settlers*, 102–103; Sykes, *Dar-ul-Islam*, 286.

officials had the same concerns about the health impact of the environment, and they prompted the district governor of Zor to write to the interior ministry in January of 1911 with the message that the province could no longer use Ras al-Ayn as an administrative center because of the "insalubrity of the water and air."[73] Officials appointed to serve there would often resign the position without ever showing up. If they did show up, they sought a transfer from the day of their arrival. In response, officials called for the county administrative seat to be moved from Ras al-Ayn to al-Hasakah. They did so not only because of the better quality of its water and air, but also because of the commanding place it occupied in "the country of the Jazira."[74] Part of what made the place commanding was that it sat atop one of the Jazira's many hills that concealed ancient ruins.[75] The Council of State ratified these changes in March, declaring that al-Hasakah represented "the center of the country of the Jazira."[76] In doing so, Ottoman authorities once again confirmed the importance of the Jazira, even as it did not appear on maps as a province on its own. They also passed judgment on past efforts to settle people at Ras al-Ayn. What was supposed to be the bulwark of cultivation against the wildness of the desert had proved to not even be fit for Ottoman civil servants, to say nothing of the refugees forced to live – and die – there.

The issues of motion and nomadic administration found their way into parliament itself in the same spring. There, the question of nomadic mobility transformed into an argument about ethnicity. The conversation began as Halil Bey of Menteşe explained that in the winter, nomads had lost "90 percent" of their animals in some cases, and as a result they were "demanding to settle."[77] He requested 60,000 lira for the purpose of settling the groups in Diyarbekir. Where the problem began was when Halil specified where the nomads in question were located: they were, he noted, "on the border" (*hudutta*). He meant the provincial border between Diyarbekir and Zor. The border had long been a flash point in conflicts between the Millî and the Shammar and would become one once again for

[73] BOA, DH.İD 92–1/45, Zor District Governor to Interior Ministry, 4 Kanunusani 1326 (January 17, 1911). Complaints about insalubrity of air typically referred to malaria. See Gratien, *The Unsettled Plain*, 36.

[74] BOA, DH.İD 92–1/45, Zor General Council to Interior Ministry, 24 Kanunusani 1326 (February 6, 1911).

[75] "John Hugh Smith's Diary of a Journey from Aleppo to Urfa by Way of Deir Zor and the Khabur" in Sykes, *Dar-ul-Islam*, 281.

[76] BOA, ŞD 30/18, 24 Şubat 1326 (March 9, 1911).

[77] MMZC, İ:114, C:2, 21 Mayıs 1327 (June 3, 1911), 511–512.

Ottoman deputies. Nafi Pasha of Aleppo objected that "these tribes are not just exclusive to Diyarbekir," adding that aside from "some of the Kurdish tribes" that remained only in Diyarbekir, many of these pastoralists moved much more broadly between places such as "Mosul, Baghdad, Syria, and Aleppo." At this point the deputy of Zor interjected, "Do not forget Zor, gentlemen."[78] It should have been very difficult to forget Zor in such a conversation, seeing as the origins of the district of Zor derived from a desire to include all nomadic groups within one border. But as the history of Zor revealed, even its borders – ostensibly drawn in line with the desert – had not encompassed the full extent of nomadic motion. As the deputy of Mosul Muhammad ʿAli Fadil argued, "these tribes . . . are not native to any one province. They are the tribes that wander the Jazira."[79]

Yet many deputies were keen on going even further than Fadil. One was the representative from Kozan, Hampartsum Boyajian, the revolutionary activist who, after a stint as a medical student in Istanbul, helped lead the Armenian resistance at Sasun in 1895.[80] He claimed that if one asked nomads in his home region where they were from, they would likely say, "Our home is on top of a camel."[81] This was so, Boyajian insisted, because nomads moved easily, from Aydın on the Aegean Sea to Adana in southern Anatolia to places further east such as Diyarbekir and Mosul. When others – following Boyajian's comments – called for an expansion of the sedentarization program, Krikor Zohrap Efendi – the deputy from Istanbul – suggested implementing the programs in "Anatolia."[82] But he was quickly shouted down by those who called for something more expansive: "Ottoman Asia." The debate in parliament began with a question about where a moving group should be settled. As the debate's expansion from one about the border of Diyarbekir and Zor to one about all of Ottoman Asia underscored, the issue was relevant for much of the empire.

But the debate also became fused with an issue that had largely remained unspoken in previous years: ethnicity. It had appeared, of course, in cases such as the description of Kurdish groups as *mütearrib* (literally "Arabized") to denote that they were seminomadic, or in the

[78] Ibid., 510. [79] Ibid., 515. [80] Chalabian, *Revolutionary Figures*, 76–77.

[81] MMZC, İ:114, C:2, 21 Mayıs 1327 (June 3, 1911), 511. The phrase prefigured one that described nomads in different circumstances nearly a century later: "wherever the camel goes, that is Somalia." Weitzberg, *We Do Not Have Borders*, 1.

[82] MMZC, İ:114, C:2, 21 Mayıs 1327 (June 3, 1911), 514.

racist language of the governor of Kurdistan, describing disorder by the Shammar with an idiom literally meaning "Arab hair." The parliamentary debates, however, connected the issue of ethnicity to the question of territory in a different way. It began when the deputy of Zor, Hızır Lütfi Efendi – who had earlier in the debate implored people not to forget Zor – brought up several problems with the proposal. The first was that the nomads in question were Arab, whereas nomads linked to Diyarbekir were Kurdish. He continued, explaining that the Shammar lived in Aleppo, Baghdad, and Zor. It did not make sense then for them to settle in another province. But even beyond this problem, Hızır Lütfi Bey pointed out that the place in which Diyarbekir province planned to settle the nomads – Ras al-Ayn – was not even in Diyarbekir but was rather part of Zor. "Look at the map," he insisted. "That is how it is." In response, Feyzi Bey of Diyarbekir argued that the Shammar were not simply "native to Zor" (*Zor'a tabidir*) but were also, indeed, present in Diyarbekir.[83] Moreover, regardless of which province they were settled in, it was "the Ottoman domains" and would contribute to "the well-being of the country" and "the prosperity of Ottomans."[84] Feyzi concluded, "Why are you getting upset about this?"

What was upsetting Hızır Lütfi Bey and others was the way that the nomadic settlement program seemed to take population from a district identified as Arab, thereby weakening the empire's Arab minority. But they were reluctant to say so explicitly. As ʿAbd al-Hamid al-Zahrawi of Hama put it, "There is a plot . . . the plot is to improve [*imar etmek*] Diyarbekir, and destroy Zor."[85] Al-Zahrawi was something of an activist for Arab causes in parliament.[86] In this case, he went on to intimate that there were "political motives" behind the plan, without specifying what precisely they were. Then another deputy intervened and spelled it out: "Have you still not understood?" declared Rıza Pasha. "They are mixing the Arab element [*unsur*] with the Kurdish element." Al-Zahrawi went on to say that "plots like this will give birth to a huge disaster for the future of this nation."[87] Feyzi Bey insisted that if anyone was to blame for the problem, it was al-Zahrawi, for making it an issue in the first place. Another deputy called for al-Zahrawi to take back what he said. Al-Zahrawi testily responded, "Come over here and say as much as you want." The sniping continued through the session, but it seemed to reach a pause when Halil Bey tried to minimize

[83] Ibid., 512. [84] Ibid., 513. [85] Ibid., 516. [86] Kayalı, *Arabs and Young Turks*, 103.
[87] MMZC, İ:114, C:2, 21 Mayıs 1327 (June 3, 1911), 517.

the question of ethnicity: "In the 600-year history of this state ... is it now that we are inventing such things?" His words aimed to defuse an argument. But the comment pointed to the new ways people conceived of ethnicity and politics in the period, and the way that the Jazira's provincial borders brought these ideas to the fore. Al-Zahrawi managed one more jab, suggesting that Halil Bey had still yet to explain how both Diyarbekir and Zor were equally Ottoman, and yet one was being privileged over the other. Halil Bey acquiesced: "Absolutely, both of them are the Ottoman nation [*Osmanlı ülkesidir*], an indivisible part."[88]

The entire exchange fits with what historian Hasan Kayalı has described as "the new game of politics" concomitant with parliament, in which "opposition" almost always took the form of opposition to policies seen as Turkish nationalist.[89] But the exchange also had special meaning for the Jazira and its people. The Jazira was a part of the Ottoman domains, albeit not a province in its own right. It may have been "inseparable" in Halil Bey's words, but some of its particular challenges stemmed from the fact that the region stretched across different provinces and its jurisdiction was thus fragmented. Indeed, these particular challenges made the region seem decidedly separate. Despite the unique dilemmas of the Jazira, people inside and outside of it were able to enlist these challenges for their own politics. The notables of Aleppo used the seemingly intractable locusts as a way to fend off a reformist governor. Arab deputies in parliament, meanwhile, claimed nomads to be settled in Diyarbekir – which they viewed as a Kurdish province – as their own, based on bonds of ethnicity. The parliamentary debates revealed a shift then, one from *urbân* – meaning Arab nomad – to Arab in an ethnic sense. For a long time, it may have seemed that educated city dwellers, whether in Hama or Istanbul, had much in common, perhaps including a disdain for those wandering populations of marginal ecologies. But the parliamentary debates revealed new kinds of solidarities. It is unclear what these solidarities meant to groups such as the Shammar. Even before the discussions in Istanbul, reports suggested that "the Shammar ... have categorically refused, already, to subsequently occupy the villages whose construction is projected."[90]

And there seem to have been good reasons for the Shammar to have these doubts, given the resilience that their motion afforded them in the Jazira's political ecology. Drought had followed the harsh winter of 1910-1911.

[88] Ibid., 518. [89] Kayalı, *Arabs and Young Turks*, 113.
[90] CADN, 166PO/D/22/2, French Vice Consul in Diyarbekir, May 29, 1911.

As one American missionary put it, "the previous snow-storms seemed to have exhausted heaven's store of moisture."[91] The weather had an impact on the Shammar, whose pastures were lacking. But so too did the drought hurt cereals. And when it seemed like things could not get worse, "clouds of locusts from the southeast bore down upon these remnants and made sad havoc," with some farmers left with nothing more than stalks to be used as winter straw. In such conditions, being a settled farmer must not have seemed appealing. As state officials described their interminable fight against locusts in the same year to the west of Mardin in the region of Suruç, they even used the same language to describe locust-destruction officials as they did to describe nomadic motion, noting that officials had to be in a state of "wandering" (*geştügüzar*) in order to find the geographically scattered eggs.[92] If even state officials had to move like the Shammar, it may well have seemed to the Shammar that they ought to continue to move in their own way. It seems they did so. As surveying work on the Baghdad Railway took place amidst temperatures of over 120° Fahrenheit (48° Celsius) in the shade, some suspected the Shammar of pilfering the wooden posts set up to guide the railway.[93] Many had hopes that the railway would force the Shammar to take up agriculture, and end their role, in the words of one governor of Mosul, as "an annoying thorn in the desolate deserts of the Jazira."[94] But at this point at least, the infrastructural project that was supposed to transform the Jazira and, as a corollary, the lives of the Shammar may have ensured that the Shammar had sufficient fuel for their campfires.

A GRAMMATICAL DISPUTE IN THE OTTOMAN PARLIAMENT AND CHANGING GRAMMARS OF RULE

In the wake of such setbacks, some Ottoman officials took a pragmatic view of their successes and failures in the region. These opinions emerged in response to an uncommonly forthright memo on the history of Zor dispatched in early February of 1912. Written by Lütfü, the memo

[91] ABCFM 76, "1911 in Mesopotamia," December 31, 1911.

[92] BOA, ŞD 550/26, Minister of Forests, Mines, and Agriculture, 2 Haziran 1327 (June 15, 1911).

[93] TNA-UK, FO 406/39, Lowther to Grey, October 4, 1911, Enclosure: Vice-Consul Greig to Lowther, September 12, 1911.

[94] BOA, DH.EUM.2.Şb 1/4, Mosul Governor to Interior Ministry, 16 Temmuz 1330 (July 29, 1914).

declared that, though Zor had been established fifty or sixty years before as a "center of tribes," there was not a "trace of improvement" (*eser-i umran*) nor any progress in settling the nomads of the region.[95] Lütfü described the history of efforts at settlement, referring even to plans to form a "migratory district" (*seyyar mutasarrıflık*) like that formed for the Muntafiq in Iraq. Like so many before him, Lütfü saw in the Jazira a solution to the Ottoman dilemma of resettling refugees from the Balkans and elsewhere, often in places on the ecological margin that ended up causing problems among the local population.[96] But no need remained. "Even if refugees are sent by the millions inside the district of Zor," he wrote, "more than enough land remains in the huge desert for the Arabs [*urbân*]."

Not everyone agreed. The district governor of Zor echoed many in March of 1912 when he declared that improving regions like "the Jazira" through irrigation "has been known for a long time."[97] He echoed many too when he blamed the failure to enact these plans on lack of resources and security. But he departed from others when he discussed refugee resettlement. "The refugees who have been able to be settled in Anatolia," he wrote, "cannot be sent here." He blamed "the climate" and "weather conditions" (*ahval-ı havaıye*) and cited the "abominable disaster" of Ras al-Ayn, where the settlement of 12,000 Chechens had dwindled to fewer than 1,000. If the plan to use refugees to improve the land did not work, however, the reorganization of nomadic administration seemed to hold out promise. Extensive correspondence in 1912 concerned the matter of whether the administration of the three branches of the Shammar might be streamlined by placing them all under the authority of one chief. Yet again, however, the environment of the Jazira made such an administrative arrangement implausible. The district governor of Zor responded to the plans by noting that "even if a single chief is recognized … by the government," there was still the fact that the Shammar "were spread out in the desert and far from" any such chief.[98]

In addition to Ottoman officials, Ottoman citizens also imagined how borders might be changed in the Jazira. In April of 1913, it was the

95 BOA, DH.İD, 45/12–53, Lütfü to Interior Ministry, 25 Kanunusani 1327 (February 7, 1912).
96 Fratantuono, "Producing Ottomans"; Gratien, "Ottoman Quagmire."
97 BOA, DH.İD, 45 12–54, District Governor of Zor Mehmed Celaleddin to Interior Ministry, 20 Şubat 1327 (March 4, 1912).
98 BOA, ŞD 34/30, Zor District Governor to Interior Ministry, 14 Haziran 1328 (June 27, 1912).

notables of Mardin who called for a change in borders, given their connection to the Jazira and its moving people. They did so based on both language and geography. Their city was part of Diyarbekir province, but they called for Mardin to either become an independent district or become part of Zor.[99] Their region, they explained, was "the most fertile ... and most important ruling point of the region of the Jazira."[100] They added that most people in Mardin spoke Arabic, just like the nomads who populated the broader Jazira region. Like the Arab deputies of the Ottoman parliament then, the people of Mardin too were declaring a cultural affinity with nomads who had for many years been presented as constitutionally different than them. Such pronouncements, of course, recalled the history of relationships between Mardin and groups such as the Shammar, so deep that in 1901, when Ibrahim Pasha massacred the Shammar, the nomads' tents contained no shortage of Mardin merchants.

The governor of Diyarbekir answered the proposal from Mardin in a way that demonstrated the changing nature of Ottoman governance with respect to ethnicity. In response to what he described as Mardin's request for its "detachment from the eastern provinces and attachment to the Arab lands," he offered a tally of population figures according to ethnicity in a coded telegram to the interior ministry.[101] By his count, Mardin district contained 45,715 Arabs, 124,915 Kurds, and 34,196 Suryani. It is unclear why these numbers did not include Armenians, who formed a significant part of the region's population. But it is clear where the numbers came from.

Particularly in light of territorial losses of the Balkan Wars (1912–1913), the CUP had taken a turn toward viewing Anatolia as the homeland of the Turks.[102] Accordingly, a shift occurred in statistical practices, as the state became increasingly concerned with counting population not simply in terms of religious communities – as they had previously – but instead in relation to ethnicity.[103] A key part of this process was the 1913 establishment within the Interior Ministry of the Directorate of the

[99] BOA, DH.İD 144/2–26, Notables of Mardin to Grand Vizier, 17 Mayıs 1329 (May 30, 1913).

[100] BOA, DH.İD 144/2–26, Notables of Mardin to Şura-yı Devlet, 11 Mayıs 1329 (May 24, 1913); Notables of Mardin to Şura-yı Devlet, 15 Nisan 1329 (April 28, 1913).

[101] BOA, DH.İD 144/2–26, Diyarbekir Governor Hakkı Coded Telegram to Bab-ı Ali and Interior Ministry, 3 Haziran 1329 (June 6, 1913).

[102] Akçam, *The Young Turks' Crimes Against Humanity*, 29.

[103] Ibid., 32. See also Yosmaoğlu, *Blood Ties*.

Settlement of Tribes and Refugees.[104] Its brief brought together the motion of people within and without the empire, combining efforts to "settle and civilize" nomadic pastoralists, provide aid to refugees entering the empire, and take "preventative measures against emigration" from the empire.[105] In this context, statistics functioned as mechanisms of control. One of the civil officers for the eastern region was none other than Ziya Gökalp.[106] Thus the figure who had been involved in the dethroning of Ibrahim Pasha was also a force in amassing forms of knowledge that stood in contrast to ethnic definitions of years past, when where people lived and what they did informed ethnicity as much as the language that people spoke. The narrowing vision of ethnicity and its statistical underpinnings would become especially prominent in the coming years.

At the same time as the region's people made the argument for different borders on the basis of language and geography, locusts struck. Once again, the insects revealed the vulnerability of a system involving piecemeal administration of a broader ecology. As the same governor of Diyarbekir who had offered tallies of ethnicities in Mardin wrote in December of 1913, "the locust disaster" had hit the southern portions of the province hard "for many years" and had been especially significant the previous spring, having left people in the "desert" struggling.[107] Such observations were commonplace. Less so was Hakkı's "declaration of war" (*ilan-ı harb*) against the locusts, although he was careful to emphasize that his plans would only work if, in an echo of so many observations in the previous decades, locusts did not arrive "from outside of the borders of the province" (*vilayet hududu haricinden*). The tension between administering what was within borders and managing what came from beyond was still present. But the discourses that accompanied these phenomena were changing. They were fueled by a growing certainty in the preeminence of ethnicity as an organizing principle of politics, and in the conviction that humans could control insects.

Whatever Hakkı's hopes for controlling matters in 1913, the locusts were back again in 1914. The American missionary D. L. Dewey in Mardin wrote that the main reason the constitutional government had "not made the progress that the people had hoped" was "the heavy snow of three years ago, and the two years of drought and grasshoppers after

[104] Üngör, *The Making of Modern Turkey*, 36. [105] *İkdam*, December 29, 1913, p. 3.
[106] Üngör, *The Making of Modern Turkey*, 36.
[107] BOA, DH.İD 99/23, Diyarbekir Governor Hakkı to Interior Ministry, 20 Teşrinisani 1329 (December 3, 1913).

it."[108] Dewey described how the locusts left many in the Jazira "robbed ... of the means for their existence," all the more so because they coincided with deprivation associated with the Balkan Wars. In another village near Viranşehir, he wrote, "Their village has been visited by the grasshoppers, etc. for three years so that people who were comfortably off before, are poor now."[109] For the missionary, the greatest concern was that locals could no longer afford to pay the salary of the preacher in town. But his observation also pointed to some of the deep structures of regional history.

As the Diyarbekir governor's declaration of war against locusts and the call for Mardin to be separated from Diyarbekir implied, there was a connection between managing locusts and territory, particularly as it related to nomads. The dynamics became articulated most clearly in the Ottoman parliament in June of 1914. The occasion was, of all things, a grammatical dispute. In discussions of a Locust Law outlining the procedure throughout the empire to respond to insect swarms, the representative of Antalya, Fuat Hulusi Bey, introduced what he described as "a tiny amendment."[110] Fuat Hulusi Bey's sights were set on what he deemed a grammatical mistake in the draft law. It read "in the places whose people are uninhabited" (*ahalisi meskun olmayan yerlerde*). Fuad Hulusi Bey insisted, "People are not inhabited. They inhabit." Given that the law was aimed at allowing for different regulations in different geographies, the meticulous deputy suggested a phrase free of grammatical errors that denoted the same point: "in the places inhabited by tribes" (*aşair ile meskun mahallerde*). An uproar ensued. One deputy contended that the initial phrase required no correction, another clarified that "inhabited" in the phrase functioned as a "passive participle" (*ism-i ef'ul*), and still another marveled that it was "quite a strange phrase." Finally, the floor returned to Hulusi Bey. "I did not think this topic would go on for so long," he admitted. Somewhat conciliatory, he stated that if the intention with the wording was to refer to regions where "people are nomadic," then perhaps referring to the regions as "migratory" (*seyyar*) ought to work. But he could not help but reiterate his original position. "I still see the first suggestion" – meaning his own – "as the most precise." Then, for good measure, he declared, "People definitely cannot be inhabited ... people inhabit." It seems Fuat Hulusi Bey had the last word on the

[108] ABCFM, 16.9.8, vol. 1, Eastern Turkey, 1909–1914, D. L. Dewey to friends, May 3, 1914.

[109] Ibid., March 2, 1914.

[110] MMZC, İ:13, C:2, 29 Mayıs 1330 (June 11, 1914), 226.

question of the grammar of the legislation's text. But the difficulty of even coming up with the words for discussing these regions was symbolic of the interlinked challenges of administering moving people and insects, as if the metaphorical and literal bugs of the changing Ottoman Empire challenged the grammar of language itself.

And, of course, the creation of legislation for the entire Ottoman Empire carried with it the possibility of failing to take into account local diversity. Even after discussion moved on from Fuat Hulusi Bey's grammatical objection, many echoed earlier concerns and insisted that the law "must be implemented according to place."[111] Like climate in the Habsburg Empire at roughly the same time, locusts in the Ottoman Empire forced officials to grapple with scale and the vast differences that existed across the Ottoman domains.[112] Later in the conversation, there was even a dispute about the proper way to measure distances with respect to locust swarms. Typically, in the Ottoman Empire, distance was described in terms of time on horseback necessary to traverse a distance. But as one deputy complained, the ambiguous measurement ought to be replaced. "What does three hours mean?" he asked. Depending on whether one went by horse or donkey or train, three hours could mean very different things indeed. The only solution, the deputy argued, was addressing the question in "a scientific way," which in his view meant adopting the kilometer. In an understatement, the deputy presiding over the meeting declared, "This is not an issue that will be resolved with the law that you mentioned." The management of locusts catalyzed disputes ranging from the grammar of describing regions with pastoralist populations to the question of how to measure space in the first place.

As deputies struggled to summon the grammar and language to discuss the seemingly interminable problem of locusts and regions like the Jazira that they afflicted, parliament also became a place for discussion of environmental and demographic engineering. Indeed in 1913 and 1914, Muslims of eastern Thrace had been brought into the Ottoman Empire while Greek Orthodox Ottoman citizens of Aydın were removed.[113] Foreshadowing the logic of the population exchanges of the interwar period, the population transfers did not go unchallenged. Particularly vocal were those who saw their place as loyal Ottomans slipping away in favor of national identities presented as fundamentally at odds with

[111] MMZC, İ:13, C:2, 29 Mayıs 1330 (June 11, 1914), 228.
[112] Coen, *Climate in Motion.*
[113] Akçam, *The Young Turks' Crimes Against Humanity*, 63–68.

Ottomanism. In July of 1914, just a month after the discussion of the locust law, the Ottoman deputy of Aydın, Emmanouil Emmanouilidis, declared, "I am not Greece's representative. I am an Ottoman representative, and my greatest hope is to remain Ottoman."[114] He did so in front of the Ottoman interior minister Talat, last seen hiding out during the 1909 Counterrevolution with Armenian Revolutionary Federation leader Aknuni. In Emmanoulidis's view, the expulsions were unnecessary because the people might have been settled on the "many empty lands" between Üsküdar on Istanbul's Asian shore and Basra on the Persian Gulf. Talat responded with environmental specificity: "If we had sent the refugees there and scattered them in the deserts [*çöllere serpecek olsaydık*], they would have all died of hunger and Emmanouilidis Efendi would not have been pleased about that either."[115] Talat's words reflected the Ottoman experiences of nearly the past half-century, when refugees resettled throughout the empire – such as the Chechens at Ras al-Ayn – had struggled not only with displacement but also with settlement. The pronouncements seem ironic given what was to come in the next year.

Some even had premonitions of these events. An Armenian recalled that during the same summer, she was told to return to her house in Namrun near Tarsus.[116] She initially feared imminent massacre, as had happened in 1909 in Adana or in the 1890s all across Anatolia. But it turned out that it was not violence that she was ordered to seek refuge from, but rather a locust invasion. The insects descended "like flying carpets and covered the ground." She remembered, "We watched helplessly [as] these little, dangerous soldiers [ate] the gardens and the trees, and [left] the earth barren."[117] It was a rehearsal, in a way, for the environmental alienation that would occur in the coming years, when the destination of Deir ez-Zor became "dreaded" among Armenians because people knew it represented "the Mesopotamian deserts."[118] Or perhaps the two events blurred together in retrospect.

"LOCUSTS HOVERED OVER THE CITY"

When Archduke Franz Ferdinand was assassinated in the former Ottoman city of Sarajevo in June of 1914, it set in motion a conflict that would change much of the world, the Jazira included. It was a conflict into which the Ottoman Empire limped, having witnessed "a revolution and three wars"

[114] MMZC, İ:26, C: 2, 23 Haziran 1330 (July 6, 1914), 609. [115] Ibid., 611.
[116] Bagdikian, *The Memoir of Lydia Bagdikian*, 33. [117] Ibid., 33–34. [118] Ibid., 54.

in the span of six years before World War I, not to mention the everyday challenges of governing a sprawling empire.[119] The CUP leadership took different messages from these conflicts. From their revolutionary triumph, they believed in the power of science and, perhaps, autocracy. From the Italian conquest of Libya, they had learned guerrilla tactics and the importance of irregular paramilitaries. From their defeats in the Balkans, they began to doubt the "secular patriotism" they had espoused, a realization that had a personal impact given that many CUP leaders themselves hailed from the Balkans.[120] They also were further convinced that demography was destiny in a world in which ethnicity increasingly determined polity.[121] These lessons would shape the experience of the war all across the Ottoman domains, but particularly in the Jazira, where the desert that had defied state control would become a horrific solution to what officials saw as the empire's problems. But first, there was mobilization, and for the Ottomans this was difficult. It may have been an exaggeration when the German military chief of staff Helmuth von Moltke wrote in 1914 that the Ottoman Empire was not so much the proverbial "Sick Man of Europe" as "the Dead Man of Europe."[122] But it was undeniable that the empire's mobilization could not occur along the same lines as in other empires. They still tried. And the mobilization that accompanied the conflict seared itself into the memories of all of the region's inhabitants.

The insects associated with the Jazira offered a powerful way of understanding the voracious impact of the effort. The Maronite priest Antun Yamin called Ottoman officials with their endless requisitions in Mount Lebanon tantamount to "creeping locusts."[123] The British consul in Aleppo called them "human locusts."[124] Rafael de Nogales, a Venezuelan mercenary who fought for the Ottomans during the war, even compared locusts favorably to Ottoman officials, since the insects only destroyed harvests while "those inveterate human parasites ... would certainly have sold the very locomotives off the Baghdad Railway."[125] Such comparisons have even endured into the present in the work of the Lebanese-American novelist Rabih Alameddine, who, in

[119] Rogan, *The Fall of the Ottomans*, 1.
[120] Akın, *When the War Came Home*, 35; Aksakal, *The Ottoman Road to War in 1914*, 23.
[121] Akın, *When the War Came Home*, 44. [122] Ibid., 109.
[123] Al-Qattan, "When Mothers Ate Their Children," 728.
[124] TNA-UK, FO 195/2460, Fontana to Sir Louis Mallet, August 31, 1914.
[125] Nogales, *Four Years beneath the Crescent*, 141–142.

recalling the horrors of the war, termed the Ottomans and locusts "interchangeable pests."[126]

Aside from the long history of the insects in the region, there was a special reason why the locust carried such metaphorical weight. As 1914 turned to 1915, the Ottomans found themselves facing the army of the Russian Empire in eastern Anatolia and that of the British Empire in Iraq and Sinai. But all across the eastern Mediterranean, they were facing an adversary reminiscent of many decades in the Jazira, as locusts swarmed across the coast and helped to instigate a brutal famine, particularly in Mount Lebanon. As historians Keith Watenpaugh and Melanie Tanielian have both argued, presenting the famine as a "natural disaster" obscures the responsibility of the locust-like state officials and the devastating Allied blockade.[127] However blame is apportioned, the famine that ensued was apocalyptic, killing some 500,000 people out of a population of somewhere between 3,000,000 and 4,000,000 people in greater Syria.[128]

The cataclysmic impact of the famine and the locusts on the coast has perhaps obscured how the war years also involved a less exceptional but also devastating outbreak of locusts further inland in the Jazira. The locusts that struck Palestine, Mount Lebanon, and the coast of greater Syria were the migratory or Najdi locusts. Not native to the region or the Jazira, they instead migrated from the south. They had previously appeared, for example, in Baghdad in 1890. At the same time in the Jazira, the more commonplace Moroccan locusts were also at work, and they too would have an impact during the war years. In fact, the locust-like agents of Ottoman requisitions may have had something to do with this dynamic. As the case of Hüseyin Kazim and the winter of 1911 underscored, much of the work of locust control occurred in the winter and spring, when diligent officials hoped to plow up locust eggs and to prevent any springtime swarms. But in the rush of war, it is likely less attention was paid to these efforts. And with so many animals and people requisitioned for the war effort, many fields fell fallow altogether, providing further openings for locusts to strike into the fields of wheat that had expanded into the Jazira over the preceding decades.

126 Alameddine, *An Unnecessary Woman*, 119.

127 Watenpaugh, "Introduction," in Panian, *Goodbye, Antoura*, ix; Tanielian, *The Charity of War*, 52–54. See also Pitts, "A Hungry Population."

128 Foster, "The 1915 Locust Attack in Syria and Palestine," 370.

With the Ottomans beset by military and insect adversaries, some officials questioned the loyalties of the empire's Armenian citizens. Violence toward Ottoman Armenians, of course, was nothing new. In a perhaps environmentally overdetermined comparison, one observer remarked to American ambassador Henry Morgenthau that "massacres have occurred during this century almost as regularly as the arrival of locusts."[129] This pronouncement was not entirely accurate, as for much of the nineteenth century Armenians had been viewed as "the loyal community," with Armenian bankers and other officials integral parts of Ottoman governance.[130] Yet as the prospect of imperial dismemberment became more real during the war years, the presence of Armenians came to be seen as a threat to Ottoman sovereignty, as had previously happened in 1895 to 1896 and in 1909. With rumors of Armenian support for Russian advances in eastern Anatolia and scattered resistance in places such as Van and Sasun, the fuse was lit. On April 24, 1915, Armenian notables of Istanbul were rounded up and arrested. The first convoys of deportees were told they were bound for the desert lands far from their homes in Anatolia.

There are many painful details of the early days of the deportations, both in terms of the violence of forced removal and in terms of the courage of those – Armenian and otherwise – who struggled to save themselves and their neighbors. One such document comes from the American missionary in Mardin, A. N. Andrus. He recorded the events in a weekly calendar, its pages now yellowing, with two holes at the top of the page, and three ruled lines beside each day's entry. On May 10, 1915, he wrote tersely, "Public crier announces that 3 days of grace are allowed in which to surrender arms of all kinds, and deserters to deliver themselves."[131] Unrecorded by Andrus – perhaps for fear of his words being intercepted by Ottoman officials – was the target of these searches. As in Mardin, so too in many other places had the searches for arms been a pretext for removing Armenian populations. The specter of deserters as criminal gangs also had been a pretext for the deportations. Three days later, Andrus's calendar recorded a message just as concise: "locusts hovered over the city from the south-east." The distinctive insects of the Jazira appeared at the same time as plans to use the ecology of locusts to kill certain humans intensified.

[129] Morgenthau, *United States Diplomacy on the Bosphorous*, 314.
[130] Suny, "*They Can Live in the Desert but Nowhere Else*", 48.
[131] ABCFM 76, AN Andrus, Calendar Diary of 1915.

Unspoken in his elliptical recordings, however, were the figures at work in Mardin. As part of the province of Diyarbekir, the city was under the jurisdiction of the infamous governor Dr. Reşid.[132] A native of the Caucasus, he helped found the CUP with his classmates at the Military Medical School. He drew on his medical training to justify his actions as part of the genocide. The "Armenian bandits" – as he termed them – were "harmful microbes ... in the nation's body," he declared, and it was "a doctor's duty to kill microbes."[133] Hilmi Bey, district governor of Mardin, called Reşid a "rabid bloodhound."[134] The arms searches and crackdowns on deserters that Andrus recorded were exactly the measures that Dr. Reşid had used in Diyarbekir. But in Mardin, they did not have the same effect, because Hilmi Bey refused to follow the particularly expansive orders enacted by Dr. Reşid, who directed genocidal violence not only at Armenians but also Assyrians in Diyarbekir province. As a result, Dr. Reşid needed someone he could trust to enact his plans.

He found that person in Feyzi Bey, the parliamentarian and scion of the Pirinççizade family of Diyarbekir. In the words of scholar Uğur Ümit Üngör, Feyzi had become "a Young Turk hardliner" and allegedly had his opponent in previous parliamentary elections – Mardin's Ohannes Kazazian – assassinated.[135] Feyzi set out for Mardin on May 15, just as the locusts that Andrus had observed began to bear down on the city.[136] When Hilmi continued to refuse orders to arrest Christian notables of the city, Feyzi Bey stepped in. Prior to the war in parliament, Feyzi had decried the failure of locust-control efforts and set off a firestorm of criticism by attempting to settle the Shammar in Diyarbekir in the years leading up to the war. Amidst the war, he used his local connections and knowledge of the environment in other ways, forging links with local toughs and power brokers to begin arresting local Christians and looting their properties.[137] When Dr. Reşid finally secured the intransigent Hilmi Bey's transfer to a minor position in Mosul province, he arranged for him to be assassinated on the way (Hilmi escaped).[138] With Hilmi out of the way and the irregular forces of deportation in place, Mardin's Christians began to be forced to leave, the vast majority never returning, their bodies interred in the caves and wells of the surrounding areas,

[132] Üngör, *The Making of Modern Turkey*, 106.
[133] Güngör, "Bir Canlı Tarih Konuşuyor," 2445.
[134] Kaiser, *The Extermination of Armenians in the Diarbekir Region*, 319.
[135] Üngör, *The Making of Modern Turkey*, 48. [136] Ibid., 69. [137] Ibid., 72.
[138] Ibid., 87.

from the Roman ruins of Zerzevan Castle to the ancient cisterns of Dara.[139] The ruins of antiquity that had stoked outsiders' excitement about the Jazira had become the final resting place of people who no longer fit within the empire's narrowing sense of belonging. By early June, Feyzi Bey returned to Diyarbekir, where it was he who translated Dr. Reşid's orders to slaughter 800 Armenians into Kurdish so that the local chieftain Ömer would understand, and it was he who celebrated the massacre with a feast afterward.[140]

Similar scenarios occurred throughout the empire, with hundreds of thousands of Armenian men and women separated and marched out of town. Ostensibly bound for the south, most were robbed of what they owned before being sexually assaulted, abducted, or slaughtered along the way. Among the forces most responsible for the violence was a group known as the Special Organization, a paramilitary of bandits, toughs, and itinerant figures formed to stave off Russian invasion but primarily known for their involvement in the violence of the genocide. Some Armenians managed to avoid their depredations and were taken in by local families, whether as a form of protection or a form of enslavement or some combination of the two. The few who made it through this gauntlet faced what seemed to be an environmental death sentence, with the harsh landscape of the Jazira looming.

Provincial borders shaped the violence of the genocide. By May 27, local actions gained legal grounding with the passing of the Deportation Law (Tehcir Kanunu), which mandated the evacuation of lands threatened by Russian advances, a geographic designation that encompassed much of the Ottoman East. Interior minister Talat Pasha specified that the Armenians of the eastern Anatolian provinces of Erzurum, Van, and Bitlis be deported to Deir ez-Zor and Mosul, while those in western portions of Anatolia be sent to southern Syria and what is now Jordan. He also specified – in acknowledgement of the politics of population that had torn the empire asunder – that Armenians should not constitute more than 10 percent of any province, and his personal notebooks contained color-coded maps denoting this plan.[141] Indeed, perhaps in connection with these questions, Diyarbekir governor Dr. Reşid had even been queried as to whether Viranşehir ought to be considered part of Diyarbekir or Zor, long located on the edge of the two districts as it was. Reşid requested a postponement of deciding on the matter "because," he wrote, "I have

[139] Ibid., 88. [140] Ibid., 72–73.
[141] Bardakçı, *Talât Paşa'nın Evrak-ı Metrûkesi*, 82–83.

been busy with very important matters, and have not been able to find time to consider this issue of attachment and detachment."[142] Amidst the rupture of the genocide, then, the question remained of what Zor's borders were, borders that had been initially intended to be in line with the desert but over the course of the subsequent decades had proved challenging to align with the environment. The districts that had for so long been difficult to manage because of moving human populations would become the final destination of deportees, whose population was to be statistically monitored in accord with ethnicity (Figure 12).

In protesting the deportations, some made the familiar environmental dynamics of the Jazira quite explicit. The Armenian patriarch Zaven Der Yeghiayan utilized distinctive imagery as he made entreaties to the Ottoman minister of justice in June of 1915. As a native of Mosul who had grown up in Baghdad, Der Yeghiayan lamented "if you only knew about the places where they are being sent!"[143] He pointed out how the government had tried to settle the nomads many times before to no avail. Still, Der Yeghiayan explained, "one or more times during the year Arab *ashirets* [tribes] like Anazah, Shammar, and others pass through these areas." To describe their impact, he invoked the humble yet omnipotent insect of the Jazira. The nomads, he stated, "wipe out like locusts everything on their way."

With so many Armenian deportees bound for the Jazira, it is perhaps not surprising then that Armenians found themselves materially and symbolically linked with locusts. As in the case of Vahram Dadrian, many experienced the horrors of the insect invasion as they were deported to an unfamiliar environment. The locusts offered a clear point of reference for Aliza Harb too, an Arabic-speaking Armenian native of Viranşehir, Ibrahim Pasha's former domain and the town whose provincial provenance Dr. Reşid declined to weigh in on, busy as he was with deportations. Harb described in her memoir how, in June of 1915, "soldiers swarmed through the streets like a plague of locusts."[144] The Assyrian priest, scholar, and native of Mardin Ishaq Armalah used a similar comparison. The "soldiers," in his words, "spread like locusts in the markets of Viranşehir and its houses and arrested the men."[145]

[142] BOA, DH.İ.UM 46–2/211, Diyarbekir Governor Reşid to Interior Ministry, 27 Nisan 1331 (May 10, 1915).

[143] Der Yeghiayan, *My Patriarchal Memoirs*, 74. [144] Harb, *Aliza*, 44.

[145] Armalah, *Al-Qusara fi Nakabat al-Nasara*, 350.

FIGURE 12 Zor and the Armenian genocide

ENVIRONMENT OF GENOCIDE

The genocide built on the rhythms of the Jazira that officials had been contending with for decades. Indeed, as Interior Minister Talat Pasha said of the Armenians in August of 1915 to Henry Morgenthau, US ambassador in Istanbul, "they can live in the desert but nowhere else."[146] The phrase proved so evocative that it became the title of Ronald Grigor Suny's authoritative work on the genocide, published to coincide with the centenary of the first deportations in 2015. Fewer historians, however, have pointed out the message Talat Pasha conveyed to Morgenthau that the ambassador left out of his published memoir.[147] In his diary, Morgenthau recorded not only Talat's promise about the desert, but also the fact that Talat "said they want to treat the Armenians like we treat the negroes." Morgenthau added, "I think he meant like the Indians," though, of course, the logic of violent removal had been used against both groups.[148] In other words, Morgenthau excised from his public account the way that Talat's logic of ethnic cleansing explicitly referred to the actions of the United States and other international powers of the day.

The fig leaf of the genocide was that Armenians would develop the Jazira in ways that settled nomads and refugees had failed to do for decades. The Armenian priest Grigoris Balakian, for example, heard that Armenians had installed themselves in the region of Deir ez-Zor and "made the desert bloom."[149] But for most, the symbolism of Deir ez-Zor was decidedly different. The intellectual Yervant Odian recalled how, even though Armenians did not yet "know the scale of this criminal action," they quickly became terrified by "the evil name of Der Zor."[150] The desert figured prominently in these anxieties. Karnig Panian asked his mother if she had "ever seen this desert," and she replied, "I've only heard about deserts in fairy tales. They're supposed to be like endless fields of sand and dirt, where it's very hot all the time."[151] When Vahram Dadrian and his family learned they were headed for the desert rather than Aleppo, they wondered, "What can all those people do in the sand?"[152] In response, he recalled, "A soldier laughed at our naiveté and explained

[146] Morgenthau, *Ambassador Morgenthau's Story*, 232.
[147] Gürel, *The Limits of Westernization*, 27.
[148] Morgenthau, *United States Diplomacy on the Bosphorous*, 298. On removal in American history, see Seeley, *Race, Removal, and the Right to Remain*.
[149] Balakian, *Armenian Golgotha*, 396. [150] Odian, *Accursed Years*, 31.
[151] Panian, *Goodbye, Antoura*, 44. [152] Dadrian, *To the Desert*, 38–39.

that we shouldn't think only of vast expanses covered with sand. A desert can be covered with stones, bushes, and thistles." The soldier also warned of nomads and Chechens in the desert, foreshadowing how past subjects of Ottoman development efforts would be implicated in the violence of the genocide.

As the Armenians were deported toward Deir ez-Zor, their bodies etched outlines around the various Ottoman projects aimed at transforming the Jazira. Bodies piled up alongside the tracks and at the stations of the Baghdad Railway, which many had envisioned as rejuvenating the Jazira.[153] Bodies filled up the region's rivers, so much so that residents of Mosul were banned from drinking from the Tigris.[154] Bodies filled up the cisterns of Dara and the tributary streams such as the Khabur, where they became food for dogs and vultures.[155] Bodies etched the borders between desert and sown, as one found corpses, in the words of an observer, "everywhere ... where the desert sand borders on inhabited districts."[156] Of 1,000 women and children from Harput who departed Viranşehir in July, only 300 would reach Ras al-Ayn.[157] Those who arrived had been stripped of their clothes, their skin "burned to the color of a green olive." One survivor even recalled how, in a convoy whose members expected to be killed "in the depths of a desert wasteland," a priest administered the litany of the dying with grains of sand for communion.[158]

While the environment of the Jazira shaped Armenian suffering, so too did the nomads and refugees who had long called the region home. The reports of the German consul in Aleppo suggested that as early as July of 1915, Armenian families in Tel Abyad felt forced to sell their daughters

[153] Politisches Archiv des Auswärtiges Amts (PA-AA), R14089; A 36213; pr. 15.12.1915 p.m. Consul in Aleppo to the Imperial Chancellor, November 30, 1915, in Gust (ed.), *Der Völkermord an den Armeniern 1915/16*, 388.

[154] Üngör, *The Making of Modern Turkey*, 98.

[155] PA-AA, R14090; A 02889; pr. 01.02.1916 a.m. Consul in Aleppo to Ambassador Wolff-Metternich in Constantinople, January 3, 1916. Enclosure 1: Hoffman to Imperial Consulate, November 8, 1915, in Gust (ed.), *Der Völkermord an den Armeniern*, 405; Kévorkian, *The Armenian Genocide*, 378; PA-AA, R14087; A 23991; pr. 14.08.1915 p.m. Report, Consul in Aleppo to the Imperial Chancellor, July 27, 1915, Enclosure: From an Official on the Baghdad Railway, July 27, 1915, 217.

[156] PA-AA, R14090, A 05498; pr. 29.02.1916 a.m. Consul in Aleppo to the Imperial Chancellor, Enclosure: Wilhelm Litten, February 9, 1916, in Gust (ed.), *Der Völkermord an den Armeniern 1915/16*, 446.

[157] NARA, RG 59, 867.4016/373, J. B. Jackson, "Armenian Atrocities," March 4, 1918, in Sarafian (ed.), *United States Official Records on the Armenian Genocide*.

[158] Kazanjian, *The Cilician Armenian Ordeal*, 146.

between the ages of eight and twelve first for the price of two mecidiye, before giving them away for free in order to "spare them the fate awaiting them in the desert as a result of the climate and of the Bedouins."[159] Aside from desperate actions such as these, there was also more direct evidence of various nomadic groups participating in massacres, as they formed a key part of the Special Organization charged with doing the dirty work of the genocide. There were even reports that Khalil Bey – son of the late Ibrahim Pasha – had personally taken part in the massacre of Viranşehir's Armenians.[160] The Chechens of Ras al-Ayn, too, were implicated in the massacres.[161]

Yet the place of nomads was not simply as executioners, as some accounts would have it. This was especially the case in late 1916, when a typhoid epidemic spreading among Armenian deportees and Ottoman soldiers prompted officials to call for a new wave of violence against the Armenians in the Jazira.[162] So many Armenians in the region of the Khabur River ended up with local nomads that the soldiers began searching their camps.[163] As a result, nomads left "that corner between the Euphrates and the Khabur." In other words, the nomadic motion that had for so long thwarted Ottoman administrative schemes also served in this case as a circuit of escape for Armenian survivors. For the thousands of Armenian women and children who lived with nomads, life typically fell somewhere on a spectrum between adopted family member and enslaved labor, in ways that perhaps overlapped with long-standing practices of enslavement of people of African descent among groups such as the Shammar.

These dynamics become clear in the entries of the Aleppo orphanage run by Karen Jeppe, the Danish missionary who had compared Arabs to locusts in 1909. Composed in the 1920s, the entries typically included a terse biography and photograph of the children who had managed to survive and make their way to Aleppo. The imperatives of the orphanage

[159] PA-AA, R14087; A 23991; pr. 14.08.1915 p.m. Consul in Aleppo to the Imperial Chancellor, July 27, 1915, in Gust (ed.), *Der Völkermord an den Armeniern 1915/16*, 214.

[160] De Courtois, *The Forgotten Genocide*, 172.

[161] PA-AA; R14091; A 12911; p.r. 16.05.1916 p.m. Consul in Aleppo to the Imperial Chancellor, April 27, 1916, in Gust (ed.), *Der Völkermord an den Armeniern 1915/16*, 465. See also Lewis, *Nomads and Settlers*, 103.

[162] Mouradian, *The Resistance Network*, 138.

[163] PA-AA; R 14091; A 25739; pr. 22.09.1916 p.m., Chargé d'affaires in Aleppo to the Embassy in Constantinople, August 29, 1916, in Gust (ed.), *Der Völkermord an den Armeniern 1915/16*, 485.

to reconstitute the Armenian nation were clear. But though the dynamics of national renewal (not to mention religious and racial distinction) inflected the entries, one of their striking aspects is how they reveal the extended period of time that many Armenian children managed to survive as nomads. For example, Kirkor Papazian of the famously rebellious mountain town of Hadjin was deported to Deir ez-Zor and thirteen years old, was sold in Maskanah on the Euphrates for the sum of one mecidiye to "an Arab of a nomad tribe." [164] The next day, his mother was gone, and he lived out the war years as a shepherd. Garabed Khantroshian had a similar story. He was deported with his family to al-Hasakah, the allegedly more salubrious site to which local administration was moved from Ras al-Ayn in 1911.[165] His father was killed there, and his mother died "of emotion." The teenage native of Maraş found "himself taken to the tent of a nomad Arab who kept him as a shepherd," his siblings dispatched to separate families. He thought of fleeing, "but in the desert it seemed impossible." For others, like Megerditch, their own families made the heartrending decision to entrust the children to nomadic groups.[166] Megerditch's family gave him to "an Arab in order to save him from further sufferings" near Ras al-Ayn, while his family moved on "in an unknown direction." Work as a shepherd for little or no wages was the reality for many. Yet the remote nature of the labor also offered a refuge.

In addition to a modicum of security, others found an array of affective relationships among the Jazira's nomads. Sarkis of Sivas, for example, was a teenager at the time of the deportations. He was taken into one tent before being sold to another Arab, "who loved him" and "allowed him to marry an Armenian girl," whose place in the village was also described as "a slave."[167] For many women amidst the deportations, sexual violence was the precursor to death or a horrifying condition of survival. But it was different in the case of Varter.[168] A native of Mardin, she was deported along with her husband, who managed to escape execution by dressing as a woman. They found themselves taken among the nomads, for whom Varter worked as a "servant" and her husband as a "shepherd." "They

[164] United Nations Archives at Geneva (UNAG), C1602/498/617, August 26, 1924. For all of the names related to these materials, I have followed the spellings used in the orphanage records.

[165] UNAG, C1602/498/619, August 29, 1924.

[166] UNAG, C1602/498/1098, August 22, 1926.

[167] UNAG, C1602/498/692, May 5, 1925.

[168] UNAG, C1602/498/838, October 8, 1925.

had both to work hard," read their postwar account, "but were rather well treated." Accounts such as these suggest how, in the words of historian Nazan Maksudyan, these histories are not only "tragic stories of loss, but also sagas of courage and resilience."[169]

Regardless of the fact that it was the nomads of the Jazira who helped many Armenians to survive, prominent observers could only see these dynamics in racialized ways embedded in notions of civilizational progress. This list included Arnold Toynbee, the British intelligence officer who would go on to become one of the foremost world historians of the twentieth century. As early as 1915, he described how Armenian suffering was all the worse given the fact that they "were not savages, like the Red Indians who retired before the White Man across the American continent. They were not nomadic shepherds like their barbarous neighbours the Kurds. They were people living the same life as ourselves ... they were sedentary people."[170] Toynbee's declaration of different levels of suffering not only minimized the violence of settler colonialism in North America;[171] he also obscured how, when Talat described his plans to Morgenthau, he invoked a genealogy of violence implicating America.[172]

At the same time as Toynbee revealed the civilizational logic that undergirded some sympathy for Armenians, the Ottoman state specified how the genocide was supposed to solve ethnic problems more broadly within the empire. Although not on the same scale or with the same violence as with Armenians, the Ottoman state also engaged in deportations and resettlement of Kurdish populations, particularly those fleeing Russian advances in eastern Anatolia. The state's calculus was both ethnic and economic. To this end, in May of 1916, Talat Pasha declared that Kurds were not to be resettled in places like Urfa or Zor "because they would either Arabize or preserve their nationality there and remain a useless and harmful element."[173] Talat's words thus represented a variation on the parliamentary debates of 1911, when Arab deputies objected to the settlement of an "Arab tribe" in a "Kurdish province." Talat also echoed long-running understandings of the environmental underpinnings of ethnicity, in which seminomadic Kurdish tribes might

[169] Maksudyan, *Ottoman Children and Youth during World War I*, 106.
[170] Toynbee, *Armenian Atrocities*, 30. [171] Blackhawk, *Violence over the Land*, 15.
[172] Morgenthau, *Ambassador Morgenthau's Story*, 338.
[173] BOA, DH.ŞFR 63/172–173, Talat to Diyarbekir, May 2, 1916, cited in Üngör, *The Making of Modern Turkey*, 110.

be "Arabized" on the edge of the desert. With so many Armenians deported from their agricultural holdings, not to mention the impact of the locust invasions, the state could not risk Kurds becoming pastoralists, and so deportation almost always included emphasis on how imperative it was that the resettled Kurds cultivate the land in provinces such as Konya and Kastamonu.[174] Thus even in wartime and in the context of unprecedented emphasis on ethnic origin, provincial borders and the impetus to cultivate continued to guide Ottoman policy.

While Talat reformulated how ethnicity ought to look in relation to Ottoman provinces, the borders of the Ottoman Empire were being discussed in a different but also momentous context. In December of 1915, Sir Mark Sykes convinced the British prime minister that he ought to be the trusted hand on the Middle East.[175] Since his days in the Jazira enamored with Ibrahim Pasha, Sykes had presented himself as an expert. He was elected a member of parliament for Hull, and, spending more time at home, attempted to make it look like the Middle East of his fantasies, contracting with the Ottoman Armenian master artisan David Ohanessian for the task.[176] Sykes also maintained the healthy self-esteem that left 10 Downing Street under the impression that he was fluent in Arabic and Turkish when in fact he was not in either.[177] In a meeting whose notes survive in what historian James Barr has called Sykes's "muscular but juvenile handwriting," the man called for a postwar settlement in which the British took control of formerly Ottoman lands to the south of a line drawn on the map from the letter "E" in Acre on the Mediterranean to Kirkuk in what is now northern Iraq.[178]

He went on to spearhead negotiations over the next few weeks in which he met François Georges-Picot to decide on these matters. Picot also had a checkered past. While in his previous post as French consul in Beirut, he had corresponded with many Ottoman dissidents agitating for political change in the Arab provinces.[179] His decision to leave the letters behind when he departed Beirut at the onset of the war allowed the Ottomans – after they took the materials – to locate and execute many of those he had been writing with.[180] Together, Sykes and Picot agreed to the terms Sykes had sketched in his December meeting. The French would control land to the north of the Acre–Kirkuk line, with direct control on the coast near Beirut, British direct control near Basra, indirect control by both parties in

[174] Üngör, *The Making of Modern Turkey*, 111. [175] Barr, *A Line in the Sand*, 7–8.
[176] Moughalian, *Feast of Ashes*, 93–95, 166–169. [177] Barr, *A Line in the Sand*, 3–4.
[178] Ibid., 4, 7. [179] Ibid., 16. [180] Ibid., 17.

between, and international administration of Jerusalem.[181] The pact – which received assent from Russia too – remained secret, at least until the Russian Revolution and the publication of all secret treaties of the czar's government.[182] What became known as the Sykes–Picot Agreement bore only abstract relation to territorial arrangements put in place after the war, but it nevertheless would come to be shorthand for the colonial imposition of borders on the region.

THE WAR ON LOCUSTS

As the Jazira's ecology and its built environment shaped Armenian suffering, so too did locusts. During the horrors of the deportation, many suffered from the devastation caused by the insects. In an autobiography written in Turkish using the Armenian alphabet, Hagop Der-Garabedian remembered how, during the deportations, his convoy stopped near Suruç to get a drink from a well. "But the water was covered with dead locusts," he realized, "and had a pungent smell. It was repulsive but we had no choice."[183] Without any other options, they filled up their containers with the noxious water infused with locusts. "The repugnant smell stayed with us for the next several weeks," he wrote. Elsewhere, Armenians were forced to eat locusts. One German observer wrote of Deir ez-Zor that "here there is no grass, the locusts have consumed everything. I saw how the people were gathering the locusts and eating them raw or cooked."[184] American diplomats received similar reports.[185] The locusts also functioned in other ways for Armenians. They were used as a code word in secret correspondence. When Armenian Ephraim Jernazian's assistant Ibrahim Fawzil wrote of how "our vineyard in Kara Koepru was eaten by the locusts," he actually referred to a massacre of Armenians there.[186]

The Ottoman state as well as their German allies were concerned about the prospect of the locusts. As early as April of 1915, German officials

[181] Ibid., 26. [182] Ibid., 26. [183] Der-Garabedian, *Jail to Jail*, 18.

[184] People generally did not eat the Moroccan locust, but the Najdi locust was more popular. In his World War I memoir, Naci Kaşif Kıcıman recalled eating locusts in the Hijaz, asking how they were any different than lobster and shrimp. They were especially good with olive oil and lemon juice, he added. Kıcıman, *Medine Müdafaası*, 216–217; Rohner, Aleppo, June 26, 1916, enclosure to Rossler to Metternich, Aleppo, June 29, 1916, J. No. II 6126 AA-PA Konstantinopel 101 No. 1822, in Kaiser, *At the Crossroads of Der Zor*, 64.

[185] NARA, RG59/867.48/356, Philip Hoffman to Secretary of State, July 21, 1916, in Sarafian, *United States Official Records on the Armenian Genocide 1915–1917*, 519.

[186] Doumanian, *My Memoirs*, 34; Jernazian, *Judgment unto Truth*, 77.

proposed sending a scientific expedition to deal with the issue.[187] They also asked whether "poison gas" or "artificially producing diseases" might be used against the insects (replies to the contrary cited the failure of the same experiments in Guatemala that Ottoman officials had written of a few years before).[188] The Ottomans fought against the invasion, including noteworthy efforts by Cemal Pasha in Syria.[189] But by fall, the situation was dire, with predictions that the "internal enemy" of "famine" would be more of a threat than the Entente powers.[190] The Ottoman minister of commerce and agriculture Ahmed Nesimi concurred; he wrote that the locusts were spreading "like an oil stain" and posed "more of a danger to the Ottoman Army and Ottoman domains," Nesimi added, "than the rival states."[191] By November, the Ottoman state had approved a plan to dispatch three German experts to address what the German embassy in Istanbul described as the "great clouds of locusts devastating the fields of Syria and Mesopotamia" (the latter term was rendered as "the Jazira" in Ottoman translation of the French original).[192] The leader of the German experts was to be Dr. Hermann Bücher, an entomologist with experience in the German colony of Cameroon.[193] From the early stages of negotiation, Bücher was pessimistic. "I am ready to take on this task," he wrote, "but I must emphasize that it seems hopeless to me."[194] As a condition of his contract, he requested that he be able to return to Germany if he subsequently decided the task was impossible.

Upon arrival in Istanbul in early January, Bücher's diagnosis of the problem was systemic, and he indicted agriculture and the Armenian genocide as causes of the locust invasions. "It turned out that the extent of the locust plague," he wrote, "is much greater than could be assumed from reports in Berlin."[195] The culprits were two. The Najdi locust had overrun

[187] Bundesarchiv (B), R901/13932, Moltke to Foreign Office, April 4, 1915.
[188] B, R901/13932, Director of Biological State Institute for Agriculture and Forestry to Vice General Staff, September 22, 1915.
[189] B, R901/13933, Report by Dr. Bücher, January 7, 1916.
[190] B, R901/13932, Deputy Director Günther to Board of Directors of the Anatolian Railway Company, October 4, 1915.
[191] Askeri Tarih ve Stratejik Etüt (ATASE) Arşivi, BDH-1330–810–1, Ahmed Nesimi to War Ministry, 9 Kanunuevvel 1331 (December 22, 1915).
[192] BOA, HR.İD 1339/24, German Embassy in Pera to Ottoman Ministry of Foreign Affairs, November 4, 1915; BOA, HR.İD 1339/25, Ottoman Translation of German Embassy in Pera to Ottoman Ministry of Foreign Affairs, November 4, 1915.
[193] BOA, HR.İD 1339/28, Commerce and Agriculture Minister to Interior Ministry, 10 Teşrinisani 1331 (November 23, 1915).
[194] B, R901/13932, Bücher to Foreign Office, November 2, 1915.
[195] B, R901/13933, Report by Dr. Bücher, January 7, 1916.

the eastern Mediterranean coast, from Palestine to Adana. Bücher did not expect it to return, though concerningly he had found a few specimens as far north as the European side of Istanbul. The "most important enemy" and, in Bücher's view, that upon which "all forces should be concentrated," was the Moroccan locust. In contrast to the migratory locust, which did not inhabit the regions it afflicted, the Moroccan locust was native to both the Jazira and the Aegean Coast. In Bücher's view, the issue was not simply one of locusts, but also of agriculture more generally. Moreover, he expected little help from the government, dismissing minister of agriculture Ahmed Nesimi Bey as lacking "the slightest knowledge of agriculture."[196] Bücher estimated that the amount of land cultivated in the empire was perhaps a third of what had been cultivated before the war. He blamed both military conscription and "the annihilation of most of the Armenian people," which resulted in what he estimated at 500,000 to 800,000 dead, many of whom had previously been engaged in "valuable agricultural" work.[197] These fallow lands left unplowed by conscripted men or massacred Armenians, Bücher explained, were "the best breeding grounds" for locusts. In other words, Vahram Dadiran had seen in the locusts a symbol of Armenian powerlessness. But according to Bücher, the connection between Armenians and locusts was causative; the Armenian genocide helped to intensify the locust swarms of the war years.

Addressing such issues required urgency but faced logistical challenges. Bücher saw expanding cultivation as the center of his approach. Since Moroccan locust eggs were planted shallowly in the soil, plowing could easily dislodge them. But given the shortages of labor, animals, and supplies, plowing was no easy matter. Bücher estimated that releasing 360,000 men from military service and issuing them with horses and seeds would do the job.[198] But he also knew such a proposal was impractical. The ministry of commerce and agriculture had secured 10,000 oxen to help with cultivation in the fall of 1915, but as of January 1916, none had been delivered, meaning the winter wheat growing season had already been missed.[199] Accordingly, Bücher called for German steam plows to be sent from where they were operating in Poland and Galicia.[200] The effort also struggled with funding more generally. Bücher had assuaged

[196] B, R901/13934, Report by Dr. Bücher, January 20, 1916.
[197] B, R901/13933, Report by Dr. Bücher, January 7, 1916.
[198] B, R901/13934, Appendix to Report by Dr. Bücher, January 20, 1916.
[199] B, R901/13933, Report by Dr. Bücher, January 7, 1916.
[200] B, R901/13934, Appendix to Report by Dr. Bücher, January 20, 1916.

Ottoman concerns about costs while initially negotiating his position by assuring them that as a scientist all he required was "a microscope" and a "small laboratory."[201] Upon arrival in Istanbul, he found that his German colleagues required more materials, since there was in fact only "one microscope," and it was used by the Ottoman entomologist Mehmed Süreyya.[202] "There is nothing in the country," he concluded.[203] To avert another devastating year of locusts, however, action was required immediately, and Bücher could not risk further delays in money and supplies. So he secured a loan from the German government to fund the expedition's activities "as is customary in colonial expeditions."[204] Then he and Ottoman officials began ordering materials such as spades, clamps, zinc sheeting, arsenic, Schweinfurt green, and petroleum jelly (for protection of the skin while spraying chemicals).[205] They also requested that they receive priority transport from Germany via military trains.[206] And when the materials did not arrive as quickly as hoped, Bücher went so far as to return to Berlin to secure their timely delivery.[207]

Despite Bücher's glum predictions, the year's activities seem to have been a stunning success. The Ottoman state – notably with the support of minister of war Enver Pasha – formed thirteen "farmer battalions" (*çiftçi taburu*) to be dispatched to provinces afflicted by locusts.[208] The battalions were reinforced with the labor of local people, who were obligated to help in accordance with the Locust Law (though it had not previously been enforced, according to Bücher).[209] They did their work by digging up eggs or plowing.[210] They also used some 150 kilometers (ninety-three miles) of zinc sheeting throughout the country. Meter-high sheets of the

[201] B, R901/13932, Bücher to Foreign Office, December 13, 1915.
[202] B, R901/13933, Bücher to German Embassy in Istanbul, January 15, 1916.
[203] B, R901/13933, Report by Dr. Bücher, January 7, 1916.
[204] B, R901/13932, Bücher to Foreign Office, December 15, 1915.
[205] B, R901/13933, Ottoman Embassy in Berlin to Foreign Office, January 26, 1916.
[206] Cemil, the Ottoman delegate to the International Institute of Agriculture in Rome, relocated to Berlin for these purposes. B, R901/13933, Ottoman Embassy in Berlin to Foreign Office, January 14, 1916; R901/13934, Note from Bücher, February 26, 1916; ATASE, 1330–810–1, Ahmed Nesimi to the Head Commander, 11 Kanunusani 1331 (January 24, 1916); Ahmed Nesimi to the War Ministry, 23 Kanunuevvel 1331 (January 5, 1916).
[207] B, R90/13934, Germany Embassy in Istanbul to Chancellor, July 28, 1916.
[208] B, R901/13933, Report by Dr. Bücher, January 7, 1916.
[209] B, R901/13934, Appendix to Report by Dr. Bücher, January 20, 1916.
[210] B, R901/13935, Summary Report on Locust Control in Anatolia, Syria and Palestine in the Year 1916, Bücher, December 25, 1916.

zinc would stop the locusts crawling on the ground because the substance was too slick for them to climb. As the swarm continued marching, the new waves of the locusts would trample their counterparts, and people would simply bury the insects.[211] In total, 378,683 dunhams (355,696 kilometers/221,019 miles) were plowed, over 5,000,000 okka (14,150,000 pounds) of eggs destroyed, and over 90,000,000 okka (254,700,000 pounds) of locusts destroyed.[212] Bücher estimated that some 450,000 to 500,000 people were engaged in the fight against locusts throughout the empire during the spring of 1916. Cemal Pasha also once again coordinated mass efforts in greater Syria. The result was that the harvest was "saved," and moreover that people were convinced that the locusts' power was not inevitable.[213] Bücher warned that success was temporary, given that the year's effort was aimed at protecting existing crops rather than eradicating all locusts.[214] The Ottoman government requested that Bücher submit plans to do it again.

Yet achieving success in the other reservoir of Moroccan locusts – the Jazira – required, as usual, overcoming special challenges (Figure 13). There, efforts were led by Gustav Bredemann, an entomologist with previous experience in New Guinea who was referred to as "the locust general."[215] Bredemann had dissolved the locust district of Zor in 1916 because of its ineptitude and what he described but did not elaborate on as "unsettled political conditions."[216] He did go into more detail on the challenges of locating and compelling the labor of locust destruction. People apparently hated the work. Some would swim across the Euphrates so as to get out of it.[217] In other instances, the locust officials would find villages entirely empty of men, who had fled into surrounding areas in order to avoid it. Bredemann admitted that "extreme ruthlessness" was necessary in some instances to compel people to work. He also described how intersections with nomadic populations proved challenging. He specifically identified this challenge as occurring at the "border of

[211] Similar measures employing fencing to stop and kill locust swarms had become known as the Cyprus Method prior to World War I. Çelik, "Humans in *Animalscapes*," 60.

[212] B, R901/13935, Summary Report on Locust Control in Anatolia, Syria and Palestine in the Year 1916, Bücher, December 25, 1916.

[213] B, R901/13934, Bücher, August 1, 1916. [214] Ibid.

[215] Künzler, *Im Lande des Blutes und der Tränen*, 120; B R901/13935, Summary Report on Locust Control in Anatolia, Syria and Palestine in the Year 1916, Bücher, December 25, 1916.

[216] B, R901/13935, Bredemann, "Report on the Control of Locusts in Northern Syria and Northern Mesopotamia during the Control Period 1916/1917," 5.

[217] Ibid., 13.

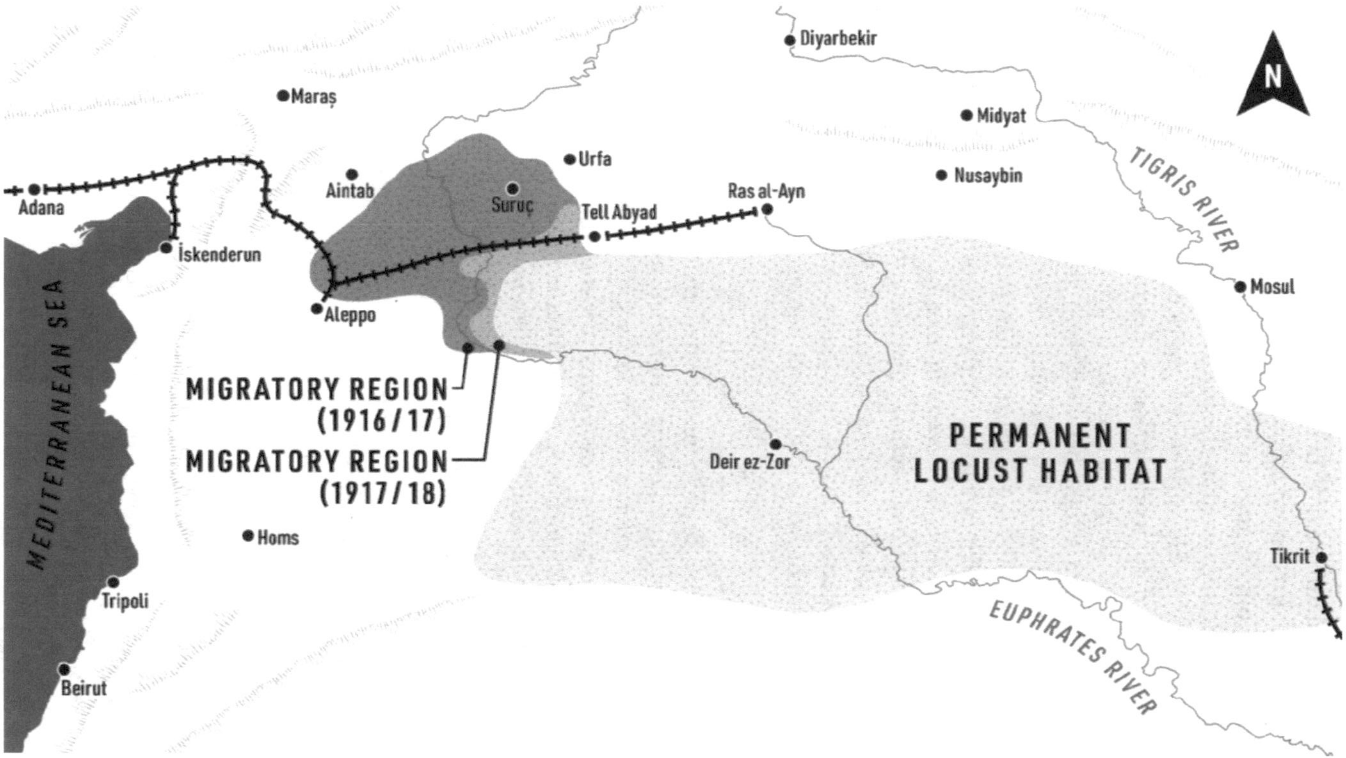

FIGURE 13 The Moroccan locust 1916–1918. Based on map in Bücher, *Die Heuschreckenplage*, 24

the cultivated land near the steppe," which lined up quite closely with the path of the Baghdad Railway.[218] In most places, locust officers would pay laborers for collecting a certain amount of locust eggs. But the policy was no good in this region, because, he explained, the nomads did not accept paper money.[219] He suggested in the future paying people with grain to avoid the problem.[220] He also faced particular difficulties with the ʿAnaza. They refused to work in locust destruction, and many non-ʿAnaza in fact claimed to be ʿAnaza so as to avoid work.[221] In one instance, there was even a fight between them and soldiers fighting the locusts.[222] Invoking a familiar comparison, he concluded, "The damage … by these Arabs is far greater than that done by the locusts, and the taming of the tribes carried out vigorously would likewise be useful, like fighting locusts." Indeed, in one of the few instances in which nomads seemed to have actually been involved in locust control, the district governor of Zor suspected that the Shammar were merely using "locust egg collection" as an "excuse" to get close enough to the Millî so that they might settle old scores against the sons of Ibrahim Pasha.[223]

Nevertheless in 1917, the locust officials seem to have achieved even more thorough success. Bredemann boasted that they were "winners across the board."[224] Egg collection amounted to some 160,000 tons, which he estimated to equal 320 billion locusts.[225] As for the locusts that did hatch, he attributed the success against them to the use of zinc walls throughout the country, which was especially effective given the challenge of finding labor.[226] Bredemann combined these tactics with a reliance on arsenic in the event that locusts appeared from the desert. In recognition of the fact that the substance was "very dangerous to man and cattle," he marked areas of application with yellow flags.[227] Even birds participated in the fight, with Bredemann mentioning "swallows, starlings, crows … and other species unknown to me by name."[228] Altogether, he estimated that they saved 95 percent of crops.[229] Bücher concluded that the locust

[218] Ibid., 9. [219] Ibid., 21. [220] Ibid., 22. [221] Ibid., 8–10. [222] Ibid., 11.

[223] BOA, DH.ŞFR 459/80, Zor District Governor Hikmet to Interior Ministry, 17 Kanunusani 1330 (October 30, 1915).

[224] B, R901/13935, Bredemann, May 14, 1917.

[225] B, R901/13935, Bredemann, "Report on the Control of Locusts in Northern Syria and Northern Mesopotamia during the Control Period 1916/1917," 50.

[226] B, R901/13935, Bredemann, May 14, 1917.

[227] B, R901/13935, Bredemann, "Report on the Control of Locusts in Northern Syria and Northern Mesopotamia during the Control Period 1916/1917," 32.

[228] Ibid., 41. [229] Ibid., 38.

plague would no longer be "an existential question for agriculture."[230] Summarizing Bredemann's work, the German missionary in Urfa Jakob Künzler wrote, "He had exterminated the enemies of mankind in the millions."[231]

Those humans targeted for extermination in the region also played a key role in Bredemann's pronouncement of victory over the locusts. Dicran Berberian, who survived in Maraş throughout the war, recalled how one day "lo and behold the whole sky was filled with billions of grasshoppers."[232] In response, the state mobilized everyone they could manage. Berberian recalled, "All that spring and summer we spent chasing and killing locusts. In the fall we swept with brooms myriads of their progeny and buried them in the ground." Kerop Bedoukian, a native of Sivas, similarly remembered being employed in locust-destruction efforts. In a camp near Birecik on the Euphrates, he and other deportees were ordered to cross the river and head toward the desert. There, they "dug ditches" in the sand without the benefit of shovels. Far from the trench warfare distinctive of the European theaters of World War I, these trenches were "three feet wide, two feet deep" and "stretched as far as the eye could see."[233] In the distance, they saw what looked like "a sandstorm" but turned out to be the locust swarm. Their drone sounded like "thunder." Bedoukian and his comrades tried to steer the insects into the ditches and bury them in sand, but it was mostly in vain. Despite their efforts, the nearby lands and even the Euphrates were covered "one inch thick with wriggling locusts," and they remained a problem for months afterward. But Bedoukian and others received food in return for their work, so it was worth it for them.

While Armenians like Berberian and Bedoukian were foot soldiers against the insects to whom people like Dadrian compared their fate, others were actually involved at a much higher level. In fact, the presiding officer of locust-destruction efforts in some parts of the Jazira was curiously enough an Armenian by the name of Hovhannes Toros Doumanian. A native of the mountain redoubt of Hadjin, Doumanian's father made money like many in the greater Adana region – thanks to the agricultural transformation of the Cilician plains, particularly through cotton cultivation. At his father's urging, Doumanian himself, born in 1894, attended

230 B, R901/13935, Bücher, "Preliminary Report on the Locust Combat in Turkey in 1917," June 15, 1917.

231 Künzler, *Im Lande des Blutes und der Tränen*, 120.

232 Kazanjian, *The Cilician Armenian Ordeal*, 118.

233 Bedoukian, *The Urchin*, 52.

the Halkalı Agricultural School in Istanbul, staying with his uncle in the Armenian enclave of Kum Kapı before returning home. It was from Hadjin that Doumanian and his family, like so many other Armenians, were deported in 1915, ending up in Aleppo beset by the omnipresent fear that they would be deported "to the Syrian desert" to "die under the scorching sun."[234] With agronomical training, Doumanian had exactly the qualifications that Bücher was seeking in locust officers.[235] In 1916, Doumanian secured his appointment as a locust-destruction officer with Dr. Bredemann, possibly under a name that concealed his Armenian identity.[236] Doumanian recalled how his family rejoiced at his appointment since it seemed to ensure his safety. Like his locust-control colleague in Palestine at the time – the Zionist Aaron Aaronsohn – Doumanian used his status as locust-control officer to enjoy freedom of movement that would normally be denied to him.[237] "Dressed in an officer's uniform," Doumanian could go anywhere in Aleppo that he liked, including neighborhoods that he had previously avoided for fear of arrest.[238]

Doumanian's memoirs of fighting locusts are interlaced with discomfort at how his work for the state afforded him a level of safety not guaranteed for his fellow Armenians, not to mention his own family. When he departed Aleppo via train with 10,000 boxes of arsenic, six soldiers, and a wagon, his parents "were very happy but they were also crying," he remembered.[239] Yet his departure for the Jazira would turn out differently than what had happened to so many before him. When Doumanian reached his first destination, he was treated with deference, staying at the "mayor's house" and having a "sheep slaughtered in [his] honor."[240] He took pride in recalling how effective his efforts were. The locust control officers under his command sprayed plants "so that when locusts ate them, they would die," and although it was not the most

234 Doumanian, *My Memoirs*, 22.

235 B, R901/13934, Report by Dr. Bücher, January 20, 1916.

236 An Ottoman list of eighty-one men with agricultural training who could lead the fight against locusts did include several Armenians, including Artin and Kevork, both graduates of the Ankara Agricultural School. Neither were assigned in 1916 to Aleppo (Halil Kamil, Ismail Hakki, and Ibrahim Halil), Urfa (Hassan Tahsin, Hüseyin Hosni, and Osman Nuri), or Zor (Ali, Yunus, and Kemal). ATASE, BDH-1330-810-1-34, 35, 159.

237 Aaronsohn, *With the Turks in Palestine*, 66.

238 Doumanian, *My Memoirs*, 28.

239 Ibid., 29. Bredemann described dispatching sixteen columns each with three fifty-kilogram barrels of Urania, three fifty-kilogram barrels of arsenic, 3,000 meters of zinc sheeting, and butterfly nets in February of 1917. B R901/13935, Bredemann, "Report on the Control of Locusts in Northern Syria and Northern Mesopotamia during the Control Period 1916/1917," 26.

240 Doumanian, *My Memoirs*, 29.

effective method, Doumanian still boasted that "not a single grain of the crop was destroyed."[241] Was Doumanian one of the locust officers – Osman or Feyzullah – whom Bredemann referred to as having done especially good work on "the Jazira side of Membij"?[242]

Though Doumanian enjoyed his reputation, the work came with pervasive reminders of the suffering of Armenians in the same region. When he described being hosted by local notables in honor of his success, he wrote that "it was obvious to me that the clothes that they were wearing were taken from the Armenians," a dynamic also observed by another locust-control officer at the time, Ahmad Wasfi Zakariyya (who would go on to become one of the preeminent historians of tribes in Syria).[243] While traveling to a meeting regarding locust destruction in Urfa, Doumanian witnessed "the heads of Armenians who had become food for the vultures."[244] At the meeting, he heard discussions of the execution of an Armenian. He recalled, "My body froze. It's as if cold water was pouring down from my head. I had to conceal my feelings. I ran to my room. I had to control myself to survive … my whole body began to shake."[245] While occupying his position of power, he also attempted to help people escape. On one occasion, an Armenian supplicated Doumanian without knowing the locust official himself was Armenian. Doumanian arranged for them to ride together to Manbij, and when they were a safe distance from Doumanian's colleagues, the Armenian-deportee-turned-locust-hunter disclosed his identity. His Armenian counterpart cried, found his way to safety, and they remained in touch after the war years.[246] While in Urfa, an Armenian woman who had been forcibly married by the mayor of the city escaped to the residence of the German locust expert Dr. Bredemann. Doumanian smuggled her to Aleppo by dressing her as a soldier and covering her head in a keffiyeh.[247]

While Doumanian survived and locusts did not in the Jazira, the region nevertheless remained a threat when it came to insects. "The locust source is undoubtedly the steppe," Bredemann wrote, and he noted that some

[241] Ibid., 30.

[242] B, R901/13935, Bredemann, "Report on the Control of Locusts in Northern Syria and Northern Mesopotamia during the Control Period 1916/1917," 13.

[243] Doumanian, *My Memoirs*, 30. Zakariyya blamed the practice of following the caravans and stripping the dead of their clothes for the spread of typhoid all across the region. He also took the opportunity to point out that it "was both funny and sad that the men would wear the clothes of dead women and the women would wear clothes of the dead men, not differentiating one from the other." Zakariyya, "Dhikriyyati ʿan Wadi al-Furat, " 51.

[244] Doumanian, *My Memoirs*, 32. [245] Ibid., 33. [246] Ibid., 32. [247] Ibid., 34–35.

40,000 dunham (37,572 kilometers/23,346 miles) of land remained unplowed on "the Jazira side" of the Euphrates.[248] If settled areas like Urfa and Aleppo were to be protected in the future, it was at "the root of the evil" in the Jazira that success would have to be achieved.[249] But any such "campaign in the desert," as Bredemann put it, would face "extraordinary difficulties" and likely require "well-armed soldiers" to protect against "extremely insubordinate Bedouins."[250] The work of the German officers involved maps detailing locust reservoirs and careful tables of the quantities of chemicals deployed, altogether offering a level of detail not to mention relying on an amount of resources absent among Ottoman authorities over the previous decades.[251] Yet in the end, the Germans came to the same conclusion that many a despairing Ottoman governor had suggested. The steppe allowed the locusts to "develop undisturbed, and from there they can always invade the cultivated land."[252] In other words, the locusts came from the places where people refused to stay in one place, and people refused to stay in one place because the locusts came.

"SMALL HILLS FORMED FROM THE BONES OF ARMENIANS"

As the architects of the genocide used the desert in an unprecedented way, as Armenians did their best to use the desert in other ways, and as Doumanian used the fight against locusts to survive, Ottoman officials themselves returned to their old concern of how to manage the Jazira more effectively through control of space and borders. In September of 1917, officials noted that the Shammar located within half an hour of construction on the Baghdad Railway between Ras al-Ayn and Arada were stealing animals, pilfering the possessions of workers, and even removing railway ties from the line.[253] Military officials called for them to be distanced from the railway to put an end to the "looting." In early December of 1917, deputies of Urfa called for southern portions of the district near Raqqa to

[248] B, R901/13935, Bredemann, "Report on the Control of Locusts in Northern Syria and Northern Mesopotamia during the Control Period 1916/1917," 48, 50.

[249] Ibid., 50. [250] Ibid., 51.

[251] The report on the campaign by Bücher and his colleagues was published in German as *Die Heuschreckenplage und ihre Bekämpfung* and translated into Ottoman by Süreyya as *1332–1333 Senelerinde Anadolu'da ve Suriye'de Çekirgelere Karşı*.

[252] Bredemann, "Die Bekämpfung der Heuschrecken in Nord-Syrien und Nord-Mesopotamien," 150.

[253] BOA, DH.EUM.2.Şb 42/24, Bronsart to Interior Ministry, 4 Eylül 1333 (September 4, 1917).

be ceded to Zor. Their reasoning was familiar. The lands were "unsettled and the wandering grounds of tribes and Arabs [*urbân*]."[254] Added to the human geography was the physical one. The deputy of Urfa, Saffet Bey, explained that "the center of the administration of Urfa is far," and "the desert [*çöl*] portion of lands cannot be administered easily from a great distance." After some debate, the measures were accepted, leaving to Zor portions of land south of a line drawn from Tall al-Samn to Qalaʿat Jaʿbar. Thus, government officials continued to speak of borders and the difficult nature of the Jazira even as hundreds of thousands of people lived in the region thanks to the deportations.

By early January of 1918, parliamentary officials were once again singing the praises of the Jazira, decrying the challenge of nomadic mobility, and lamenting the difficulty of changing these dynamics. Abdullah Safi Bey of Kirkuk crowed of how the Assyrians and Chaldeans had used the Jazira as their base, as did the Abbasid caliphs. If Egypt was rich because of the Nile, he added, the Jazira ought to be twice as rich as Egypt, since the Tigris and the Euphrates were the equivalent of two Niles.[255] Sounding much like his predecessors, Mehmed Nuri Efendi of Zor himself complained of how nomads easily manipulated local administrations. Whenever faced with a challenge, "naturally, they will take their sheep and camels that they have and load their tent on their camel's back" and move on, whether to the "plains of Iraq or the Jazira" or elsewhere. Ağaoğlu Ahmed Bey of Afyonkarahisar, meanwhile, offered a fatalistic opinion of this oft-repeated administrative challenge. In contrast to Abdullah Safi Bey's optimism, he noted that the Qurʾan was filled with stories of the prophet struggling to govern nomadic groups.[256] If the prophet and those around him could not succeed, then surely Ottoman officials could not hope to do much better.

Amidst these familiar remarks that navigated between the Jazira's past and present possibilities appeared a different kind of interaction, one that seemed more pregnant than the rest with the Jazira's reality at the time. The person who carried this message was Artin Boşgezenyan, an Armenian deputy from Aintab. He had leveraged his position in the CUP to save some of his relatives in Aleppo.[257] But he also lost many. It was poignant then when he interrupted the debate on the Jazira by saying,

[254] MMZC, İ: 15, C: 1, 8 Kanunuevvel 1333 (December 8, 1917), 223.
[255] MMZC, İ: 29, C:1, 7 Kanunusani 1334 (January 7, 1918), 518. [256] Ibid., 521.
[257] Tachjian, *Daily Life in the Abyss*, 116.

"Sirs, do you want to speak more honestly?"[258] He continued, referring to the nomads, "Leave these poor things alone." Deputies in the chamber began to laugh. The nomads lived "in a constant state of springtime," Boşgezenyan added. Why, then, should the Ottoman state exert any effort to change them? Shouts of "bravo" erupted in the chamber. Boşgezenyan addressed his fellow deputies, "I don't know if you all understood ... there, in the desert ... wherever spring is, there they are, like a swallow." Perhaps with sarcasm, perhaps with condescension, perhaps with some mix of the two, Boşgezenyan concluded, "What a good life!"

Responding to Boşgezenyan's comments was Feyzi Bey of Diyarbekir – the son of Arif Pirinççizade and cousin of Ziya Gökalp – who had called for the Shammar to be settled in Diyarbekir in 1911. Feyzi declared of the Jazira that "many of our colleagues and especially many commanders and other individuals have visited these districts on account of wartime mobilization."[259] He was not wrong, as everyone from Mustafa Kemal to Indian prisoners of war passed through Ras al-Ayn during the war years.[260] With so many people circulating, related Feyzi Bey, they became acquainted with the region, which largely consisted, in his view, of people who blindly followed their tribal shaykhs.[261] Much was left unspoken in this interaction, but little likely escaped the awareness of deputies, to say nothing of Boşgezenyan and Feyzi Bey themselves. While Boşgezenyan's comments about the Jazira were weighted with the fates of his family members and friends who had been sent to the Jazira to die, Feyzi Bey knew well the Jazira's sociology, and in particular its nomadic politics. After all, he had a different connection to the region and its people. He was the scion of one of Diyarbekir's foremost landholding families, one that had made a fortune from the rice cultivation that his family name alluded to.[262] And he had done the bloody work of the genocide, ensuring that local gangs would finish off Armenian convoys dispatched from the cities of southeastern Anatolia.

Whatever the unsaid momentousness of this interaction, both Boşgezenyan and Feyzi converged in presenting the Jazira as unique. Their fellow deputies agreed. Of course, this was also a common theme in Ottoman administration of the Jazira over the decades. As in previous

[258] MMZC, İ:29, C:1, 7 Kanunusani 1334 (January 7, 1918), 518. [259] Ibid., 519.

[260] "Mardin: Cumhuriyetten önce ve sonra Halkevi Broşürü" (Istanbul: Resimli Ay Matbaası, 1938), in Öztoprak, Güler, Karataş, and Şahin, *Cumhuriyet'in XV Yılında Türkiye*, vol. 6, 3397; Ghosh, "Shared Sorrows."

[261] MMZC, İ:29, C:1, 7 Kanunusani 1334 (January 7, 1918), 519.

[262] Gratien, "The Rice Debates."

discussions, those in January of 1918 involved an awareness of the way the provincial division of the Jazira complicated its management. Muhammad ʿAli Fadil, the deputy of Mosul, stated that in forty years he had only seen one joint military operation coordinated between Baghdad, Zor, Diyarbekir, and Mosul.[263] A new provincial map that managed "the Jazira" (*El-Cezire*) – perhaps with separate districts of Cizre, Khabur, and Sinjar – would thus be welcome, he implied. Mustafa Abdülhalik Bey of the interior ministry ridiculed the measure, saying "if it becomes necessary to consider the places in which one tribe wanders to be a province or a region, then today it would become necessary to make a province of Basra, Baghdad, Mesopotamia [*Beynennehreyn*], Urfa, Aleppo, and Syria," because this was how groups such as the Shammar moved. The laughably large proportions underscored the enduring tension between state borders and nomadic mobility in the Jazira and beyond.

But even if the proposal inspired ridicule in parliament, it garnered support from other levels of the government. By March of 1918, Sultan Mehmed Reşad and the Council of State approved a measure for "three independent districts ... between the Euphrates and the Tigris in the north part of the Jazira."[264] They did so based on the judgment of a delegation that visited the region. The materials produced by the group offer an unprecedented level of detail on the theme of drawing better borders to make the Jazira more agriculturally productive and its people civilized. One undated, unsigned notebook on the proposed Khabur district suggested that the Jazira might support a population of "sixty million people."[265] The writer offered this hyperbolic estimate (roughly three times the entire Ottoman population at the time) based on how irrigation might change the region. If properly harnessed to support cultivation, the writer noted, the Khabur would essentially flow with "gold and silver." Holding the region back, however, were the region's nomads, whom the writer disparaged. They did unspeakable things such as eating "onions like apples, and watermelons along with the rind." But more concerning for the official was the relationship between the nomads and cultivation. To make this point clear, the writer resorted to a well-worn comparison: nomads "wandered on the edges of permanently cultivated regions like a swarm of locusts [*çekirge sürüsü gibi*]."

263 MMZC, İ:32, C: 1, 12 Kanunusani 1334 (January 12, 1918), 575.
264 BOA, ŞD 2840/6, 14 Mart 1334 (March 14, 1918).
265 BOA, DH.İ.UM 10–1/2/58, Report on Khabur Livası, undated.

Little could be said of the government presence either, which could barely be differentiated, in some ways, from that of the nomads themselves. Due to the absence of buildings, local officials conducted state affairs from wherever they could manage, whether a tent or a stable. But the writer of the report insisted that such an arrangement was not right. To accentuate the point, the writer played on the Arabic saying "the honor of the place is in the owner" (*sharaf al-makan bil-makin*). The phrase was a way of suggesting that outward appearances or possessions mattered less than kindness and generosity. But the writer insisted that as far as Ottoman administration went in the region, the opposite was true: "the honor of the owner is in the place" (*sharaf al-makin bil-makan*). In other words, the Ottoman state could get no respect until it comported itself more like a state, complete with permanent buildings. At the same time, the writer encouraged the government to adopt practical policies. It was unrealistic, for example, to build houses with "baths, wardrobes, offices, and ablution spaces" for "savages who had only worn one cloak ... in their life." Echoing the deputies in parliament who called for acknowledgment of the particular nature of the Jazira and its population, the writer of the report also called for the importance of "seeing the surroundings" and "studying local customs and real needs." By this accounting, the region was empty as a result of the convergence of ineffective, impractical government and wild nomads. One could move "for days and days ... and the only thing visible is space and sky." Like the conversation in January of 1918 between Boşgezenyan and Feyzi Bey, silence persisted. The region described as outside of state control was precisely where hundreds of thousands of people were sent, and where they survived much like the nomads decried as uncivilized. The region presented as empty was anything but.

With the war winding down, however, a return to normalcy became paramount for many. In parliament, this took different forms with respect to the Jazira's moving groups, including locusts. In late October of 1918, Nesip Bey, the agriculture general director of the ministry of commerce and agriculture, appeared before parliament and discussed the state of locust destruction. With much of the empire bereft of "a man to dig graves, let alone help with locusts," women had in many cases stepped in during the war. Some deputies, however, believed this state of affairs improper.[266] The end of the war ought to mean, in their view, the end of women's engagement with such "hard labor." In a mark of how the labor

[266] MMZC, İ:6, C:1, 24 Teşrinievvel 1334 (October 24, 1918), 58.

of locust-killing had changed – or at least one government official's sense of it – Nesip Bey explained how technology had made locust destruction far from "hard labor." The primary change, he explained, was the use of zinc sheeting, the method that Bücher and Bredemann had hailed as so effective. With zinc, the work for humans primarily involved waiting. Or, as Nesip Bey put it, "there is nothing" for laborers "to do but smoke a cigarette" before stubbing out the writhing locusts on the other side of the zinc barrier by covering them with dirt. At least by this account, locusts – longtime scourge of the Jazira – had become something whose control was mundane and only made endurable by the buzz of nicotine.

As parliament became home to discussions of how easy it was to kill locusts, it also – shortly after – became home to discussions of what had happened to Armenians. On November 12, 1918, Entente warships entered the Bosporus, and the occupation of Istanbul began. On November 18, Artin Boşgezenyan spoke in parliament one more time. All of the things left unsaid in his encounter in January of 1918 with Feyzi found their way into the open. "I only know that what had originally been the Armenian den, nation, home," he declared, "is now not."[267] He described how "between Zor and Ras al-Ayn, small hills formed from the bones of Armenians, left over by the birds and the wolves." His description of small hills dotting the region of the Khabur River evoked the *tall* that had long attracted the eyes of outsiders to the Jazira, whether for the archaeological treasure they contained or the possibility of rejuvenation they intimated. The mention of the skeletal hills also stood in stark contrast to the vision of emptiness invoked by the Ottoman report of the previous year on the region. Also in contrast to precedents was how Boşgezenyan responded when his speech was interrupted by a deputy from Muş, a region famously dubbed "the slaughterhouse province."[268] Boşgezenyan countered, "This is the seat of deputies. This is not Muş." He added, "Thank god [*elhamdülillah*], the sun of justice has risen."[269] Earlier that year, Boşgezenyan had described the nomads of the Jazira in terms of an eternal spring. But in Allied-occupied Istanbul, he invoked time as a way of marking change, time as a way of talking about what had been unspoken in the past.

Even as Boşgezenyan spoke, Armenian survivors struggled to return home from the Jazira, and, in the process, they experienced the deprivation

[267] MMZC, İ:5, C:1, 18 Teşrinisani 1334 (November 18, 1918), 142.
[268] Davis, *The Slaughterhouse Province*.
[269] MMZC, İ:5, C:1, 18 Teşrinisani 1334 (November 18, 1918), 144.

of the genocide one more time. Several thousand had been conscripted into work on the Baghdad Railway. Their labor and that of prisoners of war ensured that the line reached Nusaybin by the conclusion of the war. When orders arrived from Ottoman authorities for the Armenians to be returned to their homes via the same railway, local officials paid little attention. Instead, officials – including commander of the Sixth Army Ali Ihsan Pasha – forced the Armenians to pay exorbitant rates to ride the rails from Nusaybin to Aleppo and from there home.[270] One observer wrote that it seemed they "intend to condemn the emigrants to remain." But many pressed on, and after four days, a group of several thousand Armenians arrived at the other end of the railway line in Çobanbey. From there, they were forced to walk to Katma, the railway junction near Aleppo infamous for misery during the genocide. In the course of the journey, many of the young and old were left behind. There they waited for trains that did not come, without food or shelter. Another wrote that it seemed the authorities were "trying to exterminate the remaining part of our nation."[271] They were leaving the Jazira rather than being forced to go to it, but their suffering continued. Pleading for help, they wrote to a British official whom they believed might intercede on their behalf: Mark Sykes. The man who had once ventured into the Ottoman Empire with the aim of finding unmapped territory and the man whose name was most associated with the Anglo-French division of the region had been freshly set up as part of the Franco-British military occupation of formerly Ottoman lands.

CONCLUSION

The scale of the genocide was unprecedented. But older legacies in the Jazira shaped the nature of the violence. In the years leading up to the war, officials not only continued to struggle with the mobile people and locusts of the Jazira (and even the grammar for speaking about them). They also struggled to think about the proper spatial units for controlling them. The genocide witnessed a new use of the region's environment as a weapon rather than an object to be transformed. Borders figured into these questions, whether in Dr. Reşid's demurral on the question of whether Urfa was part of Diyarbekir province or Zor or the demographic limits on how many Armenians were to be found within the borders of any one province.

[270] Hull History Centre (HHC), U DDSY2/4/194, Asadoor Solakian to Sykes, December 3, 1918.
[271] HHC, U DDSY2/4/194, Meymarian to Sykes, December 6, 1918.

So did locusts. Their invasions were especially bad because of the Armenian genocide, which, along with conscription, left considerable amounts of land fallow. As a result, locusts devastated the region during the war years, taking advantage in a familiar way of being located on both the environmental and political edge. The insects also became involved in Armenian survival in unexpected ways, whether as food or as the occupation through which Toros Doumanian survived.

Altogether, the war years left the empire and its people forever changed. On an individual level, the impact was devastating. 'Abd al-Hamid al-Zahrawi – the Ottoman parliamentary deputy from Hama who had insisted in 1911 that a plot was afoot to promote Diyarbekir at the expense of Zor – met his end in the place that would later be known as Martyrs' Square in Damascus on May 6, 1916, hanged for sedition by the Ottoman governor of Syria, Cemal Pasha.[272] Hampartsum Boyajian – the Ottoman parliamentary deputy from Kozan who had asserted that nomads had no home except for a camel's back – was among the first Armenian deportees from Istanbul and was tortured and executed at Kayseri in May of 1915.[273] Dr. Reşid – the governor of Diyarbekir too busy deporting Armenians and Assyrians to be bothered with questions of provincial borders in 1915 – committed suicide after escaping prison in 1919.[274] Not much later, Mark Sykes passed away while in Paris for the peace negotiations, a victim of the influenza pandemic. The individual deaths, of course, pale in comparison to the momentous simplicity of numbers of another order of magnitude, the 600,000 to 1,500,000 dead Armenians, the 150,000 to 230,000 dead Assyrians, the 200,000 dead inhabitants of Mount Lebanon, not to mention the dead of the Ottoman Empire as a whole, perhaps approaching the staggering figure of 5,000,000.

Yet others survived, including Feyzi Bey. The Diyarbekir deputy languished in exile in Malta for two years, before escaping and returning to Anatolia.[275] He would be the first public works minister of the Republic of Turkey. Hovhannes Toros Doumanian, too, managed to survive, thanks to his status as a locust hunter. After the war, he briefly taught at an Armenian agricultural school established under the Allied occupation of Istanbul at Beylerbeyi Palace on the Bosporus. Doumanian marveled at the irony of a situation in which "the location where the annihilation of the Armenians

[272] Qadri, *Muddhakirati 'an al-Thawra al-'Arabiyya al-Kubra*, 56. While still in parliament during the war, al-Zahrawi spoke out on the issue of locusts, warning that eggs remained in place and that more money was required to take care of the matter. Özbilge, *Çekirgeler, Kürtler ve Devlet*, 59.

[273] Kévorkian, *The Armenian Genocide*, 259.

[274] Üngör and Polatel, *Confiscation and Destruction*, 152. [275] Ibid., 153.

was plotted" had become "a school for orphaned Armenians."[276] The locust expert Hermann Bücher returned to Germany, where he used his old colonial letterhead (albeit with Cameroon crossed out) to try to secure arrears from his time fighting the locusts in the Ottoman Empire.[277] He eventually became the chair of AEG, a position he held until after World War II.

And beyond Feyzi Bey and Doumanian and Bücher were many whose names few remember. The teenagers who, like many in the Jazira before and after, had turned their place on the edge into a way to survive. Indeed, through bravery and perseverance, they had ensured that Talat Pasha's famous words – "they can live in the desert but nowhere else" – came true, albeit not in the way that Talat had imagined. They had seen their loved ones perish as the land that had long defied Ottoman plans came to serve as a weapon, its rivers, caves, and soil filled with corpses. Yet somehow, they had managed to live, hiding in the desert, taking refuge in tents, wandering with sheep and camels. Whether they suffered or, in fewer cases, enjoyed life among the nomads, they endured. And all the while they kept watch, knowing that the chauffeur bent over an overheated engine in the middle of the desert might know where their uncle had ended up, or that the "Arab Bedouin girl" with a "pitcher balanced on her head" might well be their sister, an Armenian from Maraş.[278] Many may have felt like Vaham Dadrian, who saw in locust destruction the destruction of Armenians. But as with locusts, as with the nomads so often compared to locusts, and as with the Armenians, the political ecology of the Jazira incubated a population on the margins. It would continue to do so, even as the Jazira became divided again after the end of the Ottoman Empire.

[276] Doumanian, *My Memoirs*, 57.
[277] B, R901/13935, Bücher to Foreign Office, April 6, 1919.
[278] Kazanjian, *The Cilician Armenian Ordeal*, 37.

4

"Like Swarms of Locusts" (1918–1939)

When recalling the horrors of World War I, one of Turkey's most venerable novelists – the Kurdish native of Osmaniye, Yaşar Kemal – described how "the desert of Mesopotamia" as well as "the southeast and east of Anatolia" was filled with orphaned children.[1] They were of various ethnicities and sects – "Armenians, Kurds, Turkmens, Azeris, Yazidis, Nestorians, Assyrians, and Chaldeans" – and they wandered the countryside seeking food and clothing. They ravaged villages, leaving them denuded of supplies. They were, Kemal wrote, "like swarms of locusts." Although the different groups of people suffered in different ways during the war, Kemal's words gestured to the way that the war years were difficult for so many, and he did so by invoking a durable comparison. Kemal's unique attentiveness to the rhythms of the countryside and its intricate tapestry of cultures would gain him renown over the course of the twentieth century, most notably in the series of novels *Mehmed, My Hawk*.

The agrarian struggles that his novels illuminated were also a crucial part of the making of the post-Ottoman world. In the wake of World War I, the Jazira remained divided, albeit between British Iraq, French Syria, and Republican Turkey rather than Ottoman provinces. Reşat Kasaba has described this regional shift as a change from "the Ottoman imperial kaleidoscope to the rigidly defined world of the successor nation-states."[2] In foregrounding locusts, refugees, and their intertwined places on the edge, Kemal's comparison offers insight into the material nature of the

[1] Kemal, *Yaşar Kemal Kendini Anlatıyor*, 23. [2] Kasaba, *A Moveable Empire*, 136.

process described by Kasaba, which a number of scholars have started to explore with respect to the Jazira.[3]

In the post-Ottoman period, locusts continued to fly across borders. Although bureaucrats still squabbled about who was to blame for the insects, they also found chemical insecticides effective, so much so that by the early 1930s some claimed that the insects respected national borders. Substances such as sodium arsenate and sodium arsenite not only hurt locusts, but also nomads, who had been historically compared to locusts for their mobility on the edge. Meanwhile, states worked to mold refugees into both national citizens and agents of agrarian change in the lands once ravaged by locusts. As the French welcomed Assyrians, Armenians, and Kurds, the Syrian Jazira became in historian Seda Altuğ's words "a microcosm reflecting in reverse the dynamics of Turkey's nation-building."[4] As these refugees arrived, arsenics helped to change the Jazira's mobile geography, and an expanded ecology of agricultural production took root. At around the same time, there emerged an unprecedented and fraught effort to give the Jazira a place on the map in the form of a French protectorate. The struggle involved the various moving populations that had been compared to locusts over time, including Kurdish and Arab nomads, Armenian deportees, and Assyrian refugees. The push for autonomy would ultimately fail. But the possibility gestured to the ways that agrarian transformation – abetted in part by the absence of locusts and changes in nomadic pastoralism – shaped what borders meant, and what kinds of political visions were possible in the post-Ottoman Jazira.

"THE UNSETTLED STATE OF THE COUNTRY"

In the wake of World War I, borders of various kinds took on new meaning in the Jazira. On the one hand, ethnicity gained a growing force, building on the mobilizations of the late Ottoman period and the violence of the war. The power of ethnicity also derived from the era's emphasis on self-determination, a concept endorsed as the organizing principle for an allegedly postimperial world by Lenin and Woodrow Wilson alike. It was in this context that Armenian and global

[3] Altuğ, "Sectarianism in the Syrian Jazira"; Robson, *States of Separation*, 60–61, 81, 99; Tachjian, *La France en Cilicie et en Haute-Mésopotamie*; Tejel, *Syria's Kurds*; White, "Refugees and the Definition of Syria."

[4] Altuğ, "Sectarianism in the Syrian Jazira," 18.

humanitarian organizations sought to locate survivors of the Armenian genocide and reunite them with family.

Yet even those charged with finding survivors and rebuilding the Armenian nation had a degree of ambivalence about their work. Levon Yotnakhparian, for example, worked in the southern regions of greater Syria under the auspices of the Armenian General Benevolent Union. A native of Urfa, he had trained to become a tailor there, before enlisting in the Ottoman army in 1908, as did many non-Muslims suffused with the hopes of the Constitutional Revolution.[5] During World War I, as it became clear Armenians would not be trusted to fight as fellow Ottomans, Yotnakhparian fled to Jabal Druze south of Damascus. In Jabal Druze, he passed himself off as a Kurd, and in Damascus, he passed himself off as Druze.[6] It was with this background of disguise, escape, and survival that he became involved in the work to construct an Armenian nation after the war. Writing in Gaza shortly before his death in 1970, Yotnakhparian described finding children who had survived the genocide with Arab or Kurdish families. He revealed a degree of discomfort perhaps surprising for someone so invested in the cause, remarking that "the people from whom we took away these orphan children had the right to resist. They saved the children from certain death."[7] The American relief worker Stanley Kerr described similar scenes around Aleppo in 1919. He admitted that his team "had no means of distinguishing Arab from Armenian" and worried about recreating trauma for the children, since "rescue" might cause them to recall "the terror of separation" from their families during the genocide.[8] Despite these complications, money flowed to organizations charged with finding Armenians. Appeals were often accompanied by implicit or explicit racialized imagery reminiscent of historian and intelligence officer Arnold Toynbee emphasizing the need to save white Christians from non-white Arab or Kurdish Muslims (see Chapter 3), even if these same people, as Yotnakhparian and Kerr observed, had in some cases saved the orphans.[9] Aware of this calculus, some Arab families tried to exploit the differential sympathy to keep Armenian children, insisting that they were not, in fact, Armenian but rather Kurdish.[10] The argument took advantage of not only the targeted nature of

[5] Yotnakhparian, *Crows of the Desert*, 16, 19. On conscription, see Achladi, ed., *Karamanlı Rum Ortodoks Bir Askerin Seferberlik Hatıraları*; Cora, ed., *Harbiyeli Bir Osmanlı Ermenisi*; Peçe, "The Conscription of Greek-Ottomans into the Sultan's Army."

[6] Yotnakhparian, *Crows of the Desert*, 29, 30. [7] Ibid., 112–113.

[8] Kerr, *The Lions of Marash*, 46, 44.

[9] Watenpaugh, "The League of Nations' Rescue of Armenian Genocide Survivors," 1326.

[10] Kerr, *The Lions of Marash*, 45.

relief efforts, but also the ambiguous boundaries of ethnicity. As Yotnakhparian's story and, in a way, Yaşar Kemal's imagery attested to, so many Armenians and Kurds had lived alongside each other that they could easily pass for one another.

While bonds of ethnicity blurred in some ways and came into focus in others, the boundaries of states and their relation to the Jazira also retained a sense of ambiguity after World War I. Some called for the Jazira to receive special status, much in line with the Ottoman sense that the region's environment and its nomads required a particular kind of attention. The most prominent figure in this regard was Faysal, the scion of the Hashemite family who had rebelled against Ottoman rule during World War I. The Hashemites were overshadowed by the titular white savior in the film *Lawrence of Arabia*, but they still attained mythic status within the Arab world (and especially in their own minds) as those who rebelled against an empire that was viewed as nothing more than colonialism, especially in retrospect. In the midst of World War I in exchange for rebelling, the Hashemites secured the British promise of the throne in a pan-Arab kingdom. Yet even Faysal was unsure if such a polity would make sense after the conflict. He called "for the independence of all the Arabic speaking peoples ... from the line Alexandretta–Diarbekir southward."[11] But he also admitted that it would be "impossible to constrain" this vast region "into one frame of government."[12] He specifically mentioned the "Jezireh" as consisting of "large wastes thinly peopled by semi-nomadic tribes." One British map titled "Possible Settlement of Arab Countries" reflected this possibility, placing a state of the Jazira alongside other states for Armenians, Assyrians, and Kurds that never came to be.[13]

Concurrently, a number of local groups agitated for their own independence and used the language of nature to validate their visions. In 1919, a Damascus-based organization called the Syrian General Congress, for example, called for "Syria within its natural borders."[14] At least from the vantage of the Syrian General Congress, the Jazira largely fell outside of their conception of Syrian nature. They defined their state's eastern boundary as following the Khabur, and then the Euphrates, thereby leaving significant portions of the Jazira beyond their purview.[15] Meanwhile, the organization

[11] Miller, *My Diary at the Conference of Paris, with Documents*, vol. 14, 227.

[12] Ibid., vol. 4, 298. [13] Antrim, *Mapping the Middle East*, 169–171.

[14] Gelvin, *Divided Loyalties*, 152.

[15] Others in Syria later in the twentieth century would remember these regions as clearly part of Syria, lamenting how places of "pure Arab history" such as Mardin, Diyarbekir, Urfa, Nusaybin, and Harran had been separated from their "mother country" by the perfidy of

of former Ottoman military officers in Iraq known as al-ʿAhd al-ʿIraqi called for "natural borders" of Iraq that included the Jazira, with the former districts of Zor and Diyarbekir being joined to the erstwhile Ottoman provinces of Basra, Baghdad, and Mosul that form the state of Iraq today.[16] And at roughly the same time, Turkish officials and activists suggested the border between Turkey and Aleppo was not "natural."[17] In these different positions, all parties marshaled nature as a way of making claims about which people ought to be part of the same polity. In some cases, nature referred to the environment, in others to conventional wisdom about ethnicity and language, and in still others to economic linkages. Their discourses recalled the Ottomans, who for decades had struggled with how to manage the Jazira, hoping that alignment of Zor with the desert and its mobile human geography might make it easier to manage.

In response to all of these suggestions, the Ottoman delegation in Paris after World War I offered a proposal that both intersected with and departed from the past. Ottoman officials suggested that "a national dividing line" ought to go from Mosul to Ras al-Ayn to Aleppo and then Lattakia, with Turks and Kurds to the north, and Arabs to the south.[18] These borders would not be implemented, but they nevertheless formed the basis of the Misak-ı Millî (National Pact), the guiding force in the struggle for the independence of Turkey.[19] This articulation of borders in relation to ethnicity was not wholly new. It echoed the 1911 parliamentary discussions about where the Shammar might be settled. The "national dividing line" also intersected with management of borders of a different kind. Even though Ras al-Ayn had been distant from interimperial or international borders, the Chechen settlement had long occupied the border between Zor and Diyarbekir and cultivation and the desert, respectively, and had been the site of many clashes between Ibrahim Pasha and the Shammar.

Alongside continued perceptions of the Jazira and its people as different was the familiar theme of nomads as locusts. The British administration in Iraq declared the Shammar a "public pest" in 1919.[20] The comparison also persisted in the posthumously published work of the

the French. Ibrahim al-Jarad, "Al-Athar al-Baqiyya fi al-Hasakah," *Al-ʿUmran* nos. 14–24 (1972), 23.

[16] Tauber, "The Struggle for Dayr al-Zur," 379.

[17] Altuğ, "The Turkish–Syrian Border and Politics of Difference in Turkey and Syria," 60.

[18] Budak, *Misak-ı Milli'den Lozan'a*, 78.

[19] Kayalı, *Imperial Resilience*, 121–127.

[20] "Mosul Division. Annual Report for the Year 1919," in Jarman, ed., *Iraq Administration Reports*, 477.

recently renamed Mehmed Ziya. Under this name, the Diyarbekir native had been a member of the CUP and a cheerleader of Millî Ibrahim Pasha's demise, but under the new name of Ziya Gökalp he reached new heights. As he went on to become the first professor of sociology at Istanbul University, enthralled with Durkheim, the bullet from his youthful suicide attempt remained in his head. So, too, did his sense of disdain for nomadic pastoralists. Gökalp declared in one famous work that "the Shammar are quite like locusts ... the enemies of any kind of prosperity."[21] The comparison recalled those of decades past, but his words also took on new meaning. In addition to invoking the entomological, Gökalp also spoke of ethnicity, differentiating Turk from Arab, Anatolia from elsewhere. In another passage of the same study, he invoked China's Great Wall with respect to the need to separate "our tribes" from "the Arabs in the desert" (*çölde urbân*) and thereby encourage settlement.[22]

There was also the question of actual – as opposed to metaphorical – locusts, which continued to afflict the Jazira in the wake of the war. Despite the deep-seated nature of the dilemma, newly arrived experts were careful to present their approaches to the challenge as distinctive. This was certainly the case for Yelseti Ramachandra Rao, a native of Andhra Pradesh and an entomologist who came to Iraq with the British. His description of conditions resonated with many of the narratives of rejuvenation that, unbeknownst to him, had been articulated by the Ottomans with respect to the Jazira. While Iraq had been home to a glorious past as "the earliest seat of civilization known to Science," in Rao's view it had become "at best only a backward Province."[23] Locusts were indicative of this dynamic. He suspected that the Ottomans had made little effort to control the insects, and he attributed to German influence "the large stock of Paris Green and iron" – by which he likely meant zinc – "sheets" left in Mosul in the midst of the Ottoman retreat. Yet even these efforts did not suffice in his view. "It was ... only when the British Army began to advance into the country," he insisted, "that real Entomological work may be considered to have commenced."

In presenting British administration of Iraq as a harbinger of progress in contrast to "stagnant Ottoman rule," Rao shared European visions.[24] But he also recorded the particular challenges of locusts, much like German and Ottoman officials had during World War I. Rao carefully differentiated between different species of locust, distinguishing the native

[21] Gökalp, *Kürt Aşiretleri Hakkında Sosyolojik Tetkikler*, 63. [22] Ibid., 46.
[23] Rao, *A Preliminary List of Insect Pests of 'Iraq*. [24] Dodge, *Inventing Iraq*, 47.

Moroccan locust (*Dociostaurus maroccanus*) from the migratory Najdi locust (*Schistocerca gregaria*).[25] Materials from the Department of Agriculture further emphasized the importance of accurate observation as a means of locust control. One such leaflet declared flatly that "descriptions of insects are of no value."[26] The organization preferred live specimens of the pests, preferably placed in "cigarette tins" with "air holes punched" in the sides.

Yet whatever vision of order was projected by such regulations, the British found themselves facing problems similar to those of their predecessors with respect to locusts. Indeed, in April of 1920, Rao saw locusts "in the grassy steppes" south of Mosul.[27] British officials did their best to control the invasion.[28] But as Rao saw it, the ultimate British failure owed to the region's geography, as well as, in his terms, "the apathy of the tribes and the unsettled state of the country."[29] In other words, much like the Ottomans, the British had found that the mobile geography of the Jazira made locust control nearly impossible. By May of 1920, when the locusts developed wings, they wrought damage at Shirqat, the erstwhile site of the castle that the Ottomans hoped would coax the Shammar of Farhan into settlement.[30] There, the insects destroyed experimental plots of cotton. This too was another echo of locust invasions past, as the cotton crop at Shirqat recalled the cotton crops across the Jazira that were destroyed by locust swarms in the midst of the American Civil War. As ruling regimes changed, the Jazira's political ecology remained much the same, a divided environment of moving humans and nonhumans that often evaded state control. Rao himself would go on to become India's foremost locust expert and, in the process, would grapple with such dynamics several decades later. The regions that he had studied as locust reservoirs within one imperial space in the 1930s would split between Pakistan and India after partition.[31]

[25] Rao, *A Preliminary List of Insect Pests*, 11.

[26] Iraq, Department of Agriculture, *Insects and Pests and Samples for Analysis*, 1.

[27] Rao, *A Preliminary List of Insect Pests*, 11. [28] Ibid., 11–12. [29] Ibid., 12.

[30] Ibid., 12.

[31] Rao's magnum opus was *The Desert Locust in India*, an account of his work in the 1930s on the Desert Locust and "its breeding grounds in north-western India," regions that would subsequently be divided between Pakistan and India. The work on the book stretched years, with work initially expected to take nine months but then taking some two years and then stretching to ten years. The manuscript was submitted in 1951, but then lost. The work was then freshly composed again. It was only published in 1960, some twenty years after the initially planned completion. There is no reference to partition in the volume. Rao, *The Desert Locust in India*, v-vi, 1–2.

As locusts ravaged Mosul in the spring of 1920, the Jazira's political division was both challenged and codified. In April of 1920 at San Remo, the French and British agreed to a division of the former Ottoman territories, with Britain administering Iraq and Palestine and France administering Syria. According to their mandates from the League of Nations, it was the duty of "advanced nations" to administer "people not yet able to stand by themselves under the strenuous conditions of the modern world." Some of the first acts of tutelage in both places were violent campaigns to eliminate any challenges to British and French rule. In Syria, the French acted quickly to remove Faysal – the Hashemite leader who had been declared king in March – from power. In July of 1920, the mostly African soldiers of the French army advanced toward Damascus and defeated Syrian troops at Maysalun. Shortly after, the French took Damascus and expelled Faysal from Syria. In Iraq, meanwhile, opposition to British control turned into a revolt, which was put down through unrestrained use of airpower. As for the remaining territory of the Ottoman Empire, Greek forces occupied the western coast of Anatolia on the Aegean Sea, Italy landed on the Mediterranean coast near Antalya, and the French held large portions of southern Anatolia. Resistance emerged in Anatolia against the different occupations under the former Ottoman military officer Mustafa Kemal. The French holdings were part of what they hoped to be an Armenian state called Cilicia, which would include cities on the edge of the Jazira such as Aintab, Maraş, Mardin, and Urfa. After months of fighting that included brutal sieges, the French were forced to leave, and so too did many of the region's remaining Armenians.[32] Thus, the end of World War I did not mean the end of conflict in the region, as both fighting between states and challenges to colonial rule broke out in the immediate aftermath of the war.

In this context, many different kinds of polities seemed possible. One Ottoman military officer in Diyarbekir suggested that Arab provinces remain under Ottoman rule through a federal model along the lines of the United States, with a corollary change in the Ottoman flag by which crescents would represent every separate state, similar to the stars on the American flag.[33] The French made inquiries about installing Zionists in the Jazira, a policy that would have various iterations over the coming decades.[34] Meanwhile, the Grand National Assembly of Turkey – formed in opposition to the European occupation of Anatolia as well as the still-existing Ottoman state – established a Jazira corps that operated bases as

[32] White, "A Grudging Rescue." [33] Kayalı, *Imperial Resilience*, 111.
[34] Gil-Har, "French Policy in Syria and Zionism: Proposal for a Zionist Settlement," 158.

distant as al-Hasakah in service of expanding territorial claims southward.[35] The Grand National Assembly also cultivated relations with the ʿAnaza shaykh Hajim – a longtime rival of the Shammar to the southwest of the Euphrates – who gladly accepted the title of "President of the Patriotic Movement of the Jazira."[36] According to some rumors, the French had a plan to create a polity in the Jazira beholden to nomadic notables similar to what the British had created in Transjordan.[37] Under the rule of one of the sons of Ibrahim Pasha of the Millî, the state was to encompass Urfa, Viranşehir, Ras al-Ayn, and Nusaybin.

Of the most significance with respect to borders in the Jazira was the Ankara Agreement of 1921, which aimed to draw lines between French Syria and territory controlled by the Grand National Assembly of Turkey. Yet the text of the treaty itself also gestured to interconnections between the regions. Most notably, the border of some 400 kilometers (248.5 miles) between Çobanbey and Nusaybin would be formed by the path of the Baghdad Railway. The final stretch of the line had been built by POWs and Armenians in the waning days of World War I, and movement along it had been yet one more space in which Armenians experienced the violence of the genocide (see chapter 3). After the war, the railway would become implicated in politics in another way: the preeminent infrastructural project of the late Ottoman Empire would divide post-Ottoman states. "This border is not a border" emerged as a rallying cry among Turkish opponents of the measure.[38] Controversy would persist on other parts of the border too. East of Nusaybin – where construction of the railway had stopped at the end of World War I – negotiators suggested that the border follow another infrastructure. This one was an even older imperial relic: "the old road" dating to Roman times that went to Cizre on the Tigris. To defend their claims of where precisely the old road was, Turkish authorities would end up encouraging peasants to plant crops to cover paving stones.[39] Meanwhile, the French charged the intrepid Jesuit aviator Antoine Poidebard with determining the route of the old Roman road by conducting research based on aerial photography and consultation of "the most serious scientific German materials."[40]

35 Kayalı, *Imperial Resilience*, 144. 36 Ibid., 166.

37 TNA-UK, AIR 23/256, Office of Divisional Adviser to High Commissioner, July 2, 1921.

38 Balistreri, "Revisiting *Millî*," 48.

39 CADC, Syrie 298, Antoine Poidebard, "Notes on the Upper Jazira," December 22, 1926.

40 CADC, Syrie 299, Antoine Poidebard, "The Old Road from Nissibin to Djezireh-ibn-Omar and the Table of Peutinger," undated.

While using networks of motion as a way to divide, the Ankara Agreement also mixed up the sovereignty of these regions in other ways. Article 9 concerned the tomb of Süleyman Shah, the grandfather of Osman, the founder of the Ottoman Empire. Süleyman had drowned in the Euphrates in 1236 and had been buried nearby west of Raqqa at Qalaʿat Jaʿbar. One Turkish writer described it as "the first seed of our Ottoman Empire."[41] The burial site was to remain under Turkish control, complete with the right to fly the Turkish flag, even though it was located over 100 kilometers (sixty-two miles) south of the border. Meanwhile, Article 13 of the Ankara Agreement provided for border crossing by those Syrian citizens owning land in Turkey and vice versa, in addition to pasture rights for "semi-nomadic groups."[42] Perhaps the greatest evidence of the continued connections between the regions was that according to the treaty – which deemed all territory south of the railway as Syria – the town of Nusaybin was supposed to be in Syria. But the map used during the negotiations was faulty and placed Nusaybin to the north of the railway line.[43] And so the town ended up in Turkey.

"WAS IT POSSIBLE TO 'GAS' THEM?"

Locusts were not party to any of these agreements. In initial reports on locusts in Iraq, Ramachandra Rao spoke as if "Science" would do the job that the backward Ottomans had failed to do. But early reports on locust destruction under the British revealed none of this confidence. In 1923 in Iraq, the British complained of how winter weather kept them from inspecting territory for locust eggs.[44] Having had little impact on the locust eggs while they were buried in the ground, the British tried to fight locusts once they hatched with one of the only substances they had in great quantities: oil. In the twentieth century, Iraq would be home to no shortage of disputes over the country's oil resources, and so it is perhaps fitting that the valuable substance was implicated in the destruction of the insect that had for so long afflicted the region.[45] As one report put it, "burning of

[41] İsmail Habib, "Fıratın hatıraları," *Cumhuriyet*, February 11, 1936, p. 3.
[42] On the illusory nature of these grazing rights, see Dolbee, "Borders, Disease, and Territoriality in the Post-Ottoman Middle East."
[43] CADN, Syrie 298, Weygand to Poincaré, June 13, 1923.
[44] TNA-UK, AY 4/1083, Austin Eastwood to Imperial Institute, Enclosure: Report from Kinch on "Report of Anti-Locust Operation 1923," September 29, 1924.
[45] On oil and Kirkuk, see Bet-Shlimon, *City of Black Gold*.

swarms was commenced at Shirqat."[46] The British dispatched officials throughout northern Iraq and managed to keep the locusts from the crops until late April. But once the insects made it into the crops, destruction was difficult, since – thanks to the oil – "to destroy the locusts meant destroying the crops." The entire affair left Kinch, the British officer in charge, with little of the assuredness displayed by Rao a few years earlier. Given the importance and impossibility of coordinating efforts across the region to have success, Kinch went so far as to suggest that "the problem should be left entirely alone." Far from redeeming the promise of science through colonial occupation, at least initially some British officials were explicitly calling for the locust dilemma to be ignored.

The disputed status of the Jazira complicated the management of locusts there. As early as 1922, the French had observed Turkish officers in territory they believed to be theirs. They complained not that the Turkish officers were fomenting rebellion but rather that they were engaged in something more threatening: the mundane activities of governance.[47] They worried about reports that a Turkish officer was near Sinjar – some 100 kilometers (sixty-two miles) south of the border – counting sheep and assessing taxes on nomadic groups.[48] In 1923, matters like this became violent. A detachment of soldiers under a French flag – most of them local Syrian recruits or North African colonial fighters – were attacked near the town of Behendour. The French charged that the assailants were Kurdish tribes encouraged by Turkish officers.[49] The soldiers managed to take shelter with some Shammar, but still close to thirty were killed near Qubur al-Bayd (now Qahtaniyya).[50] The incident would be a French rallying cry for decades.[51] But the event also demonstrated the limits of French control in the region.

Given the political and environmental challenges of the Jazira, some sought new technologies for killing the locusts. When locusts made their usual "depredations" in 1924, one British official by the name of Austin Eastwood sought help from London's Imperial Institute of Entomology,

[46] TNA-UK, AY 4/1083, Austin Eastwood to Imperial Institute, Enclosure: Report from Kinch on "Report of Anti-Locust Operation 1923," September 29, 1924.
[47] BOA, HR.İM 35/5, General Weygand to Minister of Foreign Affairs, August 19, 1923.
[48] The place of Sinjar with respect to the Iraqi-Syrian border would not be definitively settled until 1932. Fuccaro, *The Other Kurds*, 114–120.
[49] BOA, HR.İM 35/5, General Weygand to Minister of Foreign Affairs, August 19, 1923.
[50] Tachjian, *La France en Cilicie et en Haute-Mésopotamie*, 314.
[51] APDF, Haute Djezireh 4, La Haute Djezireh, undated.

an outgrowth of the body Maxwell-Lefroy addressed in 1911. Eastwood wanted to know about locust-killing technologies such as "petroleum flares."[52] He was told that they – like flamethrowers – had been deemed "impractical" due to "costliness." Eastwood also wrote of a word that had largely been absent from previous efforts at killing locusts: "insecticides." The substances had appeared all around the world in the wake of World War I, as the chemicals used to kill humans were turned on the insects lodged in the interstices of human society.[53] Whatever Eastwood and the British learned about new techniques, the locusts still wrought damage in 1924.[54] However, government publications were careful to downplay the impact of the invasion, suggesting that the "loss of revenue" from destroyed crops would be "partly made up by the increased value of the crops which have survived the pests."[55] Where flamethrowers were impractical, the invisible hand, they seemed to say, proved quite effective.

As colonial officials hunted for novel techniques of killing locusts, they ignored the fact that a simple solution to locusts had emerged over many decades in the Jazira in the form of seasonal migration. As historian Chris Gratien has written of malaria, mobility was an "elegant" solution to the seasonal threat.[56] So too with locusts. After invasions of the insects, a shrub known as "zeraiza" appeared, and while locusts could not eat it, nomads' camels could.[57] In other cases, nomads simply moved. In 1924 and 1925, the destruction caused by locusts prompted some portions of the Shammar to venture out of Iraq and into Syria to find pastures and supplies.[58] Scholars have increasingly and importantly called attention to nomadic border crossing in the interwar period, but these accounts have sometimes left the impression that this phenomenon was new.[59] In fact, the Shammar division between Iraq and Syria was reminiscent of their division between Zor and Mosul in the late Ottoman period, when they similarly manipulated borders. So too was their motion in response to locusts part of a longer history. The British and French allowed some

52 *Times of Mesopotamia*, July 9, 1924, 2; TNA-UK, AY 4/1083, Austin Eastwood to Imperial Institute of Entomology, April 13, 1924; Imperial Institute to Austin Eastwood, June 2, 1924.

53 McWilliams, *American Pests*, 135.

54 TNA-UK, AY 4/1083, Austin Eastwood to Imperial Institute of Entomology, April 13, 1924.

55 *Times of Mesopotamia*, July 9, 1924, 2.

56 Gratien, "Ottoman Quagmire," 586.

57 TNA-UK, AIR 23/281, SSO Mosul to Air Headquarters, Baghdad, August 31, 1925.

58 CADN, 1SL/1/V/561, High Commissioner in Baghdad to Ponsot, January 15, 1930.

59 Fletcher, *British Imperialism and "The Tribal Question"*; White, *The Emergence of Minorities in the Middle East*, 102.

forms of motion and autonomy, even charging, for example, the Shammar to issue passes identifying nomads as being part of the Iraqi or Syrian branches of the confederation.[60] But these sorts of mobile dynamics largely did not align with British visions of political economy. As the British did in many places, they believed small-scale cultivators would form the foundation of their power in Iraq. The *Times of Mesopotamia* encapsulated this sentiment: "the clever tricks of broad-casting ... to speed-up the world to jazz rhythm are very well ... but the real people who keep the globe spinning ... are the dull inarticulate folk who plant things."[61] When the British inquired about the efficacy of flamethrowers, petroleum flares, and insecticides in an effort to protect cultivation, they envisioned their efforts as part of protecting their key constituency.

And yet locusts continued to cross borders, leaving devastated fields and quarreling officials in their wake. In a reprisal of a theme from the Ottoman period, the various officials whose states carved up the region blamed their counterparts for outbreaks of locust swarms. In March of 1924, Turkey blamed Syria for, in their words, "infecting our fields" by failing to stop a swarm of locusts that spread from Deir ez-Zor in Syria to Urfa in Turkey.[62] They added, in a further jab, that it seemed "science" had not yet given the French a way to "fight them effectively." Later in the year, Turkish officials reasserted their complaints with respect to locusts. Even though Urfa had destroyed all of the locusts in its region, "locusts emigrating from Syria [*les sauterelles émigrées de la Syrie*] have caused significant damage among the crops of this region, neutralizing almost all of the impact of the measures adopted."[63] Having observed the effect on Turkish territory of actions (or inaction) in Syrian territory, the officials in Turkey called for cooperation across borders against the pests. In October, the French responded, insisting that they were hard at work so as to prevent "access to the Turkish zone of locusts born in Syrian territory."[64] The shared dilemma of an interlinked ecology of locusts was not new. But the language of control was. It went so far as to position

[60] TNA-UK, AIR 23/260, SSO Mosul, June 16, 1924.

[61] "Boosting Agriculture," *Times of Mesopotamia*, January 25, 1924, 3.

[62] BOA, HR.İM 99/7, Ministry of Foreign Affairs of Turkey, Note Verbale à Mr. le Représentat de France, 9 Mart 1340 (March 9, 1924).

[63] BOA, HR.İM 111/12, Nusret Bey to French Representative in Constantinople, August 10, 1924.

[64] BOA, HR.İM 111/12, Representative of the Government of the French Republic in Constantinople to Nusret Bey, October 16, 1924.

locusts in the same grammatical terms as humans, with concern for their emigration, their movement across borders, and even their places of birth.

At the same time as locusts and nomads crossed borders, there was a considerable amount of upheaval regarding the post-Ottoman borders of the Jazira and the degree to which they reflected local peoples' national identity. In January of 1925, a delegation from the League of Nations arrived in northern Iraq, with the aim of deciding on whether Mosul ought to be attached to Turkey or British Iraq (in a land grab aimed at obviating these sorts of questions, the British had occupied the city four days *after* the end of World War I).[65] The basis of the league's decision – in line with the precepts of self-determination – was to be the race of the region's inhabitants. The experts to decide on this question were Colonel Albert Paulis, a Belgian who had experience in Congo Free State, Einar af Wirsen, a Swedish diplomat, and Count Pal Teleki, who would go on to become prime minister of Hungary during World War II (he committed suicide upon news of the Nazi advance toward his country).[66] The entire visit involved numerous performances of identity. All villages in the north near Silopi and Cizre had received Turkish flags, and orders circulated prohibiting the use of Arabic in favor of Turkish.[67] Iraqi officials arranged for similar Iraqi flag-waving exercises in Mosul in support of their case.[68]

Forces beyond these performances also ended up playing a role in the decision. After the Hungarian Teleki completed his visit for the Committee, he joined a British officer by the name of Lyon to hunt ibex in the mountains. Before they saw any ibex, however, Teleki fell ill. It was bladder trouble, and Teleki was unable to urinate, an affliction dating, he explained, to his "wilder days."[69] After he fit himself with a catheter to handle the situation, what came out of him was not a substance harkening back to his youthful dalliances, but rather an earwig. "Most alarming," commented Lyon. He advised Teleki to seek medical attention for fear that there might be "a nest of them somewhere inside him." The British High Commissioner of Iraq, Dobbs, later credited Lyon's care for Teleki's insect-driven problems for the Hungarian's eventual vote in favor of Mosul remaining with Iraq. The fact that the race of Mosul's inhabitants was supposed to decide which country the region belonged in represented the currency that race had attained as an organizing principle in

[65] Sluglett, *Britain in Iraq*, 77.
[66] Bet-Shlimon, *City of Black Gold*, 67–68; Lyon, *Kurds, Arabs and Britons*, 152.
[67] TNA-UK, AIR 23/279, SSO, March 9, 1925.
[68] Edmonds, *Kurds, Turks and Arabs*, 405. [69] Lyon, *Kurds, Arabs, and Britons*, 151.

international politics. Yet even the committee was forced to admit that such an arrangement was impractical (their findings suggested that a state based on ethnicity would require, if anything, a Kurdish state, but such a pronouncement fell outside of the bounds of their brief entirely).[70] The fact that it was European experts who were to make these decisions – including one perhaps swayed by the gracious response of his British counterpart to an earwig discharged from his penis – represented the continued ways that local people's fates were decided by outsiders and shaped by insects.

The questions about Mosul accompanied challenges to the rule of the Republic of Turkey in the southeast. Beginning in March of 1925, a revolt emerged led by the Zaza Kurdish Naqshbandi Shaykh Said of Palu, loosely organized under the goals of securing local autonomy and reinstating the caliphate, which had been abolished in 1924.[71] By early March, Mardin had been seized by the rebels, and Diyarbekir surrounded, though not ultimately overtaken.[72] In a mark of the isolated nature of southeast Anatolia and the interconnected dimensions of the region's infrastructure, the Republic of Turkey had to receive permission from France to move troops to the embattled region, seeing as the railway required that they move through Syria and then on the Turkish-Syrian border line to get from western Turkey to the southeast.[73] Once Turkey secured this permission, they flooded the region with soldiers and dealt with the rebellion harshly. The state deported Kurdish leaders to western Anatolia and seized and auctioned off sheep.[74] When Shaykh Said was captured, he was imprisoned in Diyarbekir Prison, the same structure in which Diyarbekir's Armenian notables had awaited their fate just a decade before.[75] He was executed by hanging on June 29, 1925; a statue of Mustafa Kemal, first president of the Republic of Turkey, was subsequently built on the spot.[76] The measures marked a shift in the place of Kurds within the Republic of Turkey, as they became branded an ethnic threat to be made diffuse through their dispersal throughout the state. But the policies of deportation and assimilation also echoed Talat's plans during World War I for Armenians and Kurds both.

[70] McDowall, *A Modern History of the Kurds*, 144.
[71] Ibid., 194, 197; Üngör, *The Making of Modern Turkey*, 123.
[72] McDowall, *A Modern History of the Kurds*, 195; Üngör, *The Making of Modern Turkey*, 125.
[73] McDowall, *A Modern History of the Kurds*, 195. [74] Ibid., 196.
[75] Üngör, *The Making of Modern Turkey*, 131. [76] Ibid., 132.

The definitive decision on Mosul and the brutal response to the Shaykh Said Rebellion appeared to settle questions on what the map of the Jazira would look like. Nevertheless, locusts continued to complicate the idea of clear separation between the states that divided the region. As the insects devastated crops, they left many feuding officials behind them. Despite collecting some twenty tons of locust eggs by mid-March, the British called the destruction "the worst in the memory of living man."[77] The locusts did not stop in Iraq, with reports that they destroyed 30 percent of crops near Mardin in Turkey and 50 percent in Ras al-Ayn in Syria.[78] Continued fights over the origins of the locusts prompted some officials to resort to disingenuous science to avoid acknowledging the interlinked nature of the divided Jazira. In the summer of 1925, officials in Turkey, in what was becoming an annual tradition, complained that the British had done nothing to stop the insects.[79] What was more, they declared, the insects had set people in motion, with groups such as the Shammar moving toward Turkey's southeastern border, where they pillaged what few crops were left over from the locusts. Turkish authorities continued to sound the alarm in the summer. The British, however, objected. To do so, they resorted to subterfuge. Of course it was the same species of locust that lived all across the Jazira. But the British claimed that the locusts in Iraq were distinct from those afflicting Turkey. And because the range of these locusts was limited to some "40 miles," they could not, in fact, be responsible for what was happening to the north.[80] The British even went so far as to deem their locusts worthy of a name: "the Iraq locust." This is to say that the British nationalized the locust as a means of escaping culpability for the shared environmental challenges of the Jazira.

Turkish complaints about French inaction in Syria spurred further deliberation on the divided connection of the region. In July of 1925, Turkish officials wrote that locusts had left Deir ez-Zor in Syria and crossed into Turkey toward the city of Aintab.[81] They vowed to take the issue up with the French authorities in Aleppo, but in the communiqué,

[77] *Times of Mesopotamia*, "Crops in Lower Iraq – Condition Poor and Immature – Livestock Still Pitiable," April 3, 1925, 2; TNA-UK, AIR 23/279, SSO Mosul, May 6, 1925. See also *Times of Mesopotamia*, "Rail Construction in Northern 'Iraq – Triumph over Enormous Difficulties," September 13, 1925, 3.

[78] TNA-UK, AIR 23/281, SSO Mosul to Air Headquarters, Baghdad, August 31, 1925.

[79] BOA, HR.İM 150/26, July 8, 1925.

[80] TNA-UK, AIR 23/145, July 30, 1925, High Commissioner for Iraq to HBM Ambassador, Constantinople.

[81] BOA, HR.İM 150/28, July 8, 1925.

Aleppo was crossed out and "in Syria" scrawled in the margin. The officials had thus replaced Aleppo – the Ottoman province that had once contained both Syria's Aleppo and Turkey's Aintab – with the post-Ottoman place name of Syria that signaled the region's changing map. In this new world, Aintab was no longer Aleppo, and Aleppo was no longer simply Aleppo, but rather Syria. Border-crossing locusts prompted this post-Ottoman copy-editing. Altogether, the Jazira incubated a population of locusts, and the Jazira's division into separate states ensured that each state could blame its neighbors for its problems. As one British official summarized it, "ideas commonly held in Iraq" suggested "that our locusts come from Syria and Turkey, whilst in Syria it is believed that the locusts come from Iraq and in Turkey that the pest comes from Iraq also."[82]

Amidst locust-driven disputes in the Jazira, fall 1925 in Syria witnessed what one historian has called the "largest, longest, and most destructive of the Arab Middle Eastern revolts."[83] Beginning among the Druze in the mountains to the south of Damascus, where Levon Yotnakhparian had survived the years of World War I, alliances and solidarity spread throughout the country, most notably via connections forged through the grain trade, linking merchants of the Maydan in Damascus to comrades in the Hawran and beyond.[84] The French response was brutal, including a bombardment of Damascus in October 1925 before which locals received no warning, leaving some 1,500 people killed and entire neighborhoods destroyed.[85] The League of Nations – envisioned as a hope for a new kind of internationalism – did little, aside from prodding the French to justify their violence in terms of the League's prevailing presentation of the people of the Middle East as civilizational adolescents thanks to "Ottoman backwardness."[86]

What the League could do, however, was raise technical questions about French rule in Syria. Violence against the revolt did not fall within these responsibilities, but locusts did. One person who brought up the question was Sir Frederick Lugard, the British representative at the League and former governor-general of Nigeria. He was one of several people who gave the Permanent Mandates Commission the feel of, in the words of historian Susan Pedersen, "a spa for retired African governors."[87]

[82] NHMA, Iraq 6213, BH Bourdillon, July 30, 1925, Enclosure: Copy of Memo 23 July 1925 from Inspector General of Agriculture to Baghdad, July 23, 1925.

[83] Provence, *The Great Syrian Revolt and the Rise of Arab Nationalism*, 12.

[84] Ibid., 13; Bailony, "Transnationalism and the Syrian Migrant Public."

[85] Provence, *The Great Syrian Revolt*, 104. [86] Pedersen, *The Guardians*, 158.

[87] Ibid., 62.

Lugard had authored the famous 1922 work *The Dual Mandate in British Tropical Africa*, which articulated British renunciation of the idea of universal progress.[88] Of locusts, however, Lugard wondered about a universal solution. He mused about the insects to the British high commissioner of Iraq, "Was it possible to 'gas' them?"[89] The British high commissioner responded negatively, suggesting that poison in a solid rather than gaseous form seemed to be the most effective method. Yet Lugard's question – echoing the questions posed by German scientists during World War I and drawing upon, perhaps, memory of the toxic clouds of trench warfare – pointed to the ways that officials sought a new kind of control over the region and its environment. This vision was not beholden to forced labor gangs charged with unearthing locust eggs and killing recently hatched creatures before they developed wings. Instead, the new idea depended on toxic clouds that could easily annihilate oncoming swarms.

While the proposal for using gas did not go anywhere immediately, the need for international coordination at the root of the League did begin to shape locust control in the Jazira. In August of 1926, the International Bureau of Intelligence on Locusts was founded in Damascus, with its signatories including the administrations of Iraq, Transjordan, Palestine, Syria, and Turkey.[90] The body had no authority to "interfere in any manner with the administration of the contracting states" but was to instead function as a clearinghouse for information on locust invasions.[91] For decades, people had called for such coordination across the space afflicted by locusts. But the organization did little for the immediate and long-standing problems state officials faced in the Jazira. Officials in Turkey observed, as usual, that locusts were coming "in a winged state from the lands of Syria and Iraq."[92] But they also faced problems within their own borders, such as dozens of villages around Nusaybin that were afflicted by locust eggs and either "uninhabited or in ruins." The absence of people would make the effort to procure labor to

[88] Pursley, *Familiar Futures*, 20.

[89] UNAG, R53/1/55200/16466, Permanent Mandates Commission, Minutes of the Tenth Session Held at Geneva from November 4th to 19th, 1926, 75.

[90] NHMA, Syria and Lebanon 6211 – 158, E. C. Hole, January 14, 1931.

[91] NHMA, Iraq 6213, Iraq Government Gazette April 30, 1927, Text of International Locust Conference, August 20, 1926.

[92] Türkiye Cumhuriyeti Cumhurbaşkanlığı Devlet Arşivleri Başkanlığı Cumhuriyet Arşivi (BCA), 30-10-0-0, 185/277/7, Mardin Locust Struggle Director to Agriculture Ministry, September 13, 1926.

fight the locusts all the more difficult. In the coming years, changing technologies would increasingly resolve this dilemma. They would also make the newly founded International Bureau of Intelligence on Locusts obsolete.

"THE VAST STEPPES . . . LURED THEM BACK"

As the continued challenge of controlling the Jazira's moving people amidst the locust invasions underscored, nomadic pastoralism remained a viable way of life for many in the Jazira. And with nomads remained a significant number of the Armenian orphans to whom Yaşar Kemal had likened the locusts. Even years after people such as Stanley Kerr and Levon Yotnakhparian struggled to find survivors among Arab and Kurdish families after World War I and the genocide, the work went on. One of the main forces behind this effort through the 1920s was the Danish missionary and humanitarian worker Karen Jeppe, who operated Aleppo's Armenian Rescue Home. She had lived in Urfa for many years as a missionary and had even compared nomads to locusts for their springtime depredations on cultivated lands in 1908. From her base on the edge of the desert in Aleppo, Jeppe opened stations to find children across the Jazira in places such as Deir ez-Zor, al-Hasakah, and Ras al-Ayn.[93] From these stations, Jeppe's agents spread across the Syrian Jazira, and even into Turkey, where, "disguised as Arab Bedouin or Kurdish tribesmen," they passed along news of the possibility of a new life in Aleppo.[94] Jeppe's organization received funding from the League of Nations, and, as one of the chairs of the League's Commission for the Protection of Women and Children in the Near East, she frequently reported on her activities to Geneva. At the onset of the 1920s, Jeppe estimated that some 30,000 Armenians remained scattered across the Jazira. Over the decade, the Armenian Rescue Home presided over the "rescue" of more than 1,500 orphans.

The stories of the orphans attest to both continuity and change in the Jazira. Nomadism afforded the isolation that allowed orphans to survive, while also giving them the autonomy to make a run for it. Setrag Hagobian, for example, worked as a shepherd with Kurdish nomads near Suruç. His orphanage entry form read simply, "Hearing that his people were still alive, one day he left the sheep in the mountains and fled to Raqqa."[95] The story of another boy – Hagop – struck a similar

[93] Jeppe, *Misak*, xliii, xliv. [94] Ibid., xliii. [95] UNAG, C1601/497/1, March 20, 1922.

note. Having lived for eleven years with a seminomadic Kurdish group, "one day, up on the mountains with his sheep he remembered his parents and began to weep for them and resolved to go to Aleppo hoping to find them."[96] Hagop Nadjarian of Urfa found a different kind of opportunity in shepherding. Despite ending up in the deserts far south of the Jazira, he "never forgot his mother tongue," because "when he was guarding the sheep in the desert he spoke Armenian with them."[97]

While nomadism enabled both survival and escape for some, the changing political ecology of the Jazira also offered others a path to a new life, as automobile routes increasingly intersected with those of sheep and the people who cared for them. Simon was tending a flock when he noticed a driver stop to "fill water in the radiator" of his car.[98] Simon approached the overheated vehicle. When he learned that the driver was Armenian, Simon went with him to the Armenian church in Mosul, and from there to Aleppo. Megerditch had a similar story. Having been given away by his family to save his life at Ras al-Ayn during the genocide, Megerditch ended up working for a "very cruel man" as a shepherd for eleven years.[99] One day he quit, and while he walked along the road, a car sped toward him. Megerditch "took his head-dress and waved it in the air" and "the auto waited for him." The driver turned out to be Armenian, and he brought Megerditch to al-Hasakah, and from there he went to Aleppo. For these young men and others, it was work with sheep that allowed them to survive, and it was the way cars – some of them with Armenian drivers – traversed the Jazira that allowed them to find their way out.

The orphanage and the entire international rescue effort rested on the notion that orphans ought to be saved. But some of the entries pointed to the ways that Armenian survivors had become integrated into nomadic pastoralist groups. One Vartivar Avakian (Figure 14), of the famously rebellious village of Zeitoun, was deported to Deir ez-Zor with his family, where they were all killed. He managed to survive with the Shammar. His entry tersely read: "Came to Aleppo to find Armenians," while on the back of the sheet someone noted, "left our care … returned to the Arabs to sell his sheep."[100] The next entry – of a young man named Mesrob – simply

[96] UNAG, C1603/499/1168, October 29, 1926.
[97] UNAG, C1603/499/1376, July 18, 1927.
[98] UNAG, C1602/498/622, September 25, 1924.
[99] UNAG, C1602/498/1098, August 22, 1926.
[100] UNAG, C1603/499/1747, June 14, 1928.

FIGURE 14 Vartivar Avakian. UNAG, C1602/498/1747. Courtesy of United Nations Archives at Geneva

read "same story."[101] While it is unclear what Vartivar and Mesrob did after having sold their sheep, the fact that they owned these creatures underscores how they had held something other than the low-status positions of enslavement that many occupied, and that humanitarian campaigns emphasized. And while Vartivar and Mesrob may have sold their sheep with the intention of returning to Aleppo, if not the orphanage, there were many who chose otherwise. Jeppe reported to the League of Nations that many of the young men "after a short stay with us, returned to the Arabs."[102] She added, in a tone recalling that of Artin Boşgezenyan's comments about the Jazira's nomads in the Ottoman parliament in 1918, that "the vast steppes and the aimless life lured them back."

It was decisions such as these that prompted Jeppe to declare in 1927 that her work was done, not because her organization had found all of the

[101] UNAG, C1603/499/1748, June 14, 1928.

[102] UNAG, R641/12/60745/4631, "Report of the Commission for the Protection of Women and Children in the Near East, Aleppo, July 1st, 1926-June 30th 1927," July 28, 1927, 4.

Armenian survivors across the Jazira, but rather because of resignation. "Those who remain," she wrote, "have chosen to stay."[103] Her pronouncement had distinct gendered implications. Those who had "chosen to stay," as she put it, were in some cases young men who had made a life where they owned sheep in the Jazira, such as Vartivar or Mesrob. But they were more often young women (Figure 15), who did not have the circumscribed freedom of tending sheep to allow them to flee. In many cases, they also had children in these communities and did not wish to lose them, a reminder of the way gender structured survival in profound ways, as historian Lerna Ekmekçioğlu has argued.[104]

FIGURE 15 Loutfié Bilemdjian. UNAG, C1602/498/1010. Courtesy of United Nations Archives at Geneva

[103] Ibid., 1.

[104] Ekmekçioğlu, *Recovering Armenia*. See also Jinks, "'Marks Hard to Erase'"; Shemmassian, "The League of Nations and the Reclamation of Armenian Genocide Survivors"; Tachjian, "Gender, Nationalism, Exclusion."

The desert had thus been central to both the destruction of Armenians and their survival. Jeppe, like many officials in the region, envisioned environmental transformation as crucial to completing her mission. If, as historian Keith Watenpaugh has argued, "the Eastern Mediterranean was where much of modern humanitarianism was born," the Jazira was where a particular kind of humanitarianism emerged, one articulated in contrast to the desert and the nomads who lived there.[105] When Jeppe spoke of returning women and children to what she called "their natural environment," she meant removing them from the desert.[106] She also imagined leaving behind the desert's people by replacing nomadism with cultivation. To this end, Jeppe called for the League of Nations to support Armenian agricultural colonies within Syria. "Everybody knows," she stated, "that the future of Syria depends upon an intense cultivation of her fertile soil."[107] Armenians, she reasoned, formed the ideal advance guard for cultivation, given what she called "the energy of their race tingling in their veins."

Jeppe favored agricultural projects because they brought together economic progress and national restoration. Her scheme entailed "utilising that which seemed their greatest obstacle, their 'Arabisation,' to build up a strong and thriving peasantry."[108] Hajim Pasha – the ʿAnaza shaykh who had previously carried the title of "President of the Patriotic Movement of the Jazira" – offered to pay the relocation expenses of Armenian refugees from near Urfa to his own lands at Tall al-Samn.[109] Using tractors and planting cotton, the settlement apparently caught the attention of surrounding peasants. When Hajim Pasha died, his grave was visited by both his fellow ʿAnaza and Armenians alike.[110] Plans existed for other communities in the region of Raqqa at Sharb Bedros, north of Tall al-Samn, as well as Tall Arman, near Khirbat al-Riz. Thus Armenians had gone from being compared to locusts during the deportations of the

[105] Watenpaugh, *Bread from Stones*, 153.

[106] UNAG, R641/12/60745/4631, "Report of the Commission for the Protection of Women and Children in the Near East, Aleppo, July 1st, 1926-June 30th 1927," July 28, 1927, 3.

[107] UNAG, R1388/26/46266/43540, Commission for the Protection of Women and Children in the Near East, "Extract from the Minutes of the Thirty-Fifth Session of the Council," Second meeting, September 5, 1925, Annex: Report of the Commission, 5.

[108] UNAG, R638/12/4631/11391, Commission for the Protection of Women and Children in the Near East, "Extract from the Minutes of the Thirtieth Session of the Council," Second meeting, August 29, 1924, 5.

[109] Jeppe, *Misak*, iii.

[110] For a detailed account of the settlement, see Abrahamyan, "Armenian Refugees, State Formation and Identity Construction in Mandatory Syria."

genocide to cultivating lands on the edge of the desert that faced regular locust depredations. Meanwhile Jeppe went from rescuing Armenians from the desert to using Armenians to rescue the desert from the desert, an approach that resonated widely. League officials and those within Syria also called for the expansion of agricultural settlements for Armenians within Syria.[111] Jeppe joined a chorus of humanitarian and colonial officials seeking to make the desert bloom, much like the Zionist project to the south, for which she had high esteem.[112] But her work – implanting refugees on marginal lands for agricultural colonies – also recalled Ottoman refugee settlement policies.

There were other echoes of Ottoman policies as refugees found homes in French Mandate Syria. Indeed, the machinery of deportation employed against Armenians in the waning years of the Ottoman Empire continued after the empire's end, albeit with different victims. Many of the same officials charged with expelling Armenians took the reins of programs aimed at deporting Kurds from southeastern to western Turkey in the wake of the Shaykh Said Rebellion. The deportations recalled those during World War I, when clear orders existed to send Kurds to western Anatolia, where they would both take the place of deported Armenians and presumably become Turkified.[113] By 1927, however, the state's treatment of Kurds took on a new level of violent urgency. In January, the British ambassador in Ankara spoke with Turkish foreign minister Tevfik Rüştü, who likened the Kurds to the "Hindus of America," which, the ambassador wrote, "presumably ... meant the Red Indians" and that "they will die out."[114] Of course, just a dozen years previously, Ottoman interior minister Talat Pasha had uttered similar words to the American ambassador Henry Morgenthau with respect to Armenians. The violent politics of population bled from the late Ottoman Empire into the Republic and maintained similes of American violence.

It was this violence that continued to push people from the Republic of Turkey toward agriculture, the Jazira, and, sometimes, national categorization.[115] Armenians from the region of Batman in Turkey, for

[111] UNAG, R696/12/46805/47754, Report by Dr. Fridtjof Nansen, July 28, 1925, Appendix 1: Report by Mr. Carle on the Present Position of Armenian Refugees in Syria, 8.

[112] Fletcher, *British Imperialism and the "Tribal Question"*; Jeppe, *Misak*, lvi.

[113] Üngör, *The Making of Modern Turkey*, 111.

[114] TNA-UK, FO 424/266, G. Clerk to Sir Austen Chamberlain, January 4, 1927.

[115] Deportation to Syria and Lebanon also served to make it a place where plots abounded. The dissident Refik Halit wrote some of his voluminous novels about his trips to the Jazira while in Syria. Öztan, "Republic of Conspiracies"; Philliou, *Turkey*, 157–171.

example, ended up working as sharecroppers in the Syrian border town of Tel Abyad.[116] Others found their way to settlements likewise founded on the region's distinctive elevations: Tell Brak, Tell Beri, and Tell Aswad. Some 10,000 in total took these journeys in the 1920s.[117] While projects such as Jeppe's enshrined a clear sense of what it meant to be Armenian, migrants from Turkey also confounded these expectations. The French struggled to categorize the peasants who fled to the Syrian Jazira, resorting to terms such as "Kurdo-Christians, Assyro-Kurds, and Kurdo-Armenians" or "Christians of the Kurdish race" that gestured toward the complexity of rural life that did not easily align with exclusivist ethnic identities.[118] Armenian nationalist organizations agreed with this assessment and saw it as a problem. The Armenian Revolutionary Federation (known as Dashnaktsutyun) stationed operatives in border towns to teach Anatolian migrants language and religion that would make them properly Armenian.[119] The Armenian General Benevolent Union similarly supported the settlement of Armenians in what one of its officials called "the waste territories of Jazira."[120] Both organizations did so in pointed opposition to other Armenian organizations – notably the Hunchakian Party – which supported settlement in Soviet Armenia.

Mobility and agrarian transformation in the Jazira bore a similar relation to Kurdish activism. Arif Abbas had been born in the southeastern Anatolian town of Maden in 1900, before receiving training at the Institute of Agriculture in Istanbul.[121] The Turkish government exiled him to Syria for his political activism. At around the same time, a group of activists – including the Bedirkhan brothers and the brother of the executed rebel Shaykh Said – founded a Kurdish nationalist group by the name of Xoybun ("Independence"). Abbas spent a short time in Damascus, where he became acquainted with other young activists associated with the Xoybun. He was quickly transferred to the Jazira town of al-Hasakah, where he drew on his agricultural training and his "great experience" with locusts in southeastern Turkey to work against the insects.[122] He no doubt also continued his political activities in the Jazira, since the French had allowed a number of Kurdish nationalist

116 Altuğ, "Sectarianism in the Syrian Jazira," 63, 191. 117 Ibid., 61.
118 Ibid., 122; Tachjian, *La France en Cilicie et en Haute-Mésopotamie*, 275.
119 Zaza, *Bir Kürt Olarak Yaşamım*, 264.
120 Abrahamyan, "Armenian Refugees, State Formation and Identity Construction in Mandatory Syria."
121 Tejel Gorgas, *Le mouvement kurde de Turquie en exil*, 134–135.
122 Zaza, *Bir Kürt Olarak Yaşamım*, 82.

leaders to reside in al-Hasakah.[123] Syria and Turkey had different political regimes, but on either side of the border Kurds organized and locusts loomed. In an echo of past invocations of nature with respect to borders, one of Abbas's colleagues described how there was only "an artificial border" that separated the Syrian Jazira from Turkey.[124] If Abbas and others saw the foundation of Kurdistan in the locust-stricken Jazira, others saw confounding syncretism. Another operative of the Xoybun detailed how his travels among Kurdish nomads in the Syrian Jazira turned up no shortage of orphaned genocide survivors describing themselves as Muslim Armenians.[125] Thus while people worked to harvest the fruits of the Jazira through locust destruction and nationalist organizing, they also noticed a kind of ambiguity resistant to these dynamics.

"THE POISON METHOD IS THE BEST AND THE CHEAPEST"

In the spring of 1927, there were so many locusts between Tell Abyad and Ras al-Ayn that they overwhelmed the railroad. One train was "brought to a standstill owing to slipping of wheels."[126] The railroad, of course, was the curious thoroughfare that also marked the border between the countries of Syria and Turkey. The locusts' feat symbolized the linked political ecologies of Turkey and Syria across the Jazira: the organisms that had shaped the Jazira also overwhelmed the relic of Ottoman infrastructure that served to divide Turkey and Syria from one another in the post-Ottoman period.

Given the power of locusts in the space, the goal for officials in the region was to devise a method of killing the locusts that compensated for the low population density and difficulty procuring labor. By 1927, the British employed some 204 tons of sodium arsenite toward these ends.[127] They mixed it with 605 tons of date syrup and 903 tons of bran as a poison bait to serve as the insects' sweet last supper. In Syria, officials used the same substance. They described its use as much like cultivation, writing that it "is planted like wheat is planted."[128] As the locusts consumed the

[123] Tejel Gorgas, "The Kurdish Cultural Movement in Mandatory Syria and Lebanon," 847.

[124] Zaza, *Bir Kürt Olarak Yaşamım*, 82.

[125] Altuğ, "Sectarianism in the Syrian Jazira," 130–131. For more on collaboration between the Xoybun and the Dashnaktsutyun, see Tejel, "The Last Ottoman Rogues."

[126] TNA-UK, AIR 23/91, SSO Mosul, May 23, 1927.

[127] UNAG, R2314/6A/655/6774, Report by His Britannic Majesty's Government to the Council of the League of Nations on the Administration of 'Iraq for the Year 1927, 166–167.

[128] NHMA, Syria and Lebanon 6211 – 157, Rapport sur la lutte anti-acridienne dans les Pays du Levant adhérent à l'Accord International du 20 Mai 1926, 1927–1928, 14.

substance, they died, and then their "cadavers" served as "poisoned bait" for the next round of locusts, which – cannibalistic – consumed their dead comrades.[129] In Syria, officials also used a substance called "cyanogas" to do the job.[130] When sprayed in its vaporized form on the locust, "the effect is instantaneous" as the insects were "suffocated by the gas," a poisonous vindication of Lugard's question at the League of Nations several years before about the use of gas to kill locusts. As officials remarked, such technologies solved the circular dilemma in which the regions overwhelmed by locusts were also the "regions deserted of inhabitants" (by which of course they meant settled people).[131] Chemical companies lost no time in attempting to promote their products to the involved parties, with one Glasgow maker of the substance Paris green boasting that their product was "the best insect exterminator known."[132] They added that there was "no evidence" that it might harm other animals like birds or, for that matter, humans.

The residents of the Jazira had varied reactions to these substances. In Iraq, the British claimed that cultivators were initially "reluctant," but "when they saw and appreciated the results ... this prejudice vanished."[133] Faysal – who, after being removed by the French in Syria had been installed by the British as king of Iraq – himself visited the region to encourage efforts at spreading the substances.[134] Nomads, however, had a different response. These moving people were largely not "sedentarized," a term that one report – in a blending of language for humans and insects – used to describe not people but rather the locusts in the region.[135] Nomads and locusts intersected not only in the language used by experts. They also took advantage of the same ecologies. The verdant seasonal grasses of the Jazira had long been the refuge of nomads and the life source of the sheep that provided wool every year. The grasses were also where, one British officer explained, the young locusts "take shelter."[136] In 1927, the British in Iraq responded to this dilemma by applying oil to the grass and setting it aflame, with some 300 barrels used in Mosul. It was literally a scorched-earth campaign, with the British high

[129] Ibid., 15. [130] Ibid., 14.

[131] NHMA, Syria and Lebanon 6211 – 157, Compte Rendu Definitif de la Lutte Contre les Sauterelles dans le Vilayet d'Alep, Wagih Jazzar, May 24, 1927.

[132] NHMA, Iraq 6213, Letter from Harvey, Wilson & Co., December 9, 1927.

[133] UNAG, R2314/6A/655/6774, Report by His Britannic Majesty's Government to the Council of the League of Nations on the Administration of 'Iraq for the Year 1927, 167.

[134] NHMA, Iraq 6213, Report on the Anti-Locust Battle in the Territory of Iraq in 1927, 8.

[135] Ibid., 1. [136] Ibid., 7.

commissioner observing that "huge ... infested areas in the desert have been burnt."[137] Nomads resented such policies because it caused their sheep to die or starve.[138] They did not go so far as pastoralists did in Somaliland in 1945, when they resisted British locust-control campaigns by force of arms.[139] But the Shammar were nonetheless "hostile" to the deployment of chemical insecticides, according to one British report.[140] From the Shammar perspective, the state seemed intent on wiping out their revenue sources. In addition to banning nomads from collecting taxes as had been their way in the past, the British were also, as one report candidly admitted, "polluting the ground with poison which will kill off their livestock."[141] In other words, the desert, with its moving people and locusts, had long defied state designs. Groups such as the Shammar were increasingly tamed, however, through the destruction of the desert.

Different perspectives on locust destruction also existed between the states whose jurisdiction split up the Jazira. The British in Iraq had deemed collaboration with Syrian officials "impossible" after one meeting in 1927.[142] As in previous years meanwhile, the Turkish government suggested in December of 1927 that the British had done "nothing" compared to all of the work done by authorities on Turkish soil.[143] In response, the British charged that it had not "proved possible to get in touch" with locust officials on the Turkish side of the Jazira.[144] They also provided an itemized list of all the material they had employed in the latest campaign against the insects, including not only 235 tons of sodium arsenate (referred to in addition to sodium arsenite) but also 350 pairs of gloves for its dispersal. The International Bureau of Intelligence on Locusts at Damascus was supposed to offer the venue for resolving these types of disputes. But even in late 1927, the entity was not operating, and key signatories had not paid their dues.[145] Chemicals were thus emerging

137 NHMA, Iraq 6213, Dobbs, May 12, 1927.

138 NHMA, Iraq 6213, Report on the Anti-Locust Battle in the Territory of Iraq in 1927, 7.

139 Mohamed, "'The Evils of Locust Bait.'"

140 UNAG, R2314/6A/655/6774, Report by His Britannic Majesty's Government to the Council of the League of Nations on the Administration of 'Iraq for the Year 1927, 167.

141 TNA-UK, AIR 23/154, SSO Mosul, April 25, 1927.

142 NHMA, Iraq 6213, Report on the Anti-Locust Battle in the Territory of Iraq in 1927, 11.

143 NHMA, Turkey 6217, Embassy of the Republic of Turkey in London, December 15, 1927.

144 NHMA, Turkey 6217, Baghdad, March 19, 1928.

145 NHMA, Iraq 6213, Extract from Economic Report no. 100, undated; Extract from Economic Report no. 110, undated.

as the prevailing solution all around the world, but they fit particularly well in a sparsely populated and politically fractured region like the Jazira.

Adding to the appeal of chemicals were their endorsements from elsewhere in the British Empire. In the spring of 1928, Iraq hosted a locust expert from South Africa by the name of R. H. Williams. He spoke to British officials in Iraq about "the more recent methods and forms of equipment," namely the use of arsenates.[146] His report had an effect, leaving officials in Iraq convinced that the "poison method is the best and the cheapest," especially if combined with accurate intelligence of the locust swarms.[147] Reports also crowed of the "remarkable success" of such measures in South Africa in terms of finances, an especially convincing piece of evidence for the ever-cost-conscious British administration in Iraq.[148] While the chemicals no doubt helped to wipe out locust swarms in South Africa, the broader political landscape also played a role. Yet British officials largely omitted these details in their technological triumphalism. A significant part of the campaign in South Africa consisted of fining landowners for locusts emerging from their property, which, in brief treatment, British officials in Iraq acknowledged was "not necessarily easily arranged in other countries such as Iraq."[149] Still, officials declared that the visit was momentous. One explained that locust policy in Iraq was "completely changed" as a result.

Despite the single-minded focus on poison that characterized the official's visit, Williams also observed some of the local beliefs that gestured toward a different understanding of the relationship between humans and the environment. When the expert from South Africa visited the village of Khanaqin, northeast of Baghdad and close to the border with Iran, officials tried to inform the local villagers about the new methods of killing locusts.[150] But they were rebuffed. The headmen of the villages declared that they had no use for "new-fangled devices" such as those advocated by Williams. For their part, they preferred what they referred to as *ab-i sar*, Persian for starling water. They believed that the blessed water attracted

146 TNA-UK, CO 730/128/5, J. Ramsay Tainsh, Adviser to Ministry of Irrigation and Agriculture to High Commissioner for Iraq, Baghdad, April 28, 1928.

147 TNA-UK, CO 730/128/5, Guy Marshall to C. C. Parkinson, Dominions Office, April 14, 1928.

148 TNA-UK, CO 730/128/5, J. Ramsay Tainsh, Adviser to Ministry of Irrigation and Agriculture to High Commissioner for Iraq, Baghdad, April 28, 1928.

149 NHMA, Iraq 6213, Inspector General of Agriculture to Adviser, Ministry of Irrigation and Agriculture, April 28, 1928.

150 Rooke, *Note on Locusts in 'Iraq*, 4.

a bird that devoured locusts. The British official leading the trip recorded the name with a slightly different spelling than typical – "the samarmad" – but clearly referred to the bird that so many had mentioned over the decades. The ritual and the bird were the same ones that the Ottomans had relied on for decades before in the Jazira, a blessed ceremony in which avian predators of migratory insects intervened to protect human beings everywhere from Cyprus to Damascus. People continued to believe in the starling across a vast geography, too. In Aleppo, they left mulberries in their gardens as offerings to the vaunted birds.[151] In Amman, they recalled how Circassian immigrants brought with them a "locust water" from Anatolia.[152] And in the Syrian countryside, they sang for their avian protectors to intercede, crying "Get out, get out, oh locust! The starling and the cold are coming for you" (*irhal, irhal, ya jarrad! Jak al-samarmar wa al-barad*).[153]

But such history meant little to British observers or the expert from South Africa. The report on the matter noted, "unfortunately it has not yet been possible to have this bird identified."[154] For at least some among the British, such beliefs in the power of mythic birds were little more than further evidence that the problem of the local population was that they were "in the last degree fatalistic."[155] Yet in reality, the shift to using chemical treatment as opposed to encouraging natural predators had been a recent one, adopted in the United States only in the interwar period, too.[156] In the years that followed, the British made clear their preference for the expert's magic water rather than the locals', and their debt to South Africa took the form of reference to sodium arsenite as "poison d'Afrique."[157]

The use of sodium arsenite stepped up in 1928, as British officials embraced the new technology as they grappled with what they described as "the largest swarm yet."[158] Yet sodium arsenite made quick work of the insects, especially when combined with the other bedrock technology of

[151] Ghazzi, *Nahr al-Dhahab fi Tarikh Halab*, vol. 1, 142.
[152] Sanger, *Where the Jordan Flows*, 265–266.
[153] Jabbur, *Al-Badu wa al-Badiyya*, 124. [154] Rooke, *Note on Locusts in 'Iraq*, 4.
[155] Webster, "Introduction" in Rooke, *Note on Locusts in 'Iraq*, ii.
[156] McWilliams, *American Pests*, 91.
[157] NHMA, Syria and Lebanon 6211 – 158, Rapport General sur les Invasions Acridiennes dans les Territoires des Etats Adherents à l'Office International de Renseignements sur les Sauterelles de Damas durant l'Année 1930, 73; Rapport General sur les Invasions Acridiennes dans les Territoires des Etats Adherents à l'Office International de Renseignements sur les Sauterelles de Damas durant l'Année 1931, 26.
[158] Rooke, *Note on Locusts in 'Iraq*, 4.

British rule in Iraq: airpower. Sometimes locusts blended in with greenery, making them difficult to see from above, but in 1928 drought ensured that the long lines of locusts could easily be spied from airplanes "as one long black line," and poison could be dispersed to stop them.[159] According to reports, the result of the efforts was "total annihilation," leaving behind "a belt of dead locusts ... twenty-one miles in length and ... ten to twenty yards in width."[160] Excitement over locusts in Iraq was so great that when American diplomats attempted to touch base with Yasin Pasha al-Hashimi – a former Ottoman officer rumored to be harboring ambitions of becoming an Iraqi Mustafa Kemal – he refused to speak of political intrigues, as the Americans had hoped, and talked only of "locusts and the lack of rain."[161]

Whatever the confidence and enthusiasm about locusts, their management in the Jazira still involved the challenge of the interconnected nature of the region. In most places there was not even a sign, to say nothing of a fence demarcating the border. There was, of course, the case of the Turkish newspaper reporter riding along the Baghdad Railway who witnessed the train become, for a moment, a wall stopping a swarm of locusts – or at least some of them – from flying into Turkey from Syria (see Introduction).[162] Other observers noted how more often than not the railway posed little obstacle for anyone, let alone locusts. Because the railway crossed numerous small valleys on bridges, the border could be safely passed by people moving underneath the railroad tracks.[163] This was even more so the case for locusts, which one British official observed in "large swarms" not far from Ras al-Ayn, the town that the Ottomans had hoped to turn into a wall against the desert through the settlement of Chechens in the 1860s. The Chechens had not served as a wall against the desert in the 1860s, and the railway that crossed through Ras al-Ayn did little to act as a wall either. It more readily symbolized the interconnected separation of Syria and Turkey. As a folksong from Dörtyol in Turkey put it, "the locust ate a grove of pistachios [*Şam fıstığı*]," and "it passed the border without showing a passport."[164]

Border-crossing locusts would seem to fall under the aegis of the League of Nations, the entity charged with ushering in a new era of international peace and cooperation. In service of this aim, the body

159 Ibid., 4. 160 Ibid., 4.
161 NARA, RG 84, vol. 166, American Consulate, Baghdad, Iraq, May 26, 1928.
162 Y.M., "Şark Mektupları: Harran Ovasından Mardin'e kadar ...," *Cumhuriyet*, June 17, 1928.
163 TNA-UK, AIR 23/290, SSO Mosul to Air Staff Intel, Baghdad, Enclosure: Report of Admin Inspector, Mosul on Tour of Mardin, Aleppo, Deir ez-Zor, Mosul, June 9, 1928.
164 Özer, *Anadolu'da Görülen Çekirge İstilaları*, 184.

concerned itself with issues such as free trade, ensuring that commodities as varied as "Swiss Gruyère cheese" and "silkworm cocoons of Greek origins" received fair treatment in Syria.[165] But the League was worried about things aside from the cheese and the worms. As early as 1929, the British began suggesting that they were interested in supporting Iraq's entry into the League of Nations. Some in the League were "horrified" at the prospect, but many came to realize the British gambit, namely, the fact that supporting Iraqi independence would allow Britain to have "a cheap client state outside the realm of international scrutiny."[166] In other words, the British supported Iraqi liberation only insofar as it allowed the British themselves to be liberated from League of Nations strictures, while still maintaining power in Iraq. Controlling locusts by chemicals offered another glimpse of a technology made for this form of indirect rule.

ZINC WALLS ACROSS BORDERS AND SMUGGLERS INTO LOCUST OFFICERS

While adoption of chemical insecticides against locusts seemed to allow for new hopes of control, the insects still posed a menacing presence in the region. In Iraq, the New Zealand engineer Archibald Milne Hamilton recalled what locust invasions looked like when removed from the managerial tone of official reports. Charged with building a road in northern Iraq as part of what was euphemistically called "the pacification of the Kurdish tribes-people," he oversaw construction on a route stretching from Erbil to Ruwanduz to Iran.[167] Hamilton was a practical man. For example, he solved the issue of whether to give his multireligious work crew Friday, Saturday, or Sunday off by declaring that they would have no days off.[168] Yet even he described the locusts as terrifying. He acknowledged that "aerial bombing may mean the destruction of the whole village," but locusts were "even more dreaded than wars."[169] Such a claim carried weight, given the unprecedented ways that the British had used airpower as a means of violent control of people. (It also spoke to Hamilton's delusions; in later correspondence, he would write, "In all the long line of conquerors down the pages of

[165] UNAG, R2293/6A/5088/486/Jacket5, Permanent Mandates Commission, Provisional Minutes of Thirteenth Session, June 26, 1928, 10:15 am, 4, 6.
[166] Pedersen, *The Guardians*, 261. [167] Hamilton, *Road Through Kurdistan*, 41.
[168] Ibid., 62. [169] Ibid., 164.

history, there has never been a more lenient hand upon this ancient land than that of the British.")[170]

Whatever his misconceptions about British rule, Hamilton described the locusts as an unstoppable, relentless army. Rivers might seem to present a barrier to their advance, but even those that drowned ended up forming "a bridge for their comrades" to continue their movement.[171] When they invade, he wrote, "the solid skin of the earth appears to be moving slowly about, and it is composed of locusts." Hamilton even imagined them speaking: "'Kill us if you like,' they seem to say; 'there are a million others behind us.'" Despite their apparent omnipotence, the invading army, Hamilton explained, had to grapple with the "constant war upon them" by humans.[172] Of the mix of arsenic and bran that was preferred, Hamilton described how "they eat it and die," and then they "come and eat these dead bodies and in their turn die."

As Hamilton's account suggested, locust control efforts relied upon chemicals to do the job of killing all across the Jazira. In Iraq, the locust fight had been influenced by the visit of the South African expert the previous year with good results. Having imported "fifty tons of finely powdered sodium arsenite and four thousand hand spray pumps from South Africa," reports suggested that "mortality was fully 100 per cent" in areas where spraying had occurred shortly after the locusts hatched.[173] In addition to the hand-sprayers, the British outfitted twenty Ford trucks with tanks carrying "three hundred gallons of solution," which could distribute the substance along "a line of vegetation from one and a half to two miles in length" and a width of "about twenty feet." In other words, the interventions driven by chemicals and trucks seemed to carry with them a promise of total control of space. The exact proportions of the poison mixture may have been different in Turkey or Syria, but the idea was the same. [174] More chemicals meant more control.

And more control was cause for national celebration. One Syrian newspaper correspondent was so moved by the locust-control efforts in northeastern Syria that he too joined the effort. He began using the car

[170] UNAG, C1533/429/20A/80237/19093/Jacket1, Kennedy to Johnson, November 11, 1933, Enclosure: "The Cases for and Against the Assyrians," by A. M. Hamilton.

[171] Hamilton, *Road Through Kurdistan*, 165. [172] Ibid., 164.

[173] Rooke, *Note on Locusts in 'Iraq*, 10.

[174] NHMA, Turkey 6217, "Rapport sur la lutte anti-acridienne effectueuse au printemps de l'année 1929 dans les départments sud du territoire de la Republique Turque," 1.

hired to transport him on his reporting for the purpose of moving supplies for locust killing. He complained of the "vile odor of the rotting locusts" killed by chemicals, but he exuded pride in the efforts by locals as a way of redeeming the Jazira, which he termed the long "forgotten" part of Syria.[175] Thus even as states engaged in their fights against locusts separately, the chemicals they used demarcated a shared space that crossed borders and prompted celebrations of a technologically driven national liberation from the environment.

There were also means of killing locusts that involved some coordination across borders. Since the time of the Ottomans, officials had deployed the method of using meter-high (three-feet tall) zinc walls as a way of stopping adolescent locusts that had yet to develop wings and could not surmount the slippery surface of the sheeting. In 1929, some 700 kilometers (434 miles) of fencing traversed all different parts of Mardin province in Turkey. But one especially interesting portion of it ran north–south, meaning that it crossed the east–west railway line that marked the border of Syria and Turkey. Beginning at Tall Eylul in Syria, crossing the railway at Darbasiyya, and stretching northwest into Turkey from there, the fence attested to the interconnected geography of the Jazira at odds with state division of it. This fence to stop locusts stood perpendicular with lines meant to mark the borders between the states, an endorsement, it would seem, of the poetic lines at the beginning of a 1929 report on the scourge by Syrian locust official Subhi Hassibi, a Damascus native who had studied at the Montpellier agricultural school before World War I.[176] Hassibi waxed poetic in this account of the insects, intoning that "history is nothing but an eternal recommencement."[177] Hassibi's words seemed accurate, given the way that locusts continued to afflict the states on all sides of the borders, much as they had in the Ottoman period when the Jazira and its locusts crossed provincial borders rather than national borders (Figure 16).

Yet zinc and chemicals did not always do the job. The Syrian locust official Hassibi, for example, described how in some years the height of the Jazira's grasses was so great that it allowed locusts to climb the blades

[175] *Al-Qabas*, May 27, 1929, 1–2; *Al-Qabas*, April 24, 1929.

[176] Williams, "Cultivating Empires," 89.

[177] NHMA, Syria and Lebanon 6211 – 158, "Report on the Anti-Locust Fight Effected in the Territories of States Adhering to the International Accord of 20 May 1926 during 1929," December 30, 1929, 1.

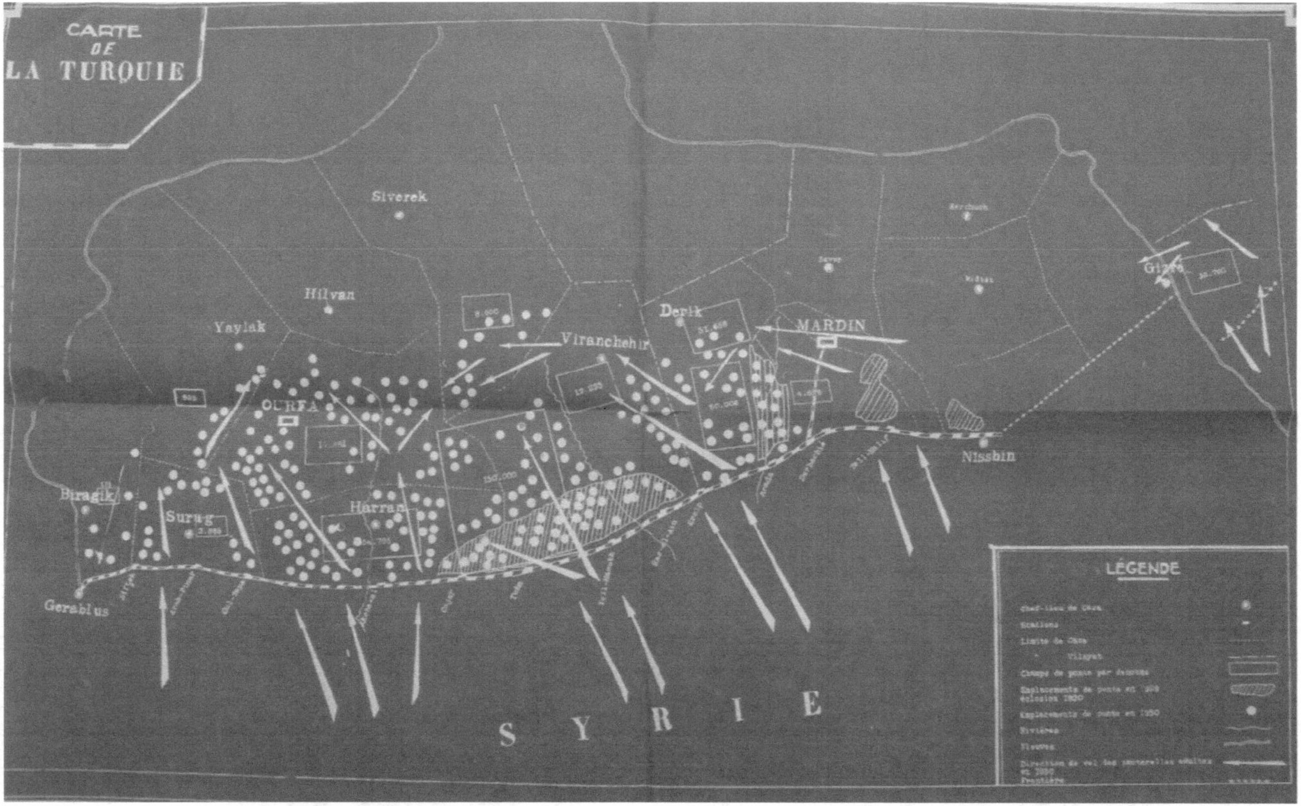

FIGURE 16 Map of locust flights into Turkey from Syria, 1929. NHMA, Turkey 6217. Courtesy of Trustees of the Natural History Museum, London

FIGURE 17 Zinc sheeting against locusts. Rooke, *Note on Locusts in 'Iraq*

and vault over the zinc sheeting meant to stop their march (Figure 17).[178] Chemicals too could only be effective where state officials could spread

[178] NHMA, Syria and Lebanon 6211 – 157, Report on 1929 Locust Struggle by Joseph Atallah, 14, 1930.

them. And even in the post-Ottoman period, some parts of the Jazira remained outside of the purview of state control, namely the areas described alternately as desert in English, *çöl* in Turkish, and *steppe* in French. These regions continued to elude management, even for the Ford trucks equipped with 300-gallon tanks of insecticides (Figure 18). The vehicles broke down in the desert, bereft of water necessary for filling up their tanks and preventing their engines from overheating.[179]

Chemical insecticides also continued to prove controversial for the people who had long had little problem moving in the region's arid expanses: nomads. For decades, it had been clear how entangled the motion of locusts and nomads were. Increasingly, locust experts took note of these dynamics in a detailed fashion, with Subhi Hassibi describing, for example, how locusts seemed to appear particularly in places where nomads had dug wells to water their animals.[180] In these areas, opposition to chemicals also emerged. Rumors of a die-off of some 20,000 sheep in Iraq had spread all across the Jazira, ensuring that in Syria, people feared for the health of their livestock. They often articulated their anxieties in religious terms. Some said that the locusts were "soldiers of God" and ought not be destroyed; that the word "soldier" might even be found written in Arabic on their wings; and that locusts laid ninety-nine eggs, but if they happened to lay 100 then they would control the world.[181] Sometimes these accounts drew on Hadith – the sayings of the prophet Muhammad – to bolster their views. By one account, the prophet had been asked what was written on locust wings and responded that they read, "I am God . . . it is I who am the master of the locust, and it is I who give them that with which they live."[182] By another account, ʿUmar, the second caliph, claimed that the prophet Muhammad said that "the first race (of living beings) who perishes at the end of the world will be that of locusts," meaning that the end of locusts was tantamount to the day of reckoning.

179 Rooke, *Note on Locusts in 'Iraq*, 10.

180 NHMA, Syria and Lebanon 6211 – 158, Subhi Hassibi, "Report on the Anti-Locust Fight Effected in the Territories of States Adhering to the International Accord of 20 May 1926 during 1929," December 30, 1929, 27.

181 *Al-Qabas*, June 25, 1929, 1, 4; NHMA, Syria and Lebanon 6211 – 157, Report on 1929 Locust Struggle by Joseph Atallah, 13; Whiting, "Jerusalem's Locust Plague," 535.

182 CADN, 166PO/D/54/7, Siouffi to Comte de Montebello, October 2, 1886.

FIGURE 18 Trucks against locusts. Rooke, *Note on Locusts in 'Iraq*

In Syria, officials took note and tried to adjust locust control measures. They worked to assure pastoralists that "poisonous chemical products" were "inoffensive" when used according to instructions.[183] But they also tried to develop different concoctions, including one that utilized soap and thus posed less of a danger to livestock. They also acknowledged – though did not necessarily encourage – older modes of resisting the locusts, among them the "bird called the samamar" and the Kandahari wheat variety, which offered a lower yield but matured earlier and with a harder glumelle than other varieties and could therefore more easily escape locust depredations.[184] Thus while new technologies seemed to offer a different kind of control of locusts, at least in Syria officials worked to adapt to a desire for locust control with less deleterious consequences.

Even as challenges remained for locust control, the expectation of deliverance from the annual nightmare increased. In Turkey, newspapers erroneously suggested that the Moroccan locust came from Morocco.[185] In obscuring how the insect was in fact native to parts of southeastern Turkey, the discussions made it seem like the affliction was in fact from a faraway place and thus could be easily avoided. The Arabic press of Damascus took a different approach and simply lampooned the repeated failure to manage locusts. In light of reports of the arrival of locusts in neighboring countries in 1930, *Al-Qabas* wondered sarcastically whether the government would once again welcome the insects with "Arab hospitality."[186] When the locusts did make their appearance in Syria, writers observed how the insects only compounded the devastation of the world economic depression set in motion the previous year. From their perspective, the protection of people from the vicissitudes of capitalism and the nonhuman world was the duty of the state. "If the government is not at least concerned with these sorts of unforeseen events, then when and with what ... is it concerned?"[187]

With little progress made, the newspaper employed even darker humor. In previous years, people had fled in fear, screaming that the locusts had arrived. But in 1930, *Al-Qabas* suggested that the situation was reversed. Because people were so hungry that they themselves were eating locusts, it was the locusts that were justified in fleeing and screaming in fear, "The Syrians have arrived."[188] The dark ripostes offered humor, but they also echoed earlier moments of suffering when people

[183] NHMA, Syria and Lebanon 6211 – 157, Report on 1929 Locust Struggle by Joseph Atallah, 13.
[184] Ibid., 22. [185] Özer, *Anadolu'da Görülen Çekirge Istilaları*, 42.
[186] *Al-Qabas*, February 12, 1930. [187] *Al-Qabas*, March 26, 1930.
[188] *Al-Qabas*, March 27, 1930.

had in fact eaten locusts, such as during the Armenian genocide. Moreover, the quips revealed the way that people's expectations of states and the nonhuman world had shifted. To be sure, locusts had been mobilized to make fun of and even topple government officials before, but the sardonic tone of the treatment of locusts in *Al-Qabas* pointed to the way that by 1930 people were no longer resigned to living with locust invasions. A similar dynamic occurred to the north in Turkey, where newspapers complained of locusts as "rude guests."[189] Writers moreover readily compared Kurdish rebels to locusts for their destructive capacities. But – in a mark too of growing human confidence toward the locusts – writers also invoked locusts to convey the simplicity with which the rebels ought to be defeated, as historian Nevcihan Özbilge has suggested.[190]

Again in 1930, the militarized fight against locusts could do little to address the local resistance that incubated the spread of the locusts. Reports from Deir ez-Zor in Syria, for example, described a city ready for battle. The insects had "entered and occupied" the city like an army, but through the diligent use of zinc panels, trenches, and soap solutions, locals had managed to prevent the insects from reaching the garden attached to the governor's house.[191] East of the city saw similarly intense efforts. There people had fashioned a wall out of "petrol tins" and dug trenches in front of the wall so as to stop the locusts' march, after which the insects were trampled.[192] Others wielded flame-throwers and arsenic amidst the "nauseating" smell of the "decaying insects." Yet despite these efforts, again local resistance complicated matters. In one telling anecdote, a Syrian newspaper reported how a man described as a Kurdish shaykh of the Jazira was given chemicals to distribute in an effort to kill locusts.[193] The shaykh agreed to complete the task but proceeded to simply bury the chemicals, rather than spreading them in the pastures around his village as he had been ordered to do. He returned the empty sacks of the substances, evidence, it seemed, of a job well done. But then his flock of sheep found the syrupy sweet concoction of bran, molasses, and sodium arsenite and ingested it. They all died. The newspaper offered this anecdote as evidence of local ignorance, based on the belief that the locust was "a soldier of god." Similar reports appeared in Turkey bemoaning "ignorant and

[189] Özbilge, *Çekirgeler, Kürtler ve Devlet*, 84. [190] Ibid., 108, 110–111.
[191] *Al-Qabas*, May 29, 1930, 1–2.
[192] NHMA, Syria and Lebanon 6211 – 158, Consul in Aleppo, June 10, 1930.
[193] *Al-Qabas*, May 29, 1930, 1–2.

fanatical religious leaders" who insisted that the locust was a soldier of god and birds like starlings could be counted on to intervene.[194]

Yet it is unclear how much these beliefs were the problem, given the way the Jazira's interconnected geography continued to facilitate the spread of locusts across borders. In 1930, British officials in Iraq claimed that the insects arrived from Turkey.[195] Turkish officials, however, claimed that a swarm of some 20,000 hectares arrived in their territory from Syria, which was subsequently, in the martial tone of one report, "reinforced by further swarms, always coming from Syria."[196] Internal government correspondence suggested the swarms struck as far "from our border with Syria and Iraq" as Erzincan, Sivas, and Kayseri.[197] The recriminations were familiar, as were the gaps in state control they underscored. As one British official put it, the desert regions of the Jazira were in the "undisputed possession" of locusts, a control only threatened by "gaily-coloured birds" – starlings, it would seem – that enjoyed eating the swarms and their eggs.[198]

Efforts at international coordination offered little hope. The International Bureau of Intelligence on Locusts was to be the prime venue to manage these efforts, but even in 1930 officials noted the great challenge posed by the compilation of statistical data.[199] What was more, the product of the efforts was dubious. As evidence of the reports' "inaccurate and unsystematic" nature, one British official in 1930 wrote "that one locality was frequently entered in reports under half-a-dozen differing names."[200] Such flaws ensured that the reports consisted of little more than "optimistic statistics of the quantities of locusts and eggs destroyed."

If international cooperation could not be secured in a formal way, state officials had to take different approaches. For example, in northern Iraq in

[194] Kaya, "Mardin İlinde Tarım Durumu," *Ulus Sesi*, October 29, 1935, 6. Another argument was that locusts were manifestations of divine displeasure with humans, such as in 1880 when a newspaper writer in Konya suggested that locusts appeared because people had not observed the fast of Ramadan adequately. TNA-UK, FO 424/107, Stewart to Goschen, August 18, 1880.

[195] NHMA, Turkey 6217, WS Edmonds in Turkey, May 23, 1930.

[196] NHMA, Turkey 6217, Report on the Locust Invasion of Turkey in 1930, Communicated Personally by Süreyya Bey, London, November 1930.

[197] BCA, 30–10–0–0, 185/277/11, Economics Minister to Prime Minister, August 25, 1930.

[198] NHMA, Syria and Lebanon 6211 – 158, Consul in Aleppo, June 10, 1930.

[199] NHMA, Syria and Lebanon 6211 – 158, Geoffrey Meade to Foreign Office, Aleppo, June 6, 1930.

[200] NHMA, Syria and Lebanon 6211 – 158, E. C. Hole to Arthur Henderson, October 11, 1930.

1930, the government appointed a man named Abdul Wahab as the locust control officer.[201] He knew like few others the local geography, since he had spent years moving tobacco between Iraq and elsewhere to avoid government taxes. This is to say that the new locust control officer of northern Iraq was qualified for the job because he was a smuggler. It was a job that many across the Jazira fell into, particularly after the Great Depression, which caused huge price differences between Turkey with its tariffs and Iraq and Syria with their League of Nations-enforced Open Door.[202] Abdul Wahab's appointment also meant that the official charged with controlling locusts was someone who had spent much of his life evading state control by crossing borders too.

LOCUSTS WITHIN BORDERS?

Over the coming years, the fight against locusts continued, as did its intended and unintended consequences. Materiel destined for the fight against the insects disappeared.[203] Poison aimed at locusts hurt sheep and, in one case, was even accidentally used in the dessert known as *muhalabiyya*, a mistake that caused the death of two humans.[204] British officials in Iraq privately admitted that they sought an alternative to sodium arsenite, seeing as it was "very destructive to livestock."[205] Despite these concerns, from the perspective of locust experts in Syria, the employment of "toxic products" proved to be "the most economical procedure," and allowed them to envision "the possibility of the progressive annihilation of Moroccan locusts in Syria."[206] Turkish officials emphasized how chemicals protected them from the negligence of neighboring countries, allowing them to turn crawling threats to their crops into "very huge hills" of locust carcasses.[207] New hills rose in the Jazira, formed not from the accumulation of dirt over ancient ruins as in the case of *tall*, but rather from the dead bodies of locusts.

201 Lyon, *Kurds, Arabs and Britons*, 175.

202 Öztan, "The Great Depression and the Making of Turkish–Syrian Border, 1921–1939."

203 *Journal Officiel de la République Syrienne*, vol. 13, no. 8, April 30, 1931; vol. 13, no. 23, December 15, 1931.

204 "Risalat Dayr al-Zur," *Al-Qabas*, February 5, 1931, 2.

205 NHMA, Iraq 6213, Economic Advisory Council, March 3, 1932.

206 NHMA, Syria and Lebanon 6211 – 157, Joseph Atallah, Rapport sur la Campagne Anti-Acridienne dans l'Etat de Syrie en 1931, 9, 12.

207 "Urfa havalisinde çekirge mücadelesi devam ediyor," *Akşam*, June 26, 1932, 6.

No matter the confidence chemicals generated, the same kind of certainty of purpose did not emerge from international cooperation on the matter. Tensions had simmered within the International Bureau of Intelligence on Locusts ever since its inception, with Anglo-French rivalries playing out at the expense of coordination across borders.[208] The Soviet Union apparently saw the body as evidence of imperialist anticommunist designs and used it as impetus for expanding its own locust-control efforts stretching from the Caucasus to Iran, Afghanistan, and Xinjiang.[209] But such assessments exaggerated the International Bureau of Intelligence on Locusts' effectiveness, as the January 1932 meeting underscored. The representative of Turkey presided over the proceedings and, faced with the charge that his country had not paid its dues for the previous year, insisted that the check was in the mail.[210] Meanwhile, the Iraqi representative circulated a letter from the French high commissioner of Syria and Lebanon that called for greater cooperation on locating reservoirs of locust eggs, sharing supplies, and assessing what to do when private property was destroyed in the course of the fight against locusts.[211] The representative of Palestine received the letter and was supportive, as was the representative of Turkey. However, the representative of Syria found himself in the awkward position of stating that he had "not received communication of this letter by his own government." In an absurd turn, the meeting minutes detail how the Syrian representative explained he "cannot therefore formulate an opinion in the name of his government on the suggestions contained in this circular." The suggestions contained in the circular were also, of course, the opinions of his government. In sum, while almost all states agreed on the importance of cooperation in order to manage an insect that easily flew across borders, their practical abilities to do so were quite limited, at least in the realm of this international body. Nevertheless, locust control remained an international issue, even as the political status of locusts' range changed – albeit with little fanfare – when Iraq gained independence and membership in the League of Nations in October of 1932, although, of course, Britain retained considerable influence.[212]

[208] TNA-UK, FO 684/5, E. C. Hole to Arthur Henderson, January 14, 1931.

[209] Forestier-Peyrat, "Fighting Locusts Together," 294, 321.

[210] NHMA, Syria and Lebanon 6211 – 158, Minutes of the Fourth Meeting, January 23, 1932.

[211] NHMA, Syria and Lebanon 6211 – 158, Minutes of the Fifth Meeting, January 23, 1932.

[212] Sluglett, *Britain in Iraq*, 155–156.

Amidst these changes, the International Bureau mobilized to facilitate scientific assessments of the region's dilemma by the world's foremost locust expert. In 1932, it arranged for the visit of Boris Uvarov. As a boy, Uvarov had collected bugs in Uralsk, located in what was then the Russian Empire, later the Soviet Union, and now Kazakhstan, before becoming an avid member of the Russian Entomological Society while studying biology in St. Petersburg.[213] He went on to supervise control of the very same species of locust afflicting the Jazira – the Moroccan locust – in Tbilisi and Stavropol, where he witnessed his own diligent locust-control efforts go to waste when – as so many Ottoman officials had claimed before him – a swarm from elsewhere swooped in.[214] In Tblisi after the end of World War I, Uvarov found himself in dire straits, selling homemade pies on the street before going off to lecture on entomology at the university.[215] He left the city and its plane trees in 1920 when he received an invitation to work in London under the auspices of the Imperial Bureau of Entomology (which would later be renamed the Imperial Institute of Entomology).[216] In the process, he developed one of the most important theories in locust biology of the twentieth century. Whereas previously locusts in swarm state had been seen as a separate species from those in solitary states, Uvarov revealed that they were actually the same species, with differences in appearance and behavior owing to what Uvarov called "phase transformation."[217] From the Imperial Bureau of Entomology's base in the British Museum, he helped coordinate locust-control policies all around the world, while also finding time to correspond, for example, with a Norwich resident curious about the bug that appeared in the bunch of bananas she purchased at the greengrocer. Uvarov would publish some 475 articles in English and Russian and receive such honors as a knighthood. His death in 1970 was met by condolence letters from places as far-flung as Ethiopia and Newfoundland, as well as one hailing him as "world's greatest authority on Locusts" from none other than Y. Ramachandra Rao, the Indian locust expert who had joined the British in their early years in Iraq.[218]

[213] Waloff and Popov, "Sir Boris Uvarov," 2–3.

[214] International Institute of Agriculture, *La Lutte contre les sauterelles dans les divers pays*, x.

[215] Waloff and Popov, "Sir Boris Uvarov," 4.

[216] Uvarov is sometimes referred to as a "White Russian," and the Bolsheviks executed his brother. But Worboys suggests that Uvarov's departure was more likely hastened by Georgian nationalism. Worboys, "Imperial Entomology," 31, 37.

[217] Waloff and Popov, "Sir Boris Uvarov," 10.

[218] TNA-UK, AY 20/16, Condolences upon the death of Uvarov, 1970.

In 1932, however, Uvarov was bound for what were described as "the main breeding areas of the Moroccan Locust ... i.e. Turkey, Syria and Iraq."[219] Yet his efforts to study the Moroccan locust there ran aground in a manner evocative of the Jazira's political ecology, which was simultaneously connected and divided. Uvarov's visit was supported by the International Bureau of Intelligence on Locusts. But though Iraq, Syria, and Turkey were all part of the organization, the body could only facilitate Uvarov's journey to Iraq and Syria, leaving Turkey outside of the purview of his research.[220] Nevertheless, Uvarov's observations were remarkable registers of the region's changing ecology and the continued challenge of controlling an insect that emerged from places where humans were not. Uvarov's consistent message was that the key to controlling locusts was not what he called "defensive" actions taken in the event of an invasion, but rather knowledge of permanent locust habitats, which he called "reservations."[221] Through proper management of these spaces, he believed that locusts could be managed in a way that obviated the elaborate systems of sodium arsenate and zinc fencing that had come to mean springtime in the Jazira just as sure as the region's bountiful greenery.

An insistence on the region's ecological unity threaded throughout Uvarov's observations. Like many, he deemed the term "desert" a "misnomer" and instead referred to "the Jezireh steppe" that transcended political borders. [222] "There is little marked change in the type of country when the frontier is crossed," he wrote, with the exception of "more cultivation in the Syrian Jezireh from the Turco-Syrian boundary to about 40–50 kilometres southwards." In other words, cultivation extended into Syria from the border rather than out toward it. The lines drawn on maps because of a railway line had become reinforced by fields of wheat.

All across the unified ecology, Uvarov moreover saw the region's distinctive *tall*, and he speculated that they served an important purpose for locusts. The "round hillocks" were typically some "30–50 feet above the general level of the plain."[223] They were modest, but they were also precisely where locusts laid eggs, as Uvarov explained, "for the simple reason that they are the only 'islands' with sufficiently hard and compact

[219] Uvarov, "Ecology of the Moroccan Locust," 407.

[220] Uvarov had visited Izmir the year before. Özbilge, *Çekirgeler, Kürtler ve Devlet*, 71–72; BCA, 30–18–1–2, 20/39/15, Decree on Uvarov's visit signed by Mustafa Kemal, June 7, 1931.

[221] Uvarov, "Ecology of the Moroccan Locust," 418. [222] Ibid., 408, 412.

[223] Ibid., 412.

soil in the midst of vast areas with soft soils avoided by the locust."[224] Particularly in the cultivated regions stretching south of the Turkish–Syrian border, these were the only areas "that remain uncultivated" and, as a result "egg-laying takes place with considerable regularity." Because of the presence of these elevations in the midst of cultivated regions, Uvarov insisted that the dynamic posed "an exceptionally great danger to the rich grain producing areas of Syria." To prevent their continued use as a stronghold of locust egg-laying behind enemy lines, Uvarov called for the *tall* to be plowed, meaning that the uncultivated dirt that had stoked dreams of agrarian plenty in the Jazira in the first place would itself come under the plow.

But alongside his specific advice, Uvarov's main call was for a map of the "natural region" of the Jazira in a way not limited "by political or administrative boundaries."[225] Uvarov was thus one more person calling for attention to "natural" borders in the region. His message was, of course, different than those who used "natural" or "artificial" to describe or decry borders. But nevertheless, Uvarov restated a dilemma that had haunted state officials in the Jazira for many decades. How were they to see its connections? Like Ottoman provincial governors who blamed their locust invasions on provinces further east, entomologists too might be harmed by state boundaries that divided up their vision of the ecologically coherent space. In fact, Uvarov's own study had been hindered in this way, given that he was not allowed to visit Turkey.

Around the same time as Uvarov's trip, Alexander Eig of the Hebrew University of Jerusalem made a trip to the Jazira, and his findings offered an even more specific understanding of locust reservoirs. A native of Minsk and a graduate of the Mikveh Agricultural School in Mandate Palestine, Eig was also the founder of what would eventually become known as the Botanic Garden of Israel, along with Michael Zohary.[226] Eig's trip was limited to Iraq, and specifically focused on Mosul. Like Uvarov, Eig suggested that the insects were hitching a ride on the ruins of ancient human infrastructures, "concentrated almost exclusively on the chains of hills and small mountains or small 'tels' which run through here

[224] Later studies disputed this claim. Bodenheimer, *Studies on the Ecology and Control of the Moroccan Locust*, 31–32.

[225] Uvarov, "Ecology of the Moroccan Locust," 417.

[226] Zohary's magnum opus was *Geobotanical Foundations of the Middle East*, an effort to build on that of Alexander von Humboldt to catalog the "phytogeographical regions" of the Middle East, a task which required, in his words, "reliable boundary lines." Zohary, *Geobotanical Foundations*, 77–78.

and there the immense plains."[227] More specifically, Eig found that reservations of locusts often coincided with the presence of three plant species, among them bluegrass (*Poa bulbosa*), buttercups (*Ranunculus asiaticus*), and another grass (*Carex stenophylla*).[228] Eig wrote that in Mosul he found no reserves of locusts without this triumvirate of plants. In other words, the group of vegetation might act as a gloss for locust presence. Reflecting on the fact that these flowers bloomed at almost precisely the time that locusts appeared, Eig wondered, "What are the natural conditions of locust development cycle outside of the artificial conditions which man has created through fields of wheat?"[229] He suspected that locusts actually preferred to consume this group of plants rather than cultivated grains, and only moved toward the cultivated lands because they had "replaced the natural vegetation."[230] Yet Eig's conceptual framework of the region was also illuminating. In viewing wheat as artificial and the blend of grasses and buttercups natural, he ignored the way that nomadic pastoralism likely made the landscape hospitable for locusts, too. Indeed, buttercups – which Eig had speculated was locusts' favorite food – had been among the vaunted saddlebag of vegetation that allegedly attracted the Shammar to the Jazira in the first place.

As scientists zeroed in on the geographical and botanical aspects of locust invasions, the movement of locusts suddenly changed. For decades they had crossed human borders. But in 1933, it seemed that they respected borders. The unexpected event took place in the wake of an especially harsh winter.[231] Nomads migrated west of Aleppo in search of pasture, and their camels were such a pest to Armenians there that they referred to the nomads' dromedaries as "the giant locusts" in Armenian.[232] And on top of the metaphorical locusts, actual locusts also struck once again in 1933. The 1933 locust invasions involved some 969 hectares filled with locust eggs in Turkey, 3,453 in Syria, and 22,430 dunams in Iraq.[233] Even in this chemical age, efforts at locust control involved typically large amounts of labor if reports are to be believed (761 day laborers in

[227] Eig, "Ecologie du criquet marocain en Iraq," 298. [228] Ibid., 296. [229] Ibid., 301.
[230] Ibid., 302.
[231] "Ghadab al-sama' ʿala al-ard," *Al-Qabas*, January 10, 1933, 3; "Zafarat fi wadi al-Furat," *Al-Qabas*, January 12, 1933, 1; "Juʿ. ʿAtash. Bard. Talaf." *Al-Qabas*, January 16, 1933, 3.
[232] "Trucks Replace Camels for Desert Transport; Nomad Breeders Suffer and Animals Starve," *New York Times*, December 25, 1932, p. E7.
[233] NHMA, Syria and Lebanon 6211 – 158, S. Hassibi, "General Report on the Anti-Locust Campaign 1933 in the States Adhering to the International Office of Intelligence on Locusts of Damascus," November 10, 1933.

Mosul, and 1,100 and 667 workers at Turkey's Kızıltepe and Derik, respectively). Nonhumans, too, figured into the fight against the insects, as storks "in considerable flocks" appeared during the invasions and "forced" the locusts "to change the direction of their march." While there were familiar concerns that the locusts would make use of "unpopulated steppes" (*steppiques dépeuplées*) across the region, there was also a new wrinkle to the fight against the locusts. "It was noticed," explained the report on the year's efforts, "that the locusts that hatched in the territory of the Turkish Republic have not crossed the Iraqi–Syrian frontiers, and inversely, the locusts that hatched in the territories of the bordering states have similarly not crossed the Turkish border." In other words, locusts – so long a creature that had unknowingly taken advantage of the Jazira's divided political ecology – suddenly seemed to respect borders that they had previously crossed.[234] Historian Cyrus Schayegh has referred to the "Ottoman twilight" that ensued after the end of the empire, as people and places remained very much connected.[235] The new-found confidence with respect to the locust gestured toward a material aspect of night falling on that Ottoman twilight.

It was perhaps fitting given this unprecedented dynamic that 1933 also witnessed the decisive breakdown of cooperative efforts between these states. The tribulations of international cooperation on locust matters emerged perhaps most poignantly in correspondence between the Turkish entomologist Mehmed Süreyya – who had authored *Haşerat-ı Ziraiye Mecmuası* before working with Bücher and the German entomologists during World War I and translating their reports on the World War I locust invasion into Turkish – and the Russian emigré at the head of the Imperial Institute of Entomology in Britain, Boris Uvarov. Through the years, an entomological and epistolary intimacy emerged between the men. Süreyya wrote that he appreciated finding a "kind letter" and "bottles of identified insects" from Uvarov waiting on his desk when he returned from a trip.[236] In another letter, he promised that should Uvarov ever visit Turkey, Süreyya would gladly arrange for a "house at a good price" on the Bosporus in Istanbul.[237]

[234] The Syrian report on the invasion declared more cynically that "the Moroccan locust is no longer to be feared" by cultivators in some parts of northeast Syria because so much land had been abandoned in response to locust invasions of previous years. S. Hassibi, "General Report on the Anti-Locust Campaign 1933 in the States Adhering to the International Office of Intelligence on Locusts of Damascus," November 10, 1933.

[235] Schayegh, *The Middle East and the Making of the Modern World*, 8.

[236] NHMA, Turkey 6217, Süreyya to Uvarov, June 3, 1929.

[237] NHMA, Turkey 6217, Süreyya to Uvarov, November 23, 1930.

By 1933, their closeness had reached such a level that Uvarov even named a species of locust after his Turkish counterpart, with a particular insect observed at Tuzlu Göl taking the scientific denomination *Sureyyaella bella*. Uvarov wrote, "It gives me a great pleasure to dedicate this remarkable and extremely pretty little insect to my friend Süreyya Bey and thus perpetuate his name in the history of entomology of Asia Minor, to which he is so largely contributing."[238]

But in the same year, the conditions for cooperation between states reached their lowest level. Uvarov complained of no progress on his efforts to send an entomologist to study the region of Urfa in Turkey.[239] Upon receiving good news of locust-control efforts in Iraq, Uvarov was still pessimistic, writing "at least one country adopted my suggestions."[240] Aside from Uvarov's own personal frustrations, there were also structural issues emerging regarding the international body formed to combat the shared threat of the locust. Süreyya apparently succeeded in keeping Iraq's allegiance to the pact. But with Palestine and Transjordan removing themselves from the agreement at the 1933 meeting in Beirut, so too did Syria.[241] The defection of Syria was a devastating blow, not least because the headquarters of the organization were located in Damascus. While conveying this disheartening news, Süreyya also carried a more personal message to Uvarov: "I think of you often. I look at your photograph. I wait [*sic*] the occasion to go to London." And then, in French, "*Hélase.*"

"IT SEEMS THAT EACH COUNTRY WILL WORK FOR ITSELF"

As the possibility and perhaps need for international cooperation on matters of locusts disappeared, another group began to move across the borders of Iraq and Syria. In August of 1933, violence broke out between demobilized Assyrian levies and the Iraqi army near the border with Syria, and it quickly gave way to what is known as the Simele massacre in the Assyrian villages of the plains surrounding Mosul.[242] The Assyrians understood their own heritage as connected to the civilizations of antiquity, the relics of which were being unearthed by archaeologists in

238 Uvarov, "Studies in Orthoptera of Turkey, Iraq and Syria," 57.

239 NHMA, Turkey 6217, Uvarov to Süreyya, April 30, 1933.

240 NHMA, Turkey 6217, Uvarov to Süreyya, October 30, 1933.

241 NHMA, Turkey 6217, Süreyya to Uvarov, December 14, 1933.

242 Donabed, *Reforging a Forgotten History*; al-Husri, "The Assyrian Affair of 1933 (I)," "The Assyrian Affair of 1933 (II)"; Zubaida, "Contested Nations: Iraq and the Assyrians."

the region.[243] They had suffered genocide during World War I, particularly in Diyarbekir province, with violence whose impact prompted Yaşar Kemal to compare Assyrian orphans to locusts in their ravenous deprivation.[244] Many Assyrians decamped to Iran and subsequently Iraq, where large numbers served in military units commanded by the British. As one British official explained, the past and present appeared to come together as the Assyrian levies protected British police stations established on the hills upon which the Assyrian empire had built forts millennia before.[245] Their role in the modern world in relation to the British, however, did little to endear them to some Arab nationalists who viewed the community at large as colonial compradors.[246] Upon Iraq's independence from Britain, questions remained with respect to how the former colonial troops would relate to the emerging Iraqi military. It was against this backdrop that the Simele massacre occurred, and it spurred the flight of thousands of Assyrians across the border into Syria and an international effort to find a home elsewhere for the refugees.

The prospect of settling the Assyrians invited no small number of schemes. Informed by the premise that Assyrians, as Christians, were white, the League of Nations considered sending them to Brazil, British Guiana, and Niger as part of what historian Laura Robson has called "the transfer solution" and its tenet that "empires . . . had both the capacity and the incentive to dole out territory for settler colonial purposes."[247] Yet these new landscapes were of course different than the environments the Assyrians would be leaving behind. Of the Brazil plan, a British official advised that sometimes these differences could be positive. He wrote, "The Assyrians should have explained to them that . . . there are no plagues in the country, such as the locusts of Iraq."[248] Thus locusts loomed even in discussions of Assyrian settlement far afield of the Jazira.

While international organizations considered these questions and dispatched missions to far-flung places, the Assyrians had to go somewhere. And they found their way to the place whose lack of population had for so long made it difficult to govern, as they settled in the Jazira on portions of

[243] Hanoosh, *The Chaldeans*, 35, 52, 66–67.

[244] Gaunt, *Massacres, Resistance, Protectors*; Gaunt, Atto, and Barthoma, eds., *Let Them Not Return*; Kaiser, *The Extermination of Armenians in the Diarbekir Region*.

[245] Browne, "The Assyrians: A Debt of Honour," 431.

[246] Pedersen, *The Guardians*, 280. [247] Robson, *States of Separation*, 103.

[248] TNA, FO 406/72, Report by Brigadier General Browne on the Settlement of the Assyrians on the Lands in Parana of Parana Plantations, Note by Brigadier-General Browne, March 26, 1934.

the Khabur River south of Ras al-Ayn. The French accepted this state of affairs, given their policy of using "non-Turkish" and "non-Arab" minorities as a "counterweight" to protect their administration from nationalist challenge throughout the mandate.[249] And the presence of the population in a place that had for so long been home to mobile human and locust populations also proved helpful. It is in this sense that historian Benjamin White has insisted that "the modern state of Syria was formed around and against refugees."[250] The Assyrian presence allowed the French to consolidate their shaky rule in the Jazira, making "the upper Khabur 'Syrian,' as it had not been before."[251] Yet their placement on the Khabur also echoed settlements of the Jazira's past. The Assyrians were just south of Ras al-Ayn, where Chechens had settled in the 1860s. As with the Chechen settlement, the Assyrians too became targets of nomads, as in August of 1933 when the Shammar – having lost many of their sheep the previous winter and spring – raided their villages.[252]

With these echoes of the past with respect to settlement of the Jazira by refugees, joint efforts to fight locusts further dwindled. Once again, the correspondence between entomological friends Uvarov and Süreyya underscored this point. In January of 1934, Uvarov fumed that the Syrian retreat from the International Locust Bureau was not surprising, given their incapacity for "understanding the value of scientific control of locusts."[253] Yet Uvarov also acknowledged that perhaps coordination was not as important as some believed, noting that "this control can be conducted by each country separately." In response, Süreyya did not seem so sure. To him, Syria's location made it "the pivot" of cooperative locust control, and without Syria, he feared little could be accomplished.[254] At the previous meeting of the body, Süreyya had requested a report from the director of locust control in Iraq, but he could not even manage to acquire a copy of it. With cooperation falling apart, Süreyya concluded, "It seems that each country will work on behalf of itself." At least in 1934, the approach did not fail. Syrian officials wrote that once again "no swarm had succeeded in crossing the borders of the neighboring states."[255]

[249] Tachjian, *La France en Cilicie et en Haute-Mésopotamie*, 17.
[250] White, "Refugees and the Definition of Syria," 143. [251] Ibid., 163.
[252] Stafford, *The Tragedy of the Assyrians*, 183.
[253] TNA-UK, AY 20/65, Uvarov to Süreyya, January 24, 1934.
[254] TNA-UK, AY 20/65, Süreyya to Uvarov, February 12, 1934.
[255] NHMA, Syria and Lebanon 6211 – 157, Rapport général sur la campagne anti-acridienne 1934 dans les états adhérents à l'Office International de Renseignements sur les Sauterelles de Damas, 1.

Meanwhile, data from Iraq's Plant Protection Department on locust damage amounted to "hardly anything to put on the graph."[256] In a mark perhaps of the way that technology had allowed states to not worry about cooperation, in Turkey, even the humble local newspaper of Urfa carried front-page news of "a new formula for killing locusts" devised by "English engineers" that had "successful results."[257]

The twin developments of refugee settlement and locust destruction also connected to questions of nomads. French policy in the region was, of course, not monolithic.[258] Particularly in the Contrôle Bedouin, some were staunchly committed to protecting nomadism, and others in the intelligence services supported agricultural settlement through immigration from Turkey, sometimes in ways at odds with the more cautious high commission.[259] In the summer of 1934, these dynamics came together, however, as one French officer in Syria called for a renewed effort to use refugees as a way of catalyzing nomadic settlement. He suggested that the French actively work to attract "Kurdish, Milli, Assyro-Chaldean, Armenian, and other emigrants" who could well be a "source of rejuvenation" since they were "eager for land" and the Jazira had plenty of it.[260] The dynamic had already had an effect, as the French deemed vast confederations like the Shammar as "nomads" in a cultural and legal sense but no longer "dignified of the full definition of the word."[261] One example of the transformation was that the Shammar had fewer camels, and they even were acquiring "mounds to install Armenian or Kurdish families or the most miserable among them, who no longer dream of anything but the time of the harvest." This is to say that even the Shammar were grabbing on to the *tall* that had been so distinctive as part of the life of the Jazira, both in fueling the dreams of its rejuvenation and enabling its dystopian realities as reservoirs of locust propagation and burial grounds for massacred Armenians. Not all welcomed these policies, with some of the romantically inclined among the French believing that

256 NHMA, Iraq 6213, *Motor Transport London*, July 14, 1934.

257 "Çekirgeleri Öldürmek İçin Yeni bir Keşif," *Yenilik*, July 12, 1934.

258 Büssow, "Negotiating the Future of a Bedouin Polity"; Neep, *Occupying Syria under the French Mandate*, 165–198; Sakatni, "From Camel to Truck?"; Stocker, "The 'Camel Dispute.'"

259 Altuğ, "Sectarianism in the Syrian Jazira," 189–190; Tejel Gorgas, "The Terrier Plan and the Emergence of a Kurdish Policy under the French Mandate in Syria," 93.

260 CADN, 1SL/1/V/991, Rapport du Lieutenant Leroya, head of Mouvance de Deir ez Zor, Summer 1934.

261 CADN, 1SL/1/V/911, "L'evolution des Populations Nomades et Semi-Sedentaires de la Haute Djezireh et l'actions des Services Speciaux," November 12, 1934.

they ought to establish "reserves of nomadism" to protect groups like the Shammar, akin to the way protection was afforded to "certain animals threatened with extinction." But change was occurring nonetheless, and one of the unspoken drivers of this dynamic was the agricultural infrastructure of pesticides that enabled a political economy of refugee resettlement and grain cultivation.

"MILES FROM NOWHERE"

By the mid-1930s, nationalism began to shape the questions of locust control and Assyrian refugees in the Jazira in decided ways. At the same time, the question of locusts across the Jazira became entangled with questions of nation on different scales. As of 1935, the International Bureau of Intelligence on Locusts only retained the membership of Iraq, Lebanon, and Turkey.[262] The final letter between Uvarov and Süreyya in the Natural History Museum Archives involved Süreyya asking if the two would meet at the fifth international conference on locusts after missing each other at meetings in Madrid and Cairo.[263] It is unclear whether they did or did not.

In Iraq, the power of nationalism asserted itself with respect to locust control, when a locust expert recommended to the ministry of agriculture was rejected due to his colonial baggage. As one Iraqi official explained the problem in a letter to Uvarov, "You know how they have a different idea of the East and the way they should be treated."[264] Uvarov advised later that there were many "who lost their jobs in Germany for racial reasons" who might be qualified for the role.[265] Nations and nationalism had changed both the Jazira and Europe, and they had consequences for the nature of locust control, whether as a way of celebrating national modernity or of leaving a colonial past behind.

As plans to settle the Assyrians abroad ran up against problems of expense and xenophobic opposition, the French endeavored to resettle them within Syria. But these plans, too, catalyzed opposition. The French initially envisioned a marshy plot of land in western Syria known as the Ghab, near the nationalist stronghold of Hama. They planned to both settle the Assyrians

[262] NHMA, Turkey 6217, Süreyya to Uvarov, July 24, 1935.

[263] NHMA, Turkey 6217, Süreyya to Uvarov, July 27, 1936, March 26, 1938. For more on the emerging international conferences on locusts, see Worboys, "Imperial Entomology."

[264] TNA-UK, AY 11/7, Director of Agriculture to B. P. Uvarov, August 22, 1936.

[265] TNA-UK, AY 20/65, Uvarov to M. Al Radi, October 2, 1936; TNA-UK, AY 11/7, Guy Marshall to Director of Agriculture, October 6, 1936.

there and reclaim land from the swamp.[266] Indeed, it was on these grounds that officials denied Assyrian leaders their request to visit and inspect the Ghab prior to settlement; any "visit would serve no useful purpose," one official explained, because "it is not proposed to settle the Assyrians in the Ghab area in its present state." To convey the extent of environmental transformation in terms that officials in Geneva with the League of Nations would understand, one map projected the settlement's total amount of territory onto Lake Geneva, the cerulean expanse at the foot of the Alps within sight of League headquarters.[267] Arab nationalists took little time to perceive the colonial purposes behind the scheme. Najib al-Rayyis, who had cheered on locust-destruction efforts in the Jazira in the pages of the Damascus daily *Al-Qabas*, wondered why, if the Assyrians had been "colonial soldiers," they could not be settled somewhere else in the British Empire, such as India, Australia, or South Africa.[268] Nationalist leader Fakhri al-Barudi took a different tack, worrying that an Assyrian state would be created in their midst much like the emerging Zionist state in Palestine to the south. This observation prompted, apparently, a letter from a young Assyrian to al-Barudi in which he declared, "You are Arabs, so go back to the Hijaz!"[269] The Assyrian continued, insisting that Barudi and his kind were "lower than flies" and that Assyrian presence was justified given their ancient roots. As the young man wrote, "The Jazira is our property alone." With Assyrians, too, movement – and its end – prompted new claims about proper alignment of territory and ethnic identity.

The Assyrians would come to remain on the Khabur rather than far to the west in the Ghab, thanks to the interface between colonial and nationalist politics. Schemes for the resettlement of the Assyrians in the Ghab persisted until early 1936, when officials made plans to move Assyrians to western Syria via the railway that etched the border between Syria and Turkey (the Turkish government stipulated that the trains would only carry Assyrians, that the trains would make no stops along the way, and that the trains carrying animals and goods would be completely sealed).[270] Yet the plan to resettle the Assyrians in western Syria fell apart as funding

[266] UNAG, R3940/4/11757/20578, FP Walters, Under Secretary-General, Director of the Political Section to Local Commission in Iraq, December 7, 1935.

[267] UNAG, R5166/13/22354/1265, *The Settlement of the Assyrians: A Work of Humanity and Appeasement* (Geneva, 1935), 28.

[268] Al-Rayyis, *Al-A'mal al-Mukhtara*, vol. 1, 314.

[269] "Hal al-Bilad al-Suriyya Watan lil-Ashuriyin?" *Al-Qabas*, May 16, 1936, 1.

[270] UNAG, R3940/4/11757/22060, Iraqi Minister for Foreign Affairs to the Chairman of the Trustee Board for the Settlement of the Assyrians, Beirut, July 5, 1936.

disappeared along with French political will, when the socialist government of Leon Blum in Paris began negotiating a treaty for Syria's independence. One of the "reputed advantages" of the Ghab scheme, in the words of one administrator, was that it was located in the ʿAlawite state, one of several distinct sectarian states that the French established during the course of their rule in Syria.[271] Thanks to the region's special administrative status, many believed that it "would remain in French hands" even after independence. But Syrian nationalists – tired of the denial of sovereignty for so long – pushed for unification, meaning the Ghab scheme would fall outside of any possible French protectorate. As a result, League of Nations officials were forced to acknowledge that perhaps there was a benefit to keeping the Assyrians on the Khabur. It was so remote, one official wrote, that it was "miles from nowhere."[272]

The Khabur may well have been remote, but Syrian nationalist officials still went there, and they carried with them yet another vision of transforming the Jazira. Fakhri al-Barudi and others visited the settlements in March of 1936. According to a letter sent by the Assyrians of the Jazira to the secretary general of the League of Nations, the message in the visit was, "Voila, we have obtained our independence like Iraq."[273] The nationalists added, according to the Assyrians, that it was necessary for the Assyrians to fight colonial machinations as well as to assimilate with the broader population by learning Arabic. The nationalists also called for agricultural development. "In a few years," the visitors from Damascus remarked, "we are going to build dams, such that this will become a paradise on earth." The Assyrians conveyed their lack of trust in what they called "the promises of a fanatical government" and fear of living "under the domination of the Muslim flag." But the promises put forward by the nationalists also pointed to the continued perception of environmental transformation of the Jazira as a means of solving political division between various groups. It also foreshadowed the massive environmental engineering projects yet to come.

As the Assyrians' temporary residence on the Khabur became permanent, officials scrambled to ensure the refugees' health in a region renowned for malaria. Both miasmatic and bacteriological understandings of malaria presented elevation as a means of prophylaxis, by increasing the distance from bad air – according to the former explanation – or parasite-bearing

[271] UNAG, R3940/R3940/4/11757/22099/Jacket1, Barcenas to Hill, March 6, 1936.
[272] UNAG, R3940/R3940/4/11757/22099/Jacket1, Barcenas to Hill, March 6, 1936.
[273] UNAG, R3940/R3940/4/11757/22099/Jacket1, Assyrians in the Jazira at al-Hasakah to Secretary General of the League of Nations, March 8, 1936.

mosquitoes – according to the latter. The names of the Assyrian villages formed in this effort bear the imprint of the attempt to raise the settlements out of the reach of malaria: Tell Tchémé, Tell Oumrane, Tell Tamer, Tell Maghas, and a dozen others (Figure 19). As in the case of Armenian

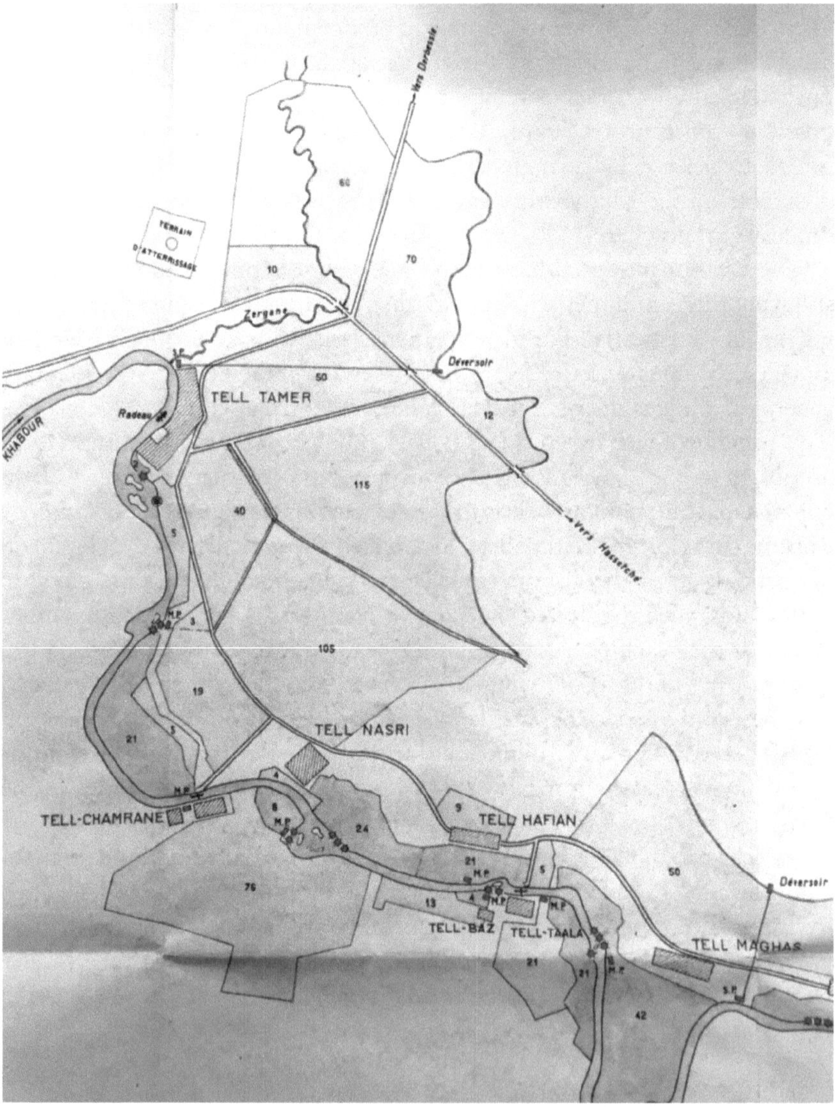

FIGURE 19 Map of Assyrian settlements, Khabur River. UNAG, R3942/4/11757/30870. Courtesy of United Nations Archive at Geneva

villages, in deference to the settlements' presence on the distinctive elevations of the Jazira, each of the villages' names began with the word *tall*. The term was so omnipresent and thus meaningless as a term of differentiation that some statistical tables on the settlements would leave out the term altogether, and refer to the villages simply by the second halves of their names.[274] There were some 8,800 Assyrians settled along the Khabur by 1937.[275] The post-Ottoman settlement scheme thus brought together the people whom Yaşar Kemal likened to locusts for their miserable hunger in the wake of World War I with the places that entomologists such as Uvarov revealed had incubated those very same locust populations.

Yet the insects would not do so for much longer. Though the Jazira remained a space apart in the nationalist imaginary, it was increasingly true that locusts did not make it so. One report declared the Moroccan locust "exterminated" and thus "no longer any danger."[276] In April of 1937, the Damascus newspaper *Al-Qabas* even mockingly complained of reports that Syria had been invited to a conference on locusts in Beirut.[277] *Al-Qabas* called instead for a conference to be held on the question of the cereal pest known as the *sunn* bug (*Eurygaster integriceps*) or of voles. The newspaper made this joke not only in light of the fact that the country seemed to suffer more from these scourges, but also, in another quip, based on the fact that the locust would continue to avoid the country because it would not "find anything in it to eat!!"

CRESCENT, CROSS, AND WHEAT

Thus by the late 1930s largely gone was the insect that had shaped the intricate choreography of nomads and, by virtue of shaping the environment so profoundly, had also greatly influenced the suffering and survival of Armenians. And largely gone too was the insect's metaphorical power, the way various moving human groups on the outside had been compared to locusts. Yet it was in this moment of liberation from the locusts and the mobility they catalyzed that the Jazira witnessed an unprecedented episode of separatism, one fertilized by colonial encouragement, nationalist agitation, and the various groups of people in the Jazira who had all been

[274] UNAG, R3940/4/11757/22099/Jacket1, État des Assyriens Installés sur le Khabour au Novembre 1936, Cap. Vuilloud.
[275] Tachjian, *La France en Cilicie et en Haute-Mésopotamie*, 327.
[276] Mudiriyat al-Zira'a, *Al-Jarad*, 3–4.
[277] "Mu'tamar al-Jarrad," *Al-Qabas*, April 11, 1937, 2.

compared to locusts in the past. Though many had seen Syria's independence as imminent in 1936, it proved stillborn. The treaty negotiated that year by the French government of Leon Blum to arrange for Syrian independence had provided for a three-year probationary period, giving French colonial interests time to regroup and torpedo the treaty.[278] Meanwhile, various regions of Syria that had enjoyed some degree of autonomy more stridently made their case for this kind of protection, and in places like the ʿAlawite coast and Jabal Druze, some flew the French flag – often with French encouragement – as a symbol of their opposition to nationalist unification.[279]

Calls for a French protectorate in the Jazira emerged too. Supporters of the policy included Kurds led by Mahmud Bey of the Millî, son of Ibrahim Pasha, along with Assyrians and Armenians in towns such as Qamishli (Figure 20). Located directly across the Turkish border from Nusaybin,

FIGURE 20 "Vive la France" and, in Arabic, "Long live the people of the Jazira." Qamishli, 1930s. Haute-Djezireh 45–2. Courtesy of Archives de la Province dominicaine de France

[278] Khoury, *Syria and the French Mandate*, 486. [279] Ibid., 487, 533.

Qamishli was, in historian Philip Khoury's words, "literally a creation of the mandate," supported by the French as a way of siphoning commerce and people from southeastern Turkey into Syria.[280] As historian Benjamin White has memorably demonstrated, the border between the two towns was permeable to the point of being nonexistent.[281] Opposing the alliance of Kurds, Armenians, and Assyrians in Jazira boomtowns were those who supported national unity with Damascus, led most notably by Daham al-Hadi, a chief of the Shammar. A great-grandson of Farhan, he had extensive landholdings in Nusaybin prior to the delimitation of the border.[282] When elected to the Syrian parliament to represent the Jazira in 1936, he circulated calling cards identifying himself as "the chief of the chiefs of the tribes of the Jazira."[283] Many of the people who supported national unity alongside Daham al-Hadi were Arabs and Muslims, but not all. The communist leader Khalid Bakdash – of Kurdish background – also dismissed the autonomist movement, later calling it "fascist."[284] Land undergirded these tensions, with the national government's apparent commitment to its supporters threatening to shake its opponents' grip on productive power through bestowing title to land and changing the flows of rivers.[285]

In other words, the tensions over a French protectorate brought together some of the crucial actors in the Jazira over the previous eighty years – Shammar, Millî, Armenians, Assyrians – all of whom had been, at one point or another, compared to locusts. The tensions would boil over in July of 1937, when the heat was described by one French missionary as "horrendous ... ideal weather for a revolution."[286] Violence broke out as disputes over parliamentary representation intersected with the nationalist–autonomist divisions.[287] Nationalism infused the rising tensions in the Jazira. Accounts in Damascus tarred the Kurdish–Christian coalition with the label of "refugees" from Turkey and suggested they ought to be sent back.[288]

[280] Ibid., 526. New Nusaybin had even been considered as a potential name. In a mark of the connections that persisted across the border, the kaymakam of Qamishli was Qaddur Bey, who previously had been kaymakam of Nusaybin, and whose own brother-in-law, ʿAbd al-Rizzaq Sahtana was a Kemalist deputy representing Mardin in Ankara. Velud, "Une experience d'administration régionale en Syrie durant le mandat français," 299, 301.

[281] White, *The Emergence of Minorities in the Middle East*, 101.

[282] Altuğ, "Sectarianism in the Syrian Jazira," 184; Williamson, "A Political History of the Shammar Jarba," 203.

[283] Khoury, *Syria and the French Mandate*, 529.

[284] Altuğ, "Sectarianism in the Syrian Jazira," 269. [285] Ibid., 261.

[286] APDF, Haute Djezireh 53, Journal of Fr. Thomas Bois, July 14, 1937.

[287] Khoury, *Syria and the French Mandate*, 530.

[288] Jalal al-Sayyid, "Al-Shaʿb fi al-Jazira," *Al-Qabas*, July 28, 1937, 1, 8.

In the charged atmosphere, on August 9, nomads affiliated with Daham al-Hadi and the nationalists attacked the village of ʿAmuda, looting its Christian quarter and killing over twenty of its inhabitants.[289] Another Shammar chief intervened at this point to stop the violence, but not before indiscriminate French reprisals caused more deaths in neighboring villages. Members of the Shammar – with Daham al-Hadi at the top of the list – quickly issued a telegram that was published in *Al-Qabas* distancing themselves from the violence.[290] The aftermath of the conflict too was refracted through the lens of nationalism. The Damascus press published translations of French accounts of the region, insisting that these revealed a "conspiracy" was at work to remove the Jazira from Syrian control.[291] Meanwhile, the Turkish press in nearby Mardin – where the flames of ʿAmuda were visible – took a different approach. They described the fighting instead as "between Armenians and Arabs."[292]

As the autonomist movement made demands on the French, they invoked the region's exceptional status in terms of its environmental transformation of the previous years. Michel Dome – the mayor of Qamishli and an Armenian Catholic born near Mardin – repeatedly made this case. He told a French representative in August of 1937 that the French ought to remain in the Jazira, given how French soldiers had "out of this desert made populated cities and fertile lands."[293] Dome was not the only one. The Syrian Catholic priest Hanna Hebbé in al-Hasakah too beseeched the French to remain in "a desert country" that became, under their rule, "the granary of Syria."[294] In another variation on the theme, an earlier petition from Kurds of the Jazira to mandate officials predicted that if France remained in the Jazira, they would continue to attract Kurds from across the border in Turkey, who would come and "develop [*mettre en valeur*] this region" that was "still desert."[295] Historian Jordi Tejel has called this particular argument and its attendant political grouping "steppe

[289] Khoury, *Syria and the French Mandate*, 531.

[290] "Barqiya min al-Jazira," *Al-Qabas*, August 10, 1937, 4.

[291] "Kayf yaktabun ʿan hawadith al-Jazira?" *Al-Qabas*, August 12, 1937, 1, 8.

[292] "Suriyeden hudutlarımız iltica edenler 1000 kişiye yaklaşıyor," *Ulus Sesi*, August 16, 1937, 1.

[293] APDF, Haute-Djézireh 45–1, Entretien de Michel Dome, Président de la Municipalité de Kamechlié avec le Comte Ostrorog, August 4, 1937.

[294] APDF, Haute-Djézireh 45–1, Mgr Hebbé, évêque des syriens catholiques, "Mon crime," July 1937.

[295] APDF, Haute-Djézireh 45–1, Note présentée par les Kurdes de la Djézireh à la Puissance Mandataire, 1930.

nationalism."[296] Their demands coalesced around the transformation of the steppe into cultivated lands. Indeed, the flag associated with the autonomist movement consisted of the French tricolor and, in its center, a crescent, cross, and two spikes of wheat, symbolizing the centrality of cultivation and French colonial rule to the multireligious coalition emerging in the region.[297]

The possibility of separatism in the Jazira derived not just from the violent nationalisms that pushed people out of countries and the cynical colonial policies that attracted refugees. It also owed its possibility to the chemical insecticides that over the previous decade had enabled states to control the region's agrarian fate in unprecedented ways. The annihilation of locusts alone did not prompt the agrarian transformation of the Jazira. It was instead the removal of the scourge alongside colonial military intervention, flows of capital, and changes in property regime. These factors together remade the Jazira such that the Jesuit Antoine Poidebard could marvel at how cultivation extended "as far as the eye can see in the valleys of the Khabur and Jaghjagh," flowing south from Nusaybin.[298] This geography helped enable the rise of the autonomist movement.

And the Jazira's status as potentially autonomous in turn led to talk of the annihilation of its people. Indeed, one of the more common reference points of residents of the Jazira in the late 1930s was the fear of massacre, seeing as so many of the region's inhabitants had survived the state violence accompanying imperial renewal and national birth. The group of Kurds that petitioned the high commissioner claimed that the nationalists did not "hide their intention of … the annihilation of elements belonging to non-Arab races."[299] The mayor of Qamishli Michel Dome echoed these sentiments, claiming that the Shammar chief Daham al-Hadi had suggested that if the nationalists could not handle the Christians, he would participate in "annihilating them all across the Jazira."[300] The rabbi of Qamishli took a different approach with respect to advocacy for his community of 2,000 Jews, almost all of whom had come from Nusaybin.[301] He beseeched the French to protect his community so they would not have to go to

[296] Tejel, *Syria's Kurds*, 33.
[297] Tachjian, *La France en Cilicie et en Haute-Mésopotamie*, 400 n. 120. [298] Ibid., 327.
[299] APDF, Haute-Djézireh 45–1, Note présentée par les Kurdes de la Djézireh à la Puissance Mandataire.
[300] APDF, Haute-Djézireh 45–1, Region de Hassetché: Rapport de Mgr Hebbé, July 25, 1937.
[301] Tachjian, *La France en Cilicie et en Haute-Mésopotamie*, 324.

Palestine to find safety there.[302] Of course, these figures had a real interest in keeping the French in the region, and they also had the very real memories of what they or their families had survived before making it to the Jazira, so it is difficult to know where fear and truth begin and end. But it is nevertheless noteworthy how once again the Jazira had become home to fears of violence. Massacres did not ultimately come to pass, nor did their dreams of a postindependence French protectorate in the region. Yet still the Jazira would inflect the histories of the various states that divided up the region in the coming years.

CONCLUSION

The post-Ottoman Jazira revealed both connection and disintegration. Trains running along the railway line that constituted the border between Turkey and Syria served fleetingly as fences stopping locust invasions, while zinc fencing intended to thwart the swarms would cross interstate borders rather than run alongside them. Refugees from the Republic of Turkey and Iraq found their way to the Syrian Jazira, where the French encouraged them to cultivate the land. Groups such as the Shammar became involved in cultivation too, as they acquired land and leased it to sharecroppers. By the 1930s, thanks to the power of chemical insecticides, the expansion of cultivation, and no small amount of luck, officials declared that locusts seemed in fact to be respecting national borders. Locusts would make appearances in the coming years, but for the rest of the twentieth century, the insect that had governed so many dynamics of the Jazira would be for the most part absent, exterminated or pushed to the margins, not unlike the various moving people with whom the locusts shared the Jazira.

While the various others of Turkish nationalism – Arabs, Armenians, Assyrians, and Kurds – contended with one another in northeast Syria in 1938, the Republic of Turkey took the fifteenth anniversary of its founding as an opportunity to celebrate both its ostensible ethnic homogeneity and the ecological transformation it had achieved. The pamphlets published for the occasion in Diyarbekir and Mardin provinces exhaustively listed the measures they had taken to remove the burden of the environment, including kilometers of zinc fencing constructed, tons of locust larva destroyed, dunhams of land plowed to prevent locust infestation, and even

[302] APDF, Haute-Djézireh 45–1, Rapport du chef de la communauté israélite de Kamechlié sur la révolte de Juillet, August 4, 1937.

the number of weasels and badgers killed.[303] A similar triumph over nature – or at least locusts – seemed apparent on the Syrian side of the border. In 1938, locusts did not mar the Assyrian settlements on the Khabur River. When one official complained of a confluence of nomads and drought, locusts loomed only as a specter, as the official deemed the destruction "even worse than the passage of locusts."[304]

In the absence of the insects that had for so long shaped the region and its people, new kinds of motion and nationalism took root. The pamphlet from Mardin celebrating Turkey's fifteenth anniversary offered a glimpse of this dynamic with a section heading that read "Mardin is Turkish" (*Mardin Türktür*), a maxim whose articulation only made sense as a reflection of profound insecurity about this fact.[305] As part of this effort to make border provinces such as Mardin sealed from Syria and presumably its many non-Turkish elements, Turkey banned its citizens from participating in seasonal agricultural work south of the border.[306] To make up for the loss, mandate officials arranged to have ʿAlawi workers from western Syria, indeed, the same region of the Ghab to which the Assyrians were supposed to be sent. "But," one official wrote, "in little time they returned, saying they were not able to work in such sweltering heat." The new circuits of labor migration pointed to a reorientation of the Jazira in accordance with it being part of Syria and disconnected from Turkey. But they also gestured to the continued exceptional nature of the Jazira, so fertile that it outpaced the local workforce yet also so hot that migrant laborers refused to work there.

303 "Diyarbakır," in Öztoprak et al., eds., *Cumhuriyet'in XV. Yılında Türkiye*, vol. 3, 1389; "Mardin: Cumhuriyetten Önce ve Sonra" in Öztoprak et al., *Cumhuriyet'in XV. Yılında Türkiye*, vol. 6, 3501–3502.

304 UNAG, R3940/4/11757/22099/Jacket2, Cuénod to Lisicky, March 23, 1937.

305 "Mardin: Cumhuriyetten Önce ve Sonra" in Öztoprak et al., *Cumhuriyet'in XV. Yılında Türkiye*, vol. 6, 3505.

306 UNAG, R3942/4/11757/30870/Jacket2, Cuénod to Celinski, May 30, 1939.

Conclusion

In 1953, a Syrian entomologist by the name of Rafeq Skaf ventured about fifty kilometers (thirty-one miles) east of the Syrian city of Raqqa.[1] There, on "an ancient hill with ruins," he saw "the population . . . copulating and lying." They were locusts. And the site at which they procreated was a *tall*, the ancient human infrastructure that gave life to swarm after swarm of insects in the Jazira. Skaf's observation of locusts belied any facile narrative of the total annihilation of the insects.[2] But his report also attested to change. For so long, locusts had appeared as if out of nowhere. Yet in Skaf's view, by 1953 there was a pattern to their range. Where there were nomads, there were locusts. As he put it, the locusts may as well have been considered a "permanent citizen" anywhere there was "grazing" or "almost permanent tents." The humans and insects had been metaphorically linked by the denigration of state officials and corporeally linked by arsenic compounds. Skaf connected them in terms of ecology and also, significantly, permanence, anathema to the mobility that haunted state efforts at control in the Jazira for so long. The use of "citizen" to refer to the locust, figurative though it may have been, also gestured to change. The locust was no longer a malevolent invader but rather a domesticated subject of states, one symbol of the political ecological transformations of the previous decades.

[1] NHMA, Syrian and Lebanon 6211 – 158, R. M. Skaf, "A Rapid View on the Moroccan Locust Problem in Syria," 1953.

[2] Indeed, one official described a zinc wall put up during the fight against the locusts in the 1940s as the "locust Maginot line," though it presumably had more success than its referent. Bodenheimer, *Studies on the Ecology and Control of the Moroccan Locust*, 118.

The change of locusts into citizens both helped cause and coincided with a number of other changes in the region. With the locusts went the mobile political geography that had deeply affected borders, environment, and identity in the region, leaving the Jazira transformed on a number of levels. It had changed from a region split between Ottoman provinces to one split between post-Ottoman nation-states. It had changed from a region of locusts and nomadic motion to one largely free of locusts and witness to rapidly expanding cultivation. It had changed from a region in which people's identities were defined by environment as much as language to one in which ethnicity and nationality – regardless of environment – prevailed as the basic political units.

The political ecology of the Jazira – a hybrid creation of borders and the desert – was the locus of power at the center of these changes. It was also the core of various attempts to remake the region and resist. While the Jazira largely did not appear on maps, many people were aware of the broader regional connections, and over the years an abbreviated list of people who explicitly used the term included Shammar shaykh ʿAbd al-Karim, Ottoman reformer Midhat Pasha, the empire's chief rabbi, numerous parliamentary deputies, Mark Sykes, and Boris Uvarov. Better borders – in the view of some – were key to managing the space, whether in its entirety or in fractured form. In the 1860s, the Ottoman state imagined how to use "lines of cordon" as a way of limiting the Shammar. They subsequently attempted to use the settlement of Chechen refugees as a means of building a bulwark of control against the desert. Finally, in 1871 they created the special administrative district of Zor, with the aim of encompassing both the desert environment and nomadic migrations within it. The entity catalyzed a revolt against borders, and the leader of the revolt – Shammar chief ʿAbd al-Karim – was executed. In his wake, the Shammar did not stop moving, but they did appear to respect the Ottoman provincial borders that had been established in the Jazira. As the Ottoman state shifted to relying on the Hamidiye brigades in the 1890s, borders emerged as a flash point for conflict once again. Provincial officials and nomadic leaders argued about who was where, with the northern border of Zor near Ras al-Ayn in particular becoming an axis of tension between the Shammar and the Millî. Far from wandering aimlessly across borders, nomadic pastoralists knew precisely where the borders were and adeptly used them to disappear, evade taxes, antagonize rivals, or otherwise do as they pleased. The district borders intended to contain nomadic motion served another purpose in the midst of World War I, when it became the deportation site of hundreds of thousands of Armenians during the genocide. As the name of the place became tantamount to a death sentence, some managed to survive with nomadic pastoralists in the region. And all the while, Ottoman officials imagined how to better draw borders around the Jazira, a task so important that a commission conducted a study on it in the waning years of the war. Thus when new borders appeared in the region after World War I, they were not altogether novel, but rather represented another layer of state efforts to control a marginal environment and mobile people within it. In fact, in some places, borders represented a continuity. This was especially the case in Ras al-Ayn. The town on the border of the French mandate of Syria and the Republic of Turkey – and in fact, floated by Ottoman officials at Versailles as a potential borderline between Turkish and Arab lands – had for decades been located on the border of both Ottoman provinces and the desert. In other places, borders took a distinctly different shape, most notably those walls of zinc sheeting that crossed the greater Jazira and, in a few cases, even crossed the border between the states of Turkey and Syria. These borders attested to the shared environment of the region, underscored by the continued threat of locusts. The longer history of borders in the region reveals repeated efforts to close the gap between the desert, human motion, and state demarcations of these spaces, as well as exploitation of these gaps. Borders thus mattered more in the Ottoman period than is conventionally understood, and they mattered in different ways in the post-Ottoman period.

The *tall* upon which Skaf witnessed the locusts offers another vantage from which to trace these transformations. It was these protuberances and their ancient ruins that had helped attract European and Ottoman interest in the Jazira in the first place in the mid-nineteenth century (Figure 21). The past plenty of Assyrian friezes in these *tall* helped to underwrite Ottoman dreams of turning the Jazira into an agricultural heartland once again. Indeed, one British consul even called for the use of the *tall* themselves to force nomads to settle by placing artillery on them. The Ottomans moreover attempted to settle the Shammar at Shirqat, a *tall* on the Tigris from which some of the first Assyrian relics had been removed. When Ibrahim Pasha fled into the Jazira in 1908, he first stopped by the castle of an ally, nestled atop a *tall* between Viranşehir and Mardin. In World War I, the Jazira witnessed new hills, forming this time not from ancient ruins but rather, in Artin Boşgezenyan's words, "from the bones of Armenians, left over by the birds and the wolves" during the genocide. By the mandate period, the *tall* saw yet another use, as they became sites for agricultural settlements of Armenian and Assyrian refugees. It was at roughly this same

FIGURE 21 *Tall* at Soğmatar, Turkey, 2014. Photo by Chris Gratien, used with permission

time that scientific experts suggested that it was also the *tall* that locusts used for procreation, just as Skaf would observe several decades later. Indeed, one locust expert who accompanied Skaf in the Jazira in 1953 presented a conflict occurring between nomads and locusts over the *tall*. He noted that "the Bedouin" had "built mud-brick farms on the tops of the old tells," rendering them "unsuitable for *Dociostaurus* breeding."[3] The humans who had been compared to locusts for their mobile challenges to state power had displaced their nonhuman referents, and they did so with farms atop the Jazira's distinctive elevations. The expansion of cultivation, changing technologies of killing locusts, and increased state power all came together to topple the locusts from their thrones atop the *tall*.

WHEAT FIELDS AND MINEFIELDS

Agricultural transformation of the Jazira in the twentieth century involved both continuity and change. Officials in the 1930s Jazira were incorrect in believing they were totally free of locusts, but for the rest of the century,

[3] NHMA, Syria and Lebanon 6211 – 158, Merton, Report on a Visit to N.E. Syria, April 1953.

the insects' control seemed to be unprecedentedly within reach thanks to the use of new kinds of chemicals. Organochlorides such as BHC and Agrocide 7 – now widely banned – became commonly used in Syria and Iraq.[4] There were complaints, such as Iraqi writer Dhu al-Nun al-Ayyub's suggestions that in fact "the real locusts are the employees of the fight against locusts," given how late they started their work.[5] But it seems efforts still proved terribly effective.[6] The Iraqi director general of agriculture described how in the course of a seventeen-mile drive in the country's north in 1947 "the earth" was "black with layer upon layer of dead locusts wherever he turned."[7] As these chemicals offered new power, the region retained its status as a place to which dispossessed populations might be sent. In 1948, there was a possibility that Palestinians would join Chechen refugees, Armenian deportees, and Assyrian refugees who had historically been sent to the supposedly empty lands of the Jazira. One Zionist official even used a charged phrase to convey the possibilities of the space, suggesting that the Jazira "in its natural boundaries" – that is, between Iraq and Syria – might even become home to a second Damascus.[8]

Even if Palestinian refugees did not end up in the Jazira, the dream of transforming the region stayed alive in part because of the labor of the children of people who – whether as nomadic pastoralists, deportees, or refugees – had ended up there. One of the largest landowners and most eager users of agricultural machinery south of Ras al-Ayn was Khalil Bey, son of Ibrahim Pasha and chief of the Millî.[9] Alongside him and especially in the village of Safeh were many of the descendants of Chechen refugees resettled during the Ottoman Empire.[10] Meanwhile, Daham al-Hadi,

[4] NHMA, Syria and Lebanon 6211 – 158, Lean, Report on Locusts; Iraq 6213, Plant Protection Ltd., Notes on Second Visit to the Middle East, 1948.

[5] al-Ayyub, *Al-Yad wa al-Ard wa al-Ma'*, 130.

[6] See, for example, the report on the use of 627 kilograms of sodium arsenate and other substances to kill 683,913 kilos of locusts in Turkey in 1945, including the Najdi locusts. Sadreddin Seyit Erkılıç and Süleyman Balamir, "Gundeydoğu Anadolu'da Çöl Çekirgesi," *Ziraat Dergisi* nos. 75–76 (April–May 1946): 41–46. See also Özer, *Anadolu'da Görülen Çekirge İstilaları*, 62–69.

[7] NHMA, Iraq 6213, Associated Press, "Stiff Defense by Government Halts Iraq Locust Onslaught," undated.

[8] Masalha, *Expulsion of the Palestinians*, 138. See also Shlaim, *Israel and Palestine*, 64, 74–75.

[9] Boghossian, *La Haute Djézireh*, 49.

[10] Ibid., 51–52. Some of the Chechens of Ras al-Ayn departed for the Turkish side of the border, where they settled in the formerly Armenian village of Tel Arman, now Kızıltepe. Adamiak, "To the Edge of the Desert," 235.

shaykh of the Shammar and great-grandson of Farhan, was among the largest holders of irrigated lands in the Syrian Jazira.[11] He, and others like him, became known as "the shaykhs of cotton" as they capitalized on the region's white gold.[12] Across the border in Iraq, the country's largest landholder was another Shammar chief, Ahmad ʿAjil al-Yawer, grandson of Farhan.[13] In other words, nearly one hundred years after the Ottomans dreamed of using land reforms to help increase peasant title to land and settle nomads to expand cultivation, nomadic chiefs had seemed to do the thing that Midhat Pasha had feared they were constitutionally incapable of: acquiring property and engaging in cultivation. Two particularly excited French observers placed this "feverish metamorphosis" in the type of world-historical terms that people had long utilized to describe the region: "the Jazira, which was four thousand years ago one of the most prosperous and civilized regions of the ancient world, is waking up with the sound of tractors and mechanical threshers."[14] Other protagonists of the Jazira's past participated in these changes too. One man who had grown up in Aleppo recalled how in the 1950s during the wheat and cotton harvests, Armenian mechanics would flood the Jazira to maintain the tractors and trucks necessary to the processes.[15] "Syria's economy depends on us, and we make it work," one mechanic told him. For its economic opportunity, some called the Jazira the California of Syria.[16]

But alongside these clear symbols of economic savvy, a perception of the Jazira and its people as backward persisted and would even come to inflect the social movements emerging in Iraq and Syria. In Turkey, one reporter scoffed that the people in a village on the Euphrates in Urfa province lived as if in the "Stone Age."[17] On the Syrian side of the border, denigrating views abounded, as one article in the military journal *Al-Jundi* declared that the prominence of tribes in the Jazira ensured that most Syrians viewed it as tantamount to "a region of unknown African jungles."[18] With the Jazira coded as backward on all of these fronts by

[11] Ababsa, "Contre-réforme agraire et conflits fonciers en Jâzira syrienne," 213.

[12] Ababsa, *Raqqa*, 7, 67.

[13] Batatu, *The Old Social Classes and the Revolutionary Moments of Iraq*, 48.

[14] Gibert and Fevret, "La Djezireh syrienne et son réveil économique," 2.

[15] Personal communication with Khachig Tölölyan, July 30, 2020.

[16] Ramzi and Hana, *Al-Jazira wa Rijalatuha*, 3.

[17] Hıfzı Topuz, "Kamış çardaklarda yaşayan 2000 kişi," *Akşam*, August 6, 1949, 4.

[18] "Ahamiyyat mintaqat al-Jazira," *Al-Jundi*, April 21, 1955. See also Siham Turjuman, "Fi tariqina ila al-sahraʾ," *Al-Jundi*, July 10, 1958, 24; "Wa raʾaytu al-mukhayamat al-ʿarabiyya ʿala haqiqatiha," *Al-Jundi*, July 17, 1958.

reporters from the country's urban centers, it was no great leap to position tribal shaykhs with their large landholdings as avatars of a different time, one that might be left behind with more radical land reforms.

Locusts remained in people's consciousness, but not always in the same ways as in the past. As the Syrian entomologist Skaf noted, it was not simply the spread of cultivation that crowded out locusts, but also the overgrazing that ensued as more and more livestock lived on more marginal lands.[19] Meanwhile, the horrors of previous invasions gave way to the wonders of technology.[20] Following the 1958 Revolution in Iraq, one of the government's first actions was the purchase of organochlorides such as BHC and Aldrin so that they could respond to a swarm of desert locusts.[21] Airplanes became increasingly involved in locust control.[22]

Yet anxieties about locusts, borders, and pesticides continued. For example, when the Turkish parliamentarian Tahsin Demiray reached for a metaphor to describe the state's financial crisis in 1962, he reasoned that the affliction he used had to convey a sense of contagion that could come from anywhere.[23] Microbes, malaria, and syphilis all did the job. But locusts, he explained, did not. After all, they came from one place and one place only: Syria. In doing so, Demiray replayed disputes about border-crossing locusts stretching back to the Ottoman period, while adding a layer of disdain for Syria. People continued to worry, too, about what killing locusts meant to god. The Turkish ministry of religious affairs issued a communiqué in 1958 requesting that Friday sermons explain that locusts – although creatures of god – were also very harmful, and thus killing them was "both a religious and a national duty."[24]

If locusts could still cross the border, people increasingly had a more difficult time of it. For many years there were no physical barriers between Turkey and Syria. It was a telling expression of the permeability of the border that one way that officials worked to catch border crossers was to

[19] Skaf, "Le Criquet marocain au Proche-Orient et sa grégarisation sous l'influence de l'homme," 314–315.

[20] DDT was used broadly across the region. In Israel, Iraqi immigrants were even sprayed with it upon arrival. Bashkin, *Impossible Exodus*, 29–30.

[21] NHMA, Iraq 6213, Sadiq Abdul Ghani, Director General of Agriculture, Report on the Locust Campaign in Iraq during 1959.

[22] "Şehrimizde çekirge mücadelesi başladı," *Yeni Mardin*, April 19, 1960; "Çekirge Savaşı," *Yeni Mardin*, April 22, 1960; "Al-Aqmar allati tanbi bi-ahwal al-taqs: silah lil-insan al-jadid fi muharibat ghazu al-jarad," *Al-Jundi*, June 25, 1963; Majid Hamwi, "Al-Jarad wa asalib mukafahatihi al-haditha," *Al-Jundi*, August 6, 1963, 17.

[23] *Millet Meclisi Tutanak Dergisi*, Dönem: 1, Cilt: 5, Toplantı: 1, May 24, 1962, 415.

[24] BCA, 051–0–0–0–0, 4/32/42, President of Religious Affairs, July 31, 1958.

plow the land alongside the border at dusk.[25] If they found footprints in the soil the next morning, they would punish any nearby villages. People with connections of kin or religiosity such as the Naqshbandi also moved back and forth across the border.[26] But motion became more difficult beginning in the mid-1950s, when Turkey "planted" land mines along the border. When they did so, they were not aiming to stop armies, but rather contraband that continued to move across the borders, a fact dramatized in Yılmaz Güney's 1966 film *Law of the Border* (*Hudutların Kanunu*). In the climactic scene, land mines cut a bloody swath through a smuggler's spooked flock of sheep.

In the absence of these explosions, a preternaturally serene landscape reigned, although, of course, the "natural" scene was only enabled by the "artificial" borders. As an article in the newspaper *Nusaybin Sesi* detailed, one might see "birds perching on barbed wire" or darting into the "green grasses of the minefield."[27] The explosive minefield and the barbed wire surrounding it seemed to convey the artificial nature of borders in the region. But they also created a natural space by enabling the grasses to grow, grasses that birds might flit in and out of, grasses that animals might feed on, and grasses that smugglers might hide in. "Artificial" and "natural" – the terms so often used to discredit or legitimate claims about borders – thus came together in the minefield. In fact, it was only because of the artificial border that a seemingly natural ecology of grasses could persist in a fertile region rapidly cordoned off into fields of crops or grazing lands.[28] Might locusts have lurked in these untilled soils, the quintessential borderland ecology of the Jazira? The insects had stopped trains on the railway that marked the border in precisely this space, so it would not be surprising. In any case, they certainly continued to swarm in cultural production, but in life they seem largely gone thanks to tractor-fueled expansion of cultivation and overgrazing.[29] Skaf – the entomologist who had stumbled upon the copulating locusts at Raqqa in 1953 – even noted that the dam on the Euphrates at Tabqa played a role in fighting locusts.

25 Aras, *The Wall*, 68.

26 Aras, "Naqshbandi Sufis and Their Conception of Place, Time and Fear on the Turkish–Syrian Border and Borderland"; Tejel, "'Des femmes contre des moutons.'" Of the relationship between the government and the local Khaznawi brotherhood, Salim Barakat wrote "successive governments like locusts sought their favor." Barakat, *Al-Siratan*, 88.

27 Ergün Doğanay, "Mayınlı Tarlanın yeşil otları," *Nusaybin Sesi*, May 6, 1969.

28 Aras, *The Wall*, 178–180.

29 Skaf, "Le Criquet marocain," 313. For appearances of locusts in literature, see, for example, Aliksan, "Al-Jarad"; Tamer, *Al-Jarad fi al-Madina*.

Completed in 1973, the dam led to the formation of Lake Assad, which not only displaced people, but also locust reservoirs.[30]

"CAN'T ARMENIAN MUSLIMS EXIST?"

The second half of the twentieth century saw Iraq, Syria, and Turkey look to the Jazira as an engine of agricultural production and a site for often violent demographic-engineering policies aimed at Kurdish populations.[31] As the instance of Lake Assad displacing locust reservoirs underscored, water was at the center of many of these efforts. In the late nineteenth and early twentieth centuries, state officials had largely worried about nomads crossing into the zone of cultivation. But by the late twentieth century, the problem was the opposite. In the Iraqi and Syrian Jazira, cultivation repeatedly pushed past limits established by state authorities aimed at protecting rangeland.[32] The Balikh River dried up in the 1990s, and by 1999, the Khabur – diverted in many places for irrigation – no longer flowed during the summer.[33] By 2001, the springs that gave Ras al-Ayn its name – meaning "Head of the Spring" in Arabic – had dried up. As these waters disappeared, other substances flowed, including pesticides and fertilizers from Turkey seeping into Syria's groundwater.[34] There was fear of more, too, as the Southeast Anatolia Project in Turkey (better known by its Turkish acronym of GAP) ramped up intensive agriculture and urban growth together, which threatened to send further runoff downstream on the Euphrates and the Tigris to Syria and Iraq.[35]

There were still persistent reminders of Armenian suffering that were difficult to ignore. When the formidable historian Raymond Kévorkian visited Markada on the Khabur River in 1974, bones remained in plain sight.[36] When representatives of the Syrian government sought oil there in

30 Skaf, "Le Criquet marocain," 320–321.

31 On agriculture, see Ababsa, *Raqqa*; Ajl, "The Political Economy of Thermidor in Syria." On Kurds and violence, see Barut, *Al-Takawwun al-Tarikhi al-Hadith lil-Jazira al-Suriyya*, 743–747; Haji, "La hidad ʿala mawt al-hashara"; Hiltermann, *A Poisonous Affair*; Marcus, *Blood and Belief*; Tejel, *Syria's Kurds*, 50–51, 60–61.

32 Wilkinson and Tucker, *Settlement Development in the North Jazira, Iraq*, 13; Hinnebusch, *Peasant and Bureaucracy in Ba'thist Syria*, 211.

33 Ababsa, "The End of a World," 205; de Châtel, "The Role of Drought and Climate Change in the Syrian Uprising," 532.

34 Hole, "Drivers of Unsustainable Land Use in the Semi-Arid Khabur River Basin, Syria," 10.

35 Yıldız, *The Kurds in Syria*, 70.

36 Kévorkian, "Earth, Fire, or Water," 107.

1990, they too found bones.[37] Beginning in 1990, the Armenian Martyrs' Church in Deir ez-Zor displayed some of these bones – "partial skulls, hip sockets, femurs, tibias, clavicles, eye sockets, teeth" – and became a site of pilgrimage for survivors and their families.[38] Alongside the bones were "vials of soil" from deportees' home cities in Anatolia.[39] The gift shop sold boxes containing soil, candles, incense, a cross, and two spikes of wheat.[40] For some, the grains evoked religious symbolism of communion. But the wheat also had another resonance in the Jazira, where violent transformations both of the land and people had for so long been tied to cultivation. Indeed, agriculture served to both cover up and reveal. When farmers tilled the ground at Ras al-Ayn, they found bones, placing them to the side of their fields out of respect.[41] Even the mounds of antiquity yielded bones. When archaeologists began to excavate Tall Fakhriya near Ras al-Ayn in 2006, they found human remains dating to the genocide.[42] Subsequent forensic archaeologists commented that the skeletons revealed few injuries, but, in a macabre detail, few skulls turned up.[43]

Alongside bones were living memories of the violence of the Jazira's past. There were many for whom memories of their own lives attested to these legacies, Hagop who had become ʿAbdallah, Hovannes who had become ʿAli.[44] "Can't Armenian Muslims exist?" one such figure in the Syrian Jazira asked a French-Armenian journalist in the early 2000s.[45] As Seda Altuğ was frequently told in the course of collecting oral history in the region, "there is hardly anyone among the Kurds and Arabs of the Syrian Jazira who does not have a Kurdified or Arabized Armenian as one of their grandmothers."[46] People remembered these connections to the past. One such testament is a humble mosque near Deir ez-Zor.[47] It was endowed in honor of a man's Armenian mother as Um Serbiyeh, the Arabicized version of the Tekirdağ native's Armenian name of Serpouhi.

The region's people and land would change in the first decade of the twenty-first century thanks to a particular intersection of natural and neoliberal disasters. Because the great expansion of cultivation in the region relied on such an intricate infrastructure of inputs, it also proved

[37] Balakian, *Black Dog of Fate*, 341. [38] Ibid., 339.
[39] Semerdjian, "Bone Memory," 64. [40] Ibid., 69–70.
[41] Kouyoumdjian and Siméone, *Deir-es-Zor*, 114.
[42] Ferllini and Croft, "The Case of an Armenian Mass Grave," 234. [43] Ibid., 239.
[44] Kouyoumdjian and Siméone, *Deir-es-Zor*, 76. [45] Ibid., 84.
[46] Altuğ, "Sectarianism in the Syrian Jazira," 103.
[47] Kouyoumdjian and Siméone, *Deir-es-Zor*, 79.

especially vulnerable to shocks. At the same time as one of the worst three-year droughts on record, the Syrian government cut subsidies to fertilizer and fuel necessary to power irrigation in the Jazira.[48] With the Jazira producing some two-thirds of Syria's wheat and cotton, the catastrophe had devastating consequences, and in 2008 the country was forced to import wheat for the first time in more than a decade.[49] Meanwhile, many of the Jazira's farmers fled, with the UN estimating that over 60 percent of villages in the districts of al-Hasakah and Deir ez-Zor were deserted, their fields overtaken by the desert.[50] The region's disappearing economy prompted political challenge. Even before the drought, the region had been Syria's poorest, its relationship to the central state akin to "an internal colony," according to the Syrian writer Yassin Haj Salih.[51] Indeed, one of the first and forgotten shows of solidarity in Syria with the political uprisings known as the "Arab Spring" took place in al-Hasakah, when Hassan ʿAli al-ʿAqleh set himself on fire in protest of the region's grinding circumstances, an homage to the Tunisian fruit vendor Mohamed Bouazizi whose own self-immolation had helped to catalyze protest movements across the region and the world.[52]

As political challenge to the Syrian regime spread across the country in the spring of 2011 and eventually gave way to civil war, the Jazira continued to inflect the nature of the struggle. Many of those who fled failing farms in the northeast ended up in cities or as agricultural migrant workers.[53] In light of the large population of desperate people, some observers took the historic drought as evidence that the conflict could be considered one of the first wars of climate change.[54] But such claims risk erasing the political choices that coincided with the drought.[55] They also risk obscuring the historical durability of the dream of turning the Jazira from a desert into something else, and the disdain for the region's residents that often accompanied these visions of plenty.

[48] Kelley et al., "Climate Change in the Fertile Crescent and Implications of the Recent Syrian Drought"; de Châtel, "The Role of Drought," 526.
[49] Ababsa, "The End of a World," 199; de Châtel, "The Role of Drought," 527.
[50] De Châtel, "The Role of Drought," 527; Robert F. Worth, "Earth Is Parched Where Syrian Farms Thrived," *New York Times*, October 13, 2010.
[51] Salih, *Al-Thawra al-Mustahila*, 127.
[52] Yassin-Kassab and al-Shami, *Burning Country*, 39.
[53] Sajadian, "Debts of Displacement"; Robert Worth, "Earth Is Parched Where Syrian Farms Thrived," *New York Times*, October 13, 2010.
[54] Serra, "Over-Grazing and Desertification in the Syrian Steppe Are the Root Causes of War."
[55] Daoudy, *The Origins of the Syrian Conflict*, 206; De Châtel, "The Role of Drought," 522.

"BECAUSE IT IS A DESERT"

In February of 2014, I drove with a friend across many parts of the Turkish Jazira. Again and again, motion across borders in the present intersected with sites of violence in the past. Outside of Midyat with its jewelry shops and winemakers, after passing a refugee camp, we ended up in Ayn Warda, an Assyrian village famous for having been a bastion of resistance to the genocide during World War I. In recent years, money from European emigrés had flowed to the city, and its stone buildings and churches stood out from the treeless terraced landscape around it. At one church, we met an older man who had fled there from the Syrian border town of Qamishli. His community had gone in the opposite direction nearly a century before. The hilltop village was once again a place of refuge in wartime. We drove south to the border at Nusaybin, where we saw the restored train station that was supposed to determine the border. We also glimpsed Qamishli across the border fence and barbed wire. We drove from there on a highway dotted with modest hotels named İpek Yolu in homage to the region's past history as a thoroughfare connected to the Silk Road. We went all the way to Urfa, over 200 km (120 miles) to the west. We traveled south amidst the fields of cotton of the Harran plains, sometimes trailing slow-moving tractors, and then from Harran to Viranşehir – erstwhile seat of Ibrahim Pasha – along a bumpy stretch of road under construction. We were stopped along the way by young boys walking along the gravel road who simply asked for water. Visiting caves and ruins according to a tourist map we had been given in Urfa, at almost every place – no matter how small – we met Syrian refugees. We visited Dara, famous for its ancient cisterns. A century before, the cisterns had been filled with Armenian bodies.

And finally, we reached Mardin. From the stone city's hilltop perch, we could sip tea and look out into the distance. In the 1860s, locusts had cut across the swath of land below, and starlings had been reported attacking the swarms. The Shammar regularly moved across these lands as well, and it was for this reason that so many merchants of Mardin were among the Shammar during Ibrahim Pasha's attack on them in May of 1901. In 1913, notables of Mardin considered their city such an integral part of the Jazira that they even called for their city to be part of its own district or a part of Zor. Armenian deportees had traversed these parts – and been killed in them – during the genocide. And in the post-Ottoman era, chemicals and expanded cultivation helped to wipe out the locusts that had for so long shaped the region. The fertile soil had brought together different kinds of motion. But on those February days in 2014, with a tulip-shaped tea cup in hand, I could only see clouds and green lands stretching to the horizon.

Suddenly, in the context of the war, the Jazira appeared once again on maps all over the world, although it was rarely identified as such. In the summer of 2014, a group calling itself the Islamic State of Iraq and Syria brought together territory stretching from Raqqa on the Syrian Euphrates to Mosul on the Iraqi Tigris. They claimed to be breaking the borders of Sykes–Picot, and many breathless commentaries repeated this rhetoric. Yet ISIS was also drawing on long-standing connections across the Jazira. They not only echoed Ottoman efforts to bring together the space, but they also took advantage of the impoverishment of the region over the course of the twentieth century. And they exploited too the chaos caused by the American-led destruction of Iraq and the smuggling networks that emerged across the Iraqi–Syrian border. The group welcomed recruits from all over, ranging from Deir ez-Zor to the Parisian suburbs to Chechnya. The land that had witnessed both ethnic cleansing and tremendous diversity in the past became home to a new totalitarian vision of purity, Muslims fleeing in fear, the doors of Christians marked with spray-paint, Yazidis massacred or taken as sex slaves. Their flight to Mount Sinjar in search of safety recalled the convoys of the past that had marched through the desert. The memorial to past suffering – the Armenian Martyrs' Church in Deir ez-Zor – was destroyed in 2016 during fighting between ISIS and Jabhat al-Nusra. Those arrayed against ISIS also carried echoes of the past, as among them were most notably the largely Kurdish People's Protection Units (Yekîneyên Parastina Gel), which later formed the Syrian Democratic Forces along with, among others, the Shammar militia (Jaysh al-Sanadid), as well as Assyrian units that emerged in the Khabur Valley and Qamishli.

As it had so many times before, the metaphor and the material blurred. The Iraqi-American poet Dunya Mikhaʾil used a familiar term to refer to the ominous approach of ISIS. They were like "locusts," she wrote, the headlights of their pickup trucks stretching across the Jazira's flat expanses.[56] American columnist Thomas Friedman described ISIS as an "invasive species," and called for the use of "herbicides" in the form of "Obama's air war."[57] In fact, as the United States moved into northeastern Syria, they established one of their first Special Forces bases at an

[56] Mikhaʾil, *Fi Suq al-Sabaya*, 173.
[57] Thomas L. Friedman, "I.S.=Invasive Species," *New York Times*, October 11, 2014.

airfield previously used for crop-dusting.[58] The apparent coincidence gestured to a broader structure of empire and environment. As Eyal Weizman and Fazal Sheikh have revealed, a plot of US drone strikes from Niger to Pakistan almost perfectly maps onto the 250 mm isohyet denoting where rain-fed agriculture is possible.[59]

Echoes of the past also remained as borders and the environment came into focus in relation to new movements of people. In 2015, Turkey went to war again with the Kurdistan Workers' Party (PKK), leaving cities such as Diyarbekir and Nusaybin in ruins. Turkey also began building a wall along its border, reinforcing the line that had first been a railway, then a minefield. In an echo of Ziya Gökalp's consideration of a wall like China's for separating Arab nomads from Turkey, the structure was hailed as "the Turkish Great Wall."[60] The US-led bombing of Raqqa, Mosul, and elsewhere left cities and even the Tabqa Dam in ruins, with untold civilian casualties and massacres all across the Jazira.[61] Then in the fall of 2019, with the blessing of US president Donald Trump, Turkey invaded northern Syria. The operation – ironically named "Peace Spring" – was justified as part of the fight against the PKK. But it was also envisioned as a way for some of the millions of Syrian refugees in Turkey to return. The latent environmental associations of ethnicity even came into play when the Turkish president Recep Tayyıp Erdoğan declared that the region was "appropriate" for Arabs and not for Kurds "because it is a desert."[62] There was, of course, another echo in the plan to send Syrians to the Jazira from the north. They were yet another unwanted group for whom the Jazira seemed to pose a solution.

As I followed news of the invasion, I was conducting research in the League of Nations Archive in Geneva, reading about Assyrian refugee settlements during the 1930s. On my computer, I saw plumes of smoke rising from and armies advancing toward and refugees fleeing from places such as Tel Tamer, Ras al-Ayn, and Qamishli, the same borderland towns being discussed in the nearly century-old yellowing papers that lay on the

[58] Clarissa Ward and Tim Lister, "Inside Syria: The Farm Airstrip That's Part of the U.S. Fight against ISIS," CNN, February 3, 2016.

[59] Weizman and Sheikh, *The Conflict Shoreline*, 12.

[60] Selçuk Adıgüzel, "Suriye sınırına Türk Seddi . . . " *Sabah*, April 13, 2018.

[61] Azmat Khan and Anand Ghopal, "The Uncounted," *New York Times Magazine*, November 16, 2017; Azmat Khan and Ivor Prickett, "The Human Toll of America's Air Wars," *New York Times Magazine*, January 2, 2022; Dave Philipps, Azmat Khan, and Eric Schmitt, "A Dam in Syria Was on a 'No-Strike' List. The U.S. Bombed It Anyway." *New York Times*, January 21, 2022.

[62] "Erdoğan: Oralara en uygun olan Araplardır, çünkü çöl," *BirGün*, October 25, 2019.

FIGURE 22 The Jazira's wildflowers, 1930s. UNAG, R34942/4/11757/30870. Courtesy of United Nations Archives at Geneva

table before me. Thinking of how violence of the present rhymed with the past, I opened a folder containing three photos of the Jazira's famous wildflowers (Figure 22). Free of any date or captions, the simplicity of the black-and-white luxuriance offered a reminder of cycles of time that turned a land long dismissed by outsiders as a desert, a land that did not even exist according to most maps, into something else. But over many years, various people have found something in these springtime blooms. The Jazira's soils and the borders that divided them up could give people a chance to escape, fight, or survive. They still do.

Bibliography

Names beginning with al- are alphabetized under the element following al-. For example, al-Barudi may be found under Barudi.

ARCHIVAL SOURCES

American Board of Commissioners for Foreign Missions, Cambridge, MA (ABCFM)
Archives de la Province dominicaine de France, Bibliothèque du Saulchoir, Paris (APDF)
Askeri Tarih ve Stratejik Etüt ve Denetleme Başkanlığı Arşivi, Ankara (ATASE)
Bundesarchiv, Berlin (B)
Devlet Arşivleri Başkanlığı Cumhuriyet Arşivi (BCA)
Devlet Arşivleri Başkanlığı Osmanlı Arşivi (BOA)
Centre des archives diplomatiques, La Courneuve (CADC)
Centre des archives diplomatiques, Nantes (CADN)
Gertrude Bell Archive, Newcastle (GBA)
Hull History Centre, Hull (HHC)
National Archives and Records Administration, College Park, MD (NARA)
Natural History Museum Archives, London (NHMA)
The National Archives, London (TNA-UK)
United Nations Archive at Geneva (UNAG)

PERIODICALS

Akşam (Istanbul)
Çekirge (Istanbul)
Cumhuriyet (Istanbul)
Diyarbekir Gazetesi (Diyarbekir)
Al-Jundi (Damascus)

Mècheroutiette (Paris)
Al-Muqtabas (Damascus)
Nusaybin Sesi (Nusaybin)
Orman ve Maaden ve Ziraat Mecmuası (Istanbul)
Al-Qabas (Damascus)
Sada Babil (Baghdad)
Servet-i Fünun (Istanbul)
Tanin (Istanbul)
Ticaret ve Ziraat Nezareti Mecmuası (Istanbul)
Times of Mesopotamia (Baghdad)
Ulus Sesi (Mardin)
Al-ʿUmran (Damascus)
Umur-ı Nafia ve Ziraat Mecmuası (Istanbul)
Yeni Mardin (Mardin)
Yenilik (Urfa)
Ziraat Dergisi (Ankara)

OTTOMAN PROVINCIAL YEARBOOKS

Bağdad Salnamesi
Diyarbekir Salnamesi
Halep Salnamesi
Musul Salnamesi

PARLIAMENTARY MINUTES

Meclis-i Mebusan Zabıt Ceridesi (MMZC) (Minutes of the Ottoman Chamber of Deputies)
Millet Meclisi Tutanak Dergisi

PRINTED PRIMARY SOURCES

Aaronsohn, Alexander. *With the Turks in Palestine*. Boston, MA: Houghton Mifflin, 1916.
Achladi, Evangelia, ed. *Karamanlı Rum Ortodoks Bir Askerin Seferberlik Hatıraları: Çanakkale ve Doğu Cepheleri, 1915–1919*. Istanbul: Kitap Yayınevi, 2017.
Alameddine, Rabih. *An Unnecessary Woman*. New York: Grove Press, 2014.
Aliksan, Jan. "Al-Jarad." In *Al-Hudud wa al-Aswar*, 15–22. Damascus: Manshurat al-Ittihad al-Kuttab al-ʿArab, 1970.
Anter, Musa. *Hatıralarım*. Vol. 1. Istanbul: Doz Yayınları, 1990.
Armalah, Ishaq. *Al-Qusara fi Nakabat al-Nasara*. Piscataway, NJ: Gorgias Press, 2011.
Asena, Ahmet Cemil. "Arkadaşım Ziya Gökalp," *Ziya Gökalp* 1.2 (April 1975): 319–325.

"Arkadaşım Ziya Gökalp," *Ziya Gökalp: Ziya Gökalp Derneğince Yayımlanır Kültür Dergisi* 2.6 (January–April 1977): 157–162.

"Arkadaşım Ziya Gökalp," *Ziya Gökalp: Ziya Gökalp Derneğince Yayımlanır Kültür Dergisi* 3.12 (January 1979): 234–235.

Atay, Falih Rıfkı. *Zeytindağı*. Istanbul: Remzi Kitabevi, 1938.

Ayyub, Dhu al-Nun. *Al-Yad wa al-Ard wa al-Ma'*. 2nd ed. Baghdad: Matba'at Shafiq, 1970.

Bagdikian, Lydia. *The Memoir of Lydia Bagdikian*. Berkeley, CA: Ben Bagdikian, 1997.

Balakian, Grigoris. *Armenian Golgotha: A Memoir of the Armenian Genocide, 1915–1918*, translated by Peter Balakian with Aris Sevag. New York: Random House, 2010.

Balakian, Peter. *Black Dog of Fate: A Memoir*. New York: Basic Books, 2009.

Barakat, Salim. *Al-Siratan*. Beirut: Dar al-Jadid, 1998.

Bardakçı, Murat. *Talât Paşa'nın Evrak-ı Metrûkesi: Sadrazam Talât Paşa'nın Özel Arşivinde bulunan Ermeni Tehciri Konusundaki Belgeler ve Hususî Yazışmalar*. Istanbul: Everest Yayınları, 2008.

Basir, Muhammad Mahdi al-. *Tarikh al-Qadiyya al-'Iraqiyya*. Baghdad: Matba'at al-Fallah, 1924.

Bedoukian, Kerop. *The Urchin: An Armenian's Escape*. London: John Murray, 1978.

Bell, Gertrude. *Amurath to Amurath*. London: William Heinemann, 1911.

Blunt, Anne. *Bedouin Tribes of the Euphrates*. New York: Harper & Brothers, 1879.

Bodenheimer, Friedrich Simon. *Studies on the Ecology and Control of the Moroccan Locust (*Dociostaurus maroccanus*) in 'Iraq. I. Results of a Mission of the 'Iraq Department of Agriculture to N. 'Iraq in Spring 1943*. Baghdad: Government Press, 1944.

Boghossian, Roupen. *La Haute Djézireh*. Aleppo: Imprimerie Chiraz, 1952.

Bredemann, Gustav. "Die Bekämpfung der Heuschrecken in Nord-Syrien und Nord-Mesopotamien." In *Die Heuschreckenplage und ihre Bekämpfung*, edited by Hermann Bücher, 131–151. Berlin: Parey, 1918.

Browne, John Gilbert. "The Assyrians: A Debt of Honour." *Geographical Magazine* 4.6 (1937): 431–448.

Bücher, Hermann. 1332–1333. *Senelerinde Anadolu'da ve Suriye'de Çekirgelere Karşı İcra Edilmiş Olan Mücadeleyi Mubeyyin Eserdir*, translated by Süreyya Bey. Der Saadet: Matbaa-i Osmaniye, 1334.

Bücher, Hermann, ed. *Die Heuschreckenplage und ihre Bekämpfung: auf Grund der in Anatolien und Syrien während der Jahre 1916 und 1917 gesammelten Erfahrungen dargestellt, und im Auftrage des Kaiserlich Osmanischen Landwirtschaftsministeriums*. Berlin: Parey, 1918.

Buck, Pearl S. *The Good Earth*. London: Methuen, 1931.

Cameron, Verney Lovett. *Our Future Highway to India*. London: Macmillan, 1880.

Carson, Rachel. *Silent Spring*. Boston, MA: Houghton Mifflin, 1962.

Césaire, Aimé. *The Original 1939 Notebook of a Return to a Native Land*, translated and edited by A. James Arnold and Clayton Eshleman. Middletown, CT: Wesleyan University Press, 2013.

Cora, Yaşar Tolga, ed. *Harbiyeli Bir Osmanlı Ermenisi: Mulazım-ı Sani Surmenyan'ın Savaş ve Tehcir Anıları*. Istanbul: Tarih Vakfı Yurt Yayınları, 2015.

Cuinet, Vital. *La Turquie d'Asie: Géographie administrative, statistique, descriptive et raisonnée de chaque province de l'Asie-Mineure*. Paris: Ernest Leroux, 1891.

Dadrian, Vahram. *To the Desert: Pages from my Diary*, translated by Agop J. Hacikyan. Princeton, NJ: Gomidas Institute, 2003.

Davis, Leslie A. *The Slaughterhouse Province: An American Diplomat's Report on the Armenian Genocide, 1915–1917*. New Rochelle, NY: A. D. Caratzas, 1989.

Demiray, Tahsin. *Buhranlarımız Temel Sebepleri*. Istanbul: Türkiye Yayınevi, 1970.

Der-Garabedian, Hagop S. *Jail to Jail: Autobiography of a Survivor of the 1915 Armenian Genocide*, translated by Aghop H. Der-Karabetian. New York: iUniverse, 2004.

Der Yeghiayan, Zaven. *My Patriarchal Memoirs*, translated by Ared Misirliyan and edited by Vatche Ghazarian. Barrington, RI: Mayreni, 2002.

Doumanian, Hovhannes. *My Memoirs* (unpublished).

Edmonds, Cecil J. *Kurds, Turks and Arabs: Politics, Travel and Research in North-Eastern Iraq, 1919–1925*. London: Oxford University Press, 1957.

Eig, Alexander. "Ecologie du criquet marocain en Iraq." *Bulletin of Entomological Research* 26.3 (1935): 293–315.

Ellis, Tristram J. *On a Raft and Through the Desert: The Narrative of an Artist's Journey Through Northern Syria and Kurdistan, By the Tigris to Mosul and Baghdad* ... London: Field and Tuer, 1881.

Evliya Çelebi. *Evliya Çelebi Seyahatnamesi*. Vol. 3. Istanbul: İkdam Matbaası, 1314 [1896/1897].

Fanon, Frantz. *Les damnés de la terre*. Paris: La Découverte, 2002. First published Paris: Maspero, 1961.

Field, Henry. *The Track of Man: Adventures of an Anthropologist*. Garden City, NY: Doubleday, 1953.

Foster, Benjamin R. , ed. and trans. *The Epic of Gilgamesh*. New York: WW Norton, 2001.

Foucault, Michel. *Surveillir et punir: Naissance de la prison*. Paris: Gallimard, 1975.

Fraser, David. *The Short Cut to India: The Record of a Journey Along the Baghdad Railway*. London: William Blackwood and Sons, 1909.

Geary, Grattan. *Through Asiatic Turkey: Narrative of a Journey from Bombay to the Bosphorous*. Vol. 1. London: Sampson Low, Marston, Searle, & Rivington, 1878.

Ghazzi, Kamal ibn Husayn. *Nahr al-Dhahab fi Tarikh Halab*. Vol. 1. Aleppo: Al-Matbaʿa al-Marwaniyya, 1923–1926.

Gibert, André and Maurice Fevret. "La Djezireh syrienne et son réveil économique." *Revue de Géographie de Lyon* 28 (1953): 1–15.
Gökalp, M. Nihat. "Ağabeyim Ziya Gökalp'ın Hayatı." *Ziya Gökalp* 1.3–4 (October 1975–January 1976): 434–438.
"Ağabeyim Ziya Gökalp'ın Hayatı," *Ziya Gökalp: Ziya Gökalp Derneğince Yayımlanır Kültür Dergisi* 2.6 (January–April 1977): 163–177.
Gökalp, Ziya. "Şaki Ibrahim Destanı." In *Son Şiirler*, edited by Yalçın Toker. Istanbul: Toker Yayınları, 2004.
Kürt Aşiretleri Hakkında Sosyolojik Tetkikler, edited by Şevket Beysanoğlu. Istanbul: Sosyal Yayınlar, 1992.
Gust, Wolfgang, ed. *Der Völkermord an den Armeniern 1915/16: Dokumente aus dem Politischen Archiv des deutschen Auswärtigen Amts*. Springe: Zu Klampen, 2005.
Hakim, Tewfik al-. *Fate of a Cockroach: Four Plays of Freedom*, translated by Denys Johnson-Davies. London: Heinemann, 1973.
Hakkı Bey, Ismail. *Irak Mektupları*. Istanbul: Kanaat Kütüphanesi, 1329 [1911].
Hamilton, Archibald Milne. *Road Through Kurdistan: The Narrative of an Engineer in Iraq*. London: Faber & Faber, 1947.
Harb, Aliza with Florence Gilmore. *Aliza: An Armenian Survivor Torn Between Two Cultures*. Belmont, MA: Armenian Heritage Press, 2003.
Harris, J. Rendel and Helen B. Harris. *Letters from the Scenes of the Recent Massacres in Armenia*. London: James Nisbet, 1897.
Hay, William Rupert. *Two Years in Kurdistan: Experiences of a Political Officer, 1918–1920*. London: Sidgwick & Jackson, 1921.
Hayik, Iskandar Yusuf al-. *Rihla fi al-Badiya*. Damascus: Dar Shahrazad al-Sham, 2004.
Herculais, M. Jules Künckel d'. *Les Sauterelles, les acridiens et leurs invasions*. Paris: Association française pour l'avancement des sciences, 1888.
Hilal, Muhammad Talab. *Dirasa 'an Muhafazat al-Jazira min Nuwahi al-Qawmiyya wa al-Ijtima'iyya wa al-Siyasiyya*. Erbil: Kawa lil-Nashr wa al-Tawzi', 2001.
Husri, Sati' al -. *Ara' wa Ahadith fi al-Wataniyya wa al-Qawmiyya*. Beirut: Dar al-'Ilm lil-Milayin, 1961.
Ibn Batuta, *Rihlat Ibn Batuta*. Beirut: Dar al-Sadir, 1964.
Ibn Hawqal. *Kitab Surat al-Ard*. Leiden: Brill, 1967.
Ibn Jubayr. *Rihlat Ibn Jubayr*. Beirut: Dar al-Sadir lil-Tiba'a wa al-Nashr, 1964.
International Institute of Agriculture. *La Lutte contre les sauterelles dans les divers pays*. Rome: Imprimerie de l' Institut International d'Agriculture, 1916.
Istakhri, Abu Ishaq Ibrahim bin Muhammad al-Farisi al-. *Kitab al-Masalik wa al-Mamalik*. Leiden: Brill, 1927.
Jabbur, Jibra'il Sulayman. *Al-Badu wa al-Badiyya: Suwar min Hayat al-Badu fi Badiyat al-Sham*, edited by Suhayl Jibra'il Jabbur. Beirut: Dar al-'Ilm al-Milayin, 1988.
Jarman, Robert L, ed. *Iraq Administration Reports, 1914–1932*. Vol. 4. Slough: Archive Editions, 1992.
Jebb, Louisa. *By Desert Ways to Baghdad*. London: T. S. Fisher Unwin, 1908.
Jeppe, Karen. *Misak: An Armenian Life*, translated and edited by Jonas Kauffeldt. London: Gomidas Institute, 2015.

Jernazian, Ephraim. *Judgment unto Truth: Witnessing the Armenian Genocide*, translated by Alice Haig. New Brunswick, NJ: Transaction Publishers, 1990.

Journal Officiel de la République Syrienne.

Kafka, Franz. "The Metamorphosis." In *The Complete Stories*, edited by Nahum N. Glatzer, translated by Willa and Edwin Muir. New York: Schocken Books, 1971.

Kazanjian, Paren. *The Cilician Armenian Ordeal.* Boston, MA: Hye Intentions, 1989.

Kemal, Yaşar. *Yaşar Kemal Kendini Anlatıyor: Alain Bosquet ile Görüşmeler.* Istanbul: Yapı Kredi Yayınları, 2015.

Kerr, Stanley. *The Lions of Marash: Personal Experiences with American Near East Relief, 1919–1922.* Albany: State University of New York Press, 1973.

Kıcıman, Naci Kaşif. *Medine Müdafaası Yahut Hicaz Bizden Nasıl Ayrıldı.* Istanbul: Sebil Yayınevi, 1971.

Köy İşleri Bakanlığı Yayınları. *Köy Envanterine Göre Mardin.* Ankara: Köy İşleri Bakanlığı, 1966.

Kouyoumdjian, Bardig and Christine Siméone. *Deir-es-Zor: Sur les traces du genocide arménien de 1915.* Arles: Actes Sud, 2005.

Künzler, Jakob. *Im Lande des Blutes und der Tränen: Erlebnisse in Mesopotamien während des Weltkrieges (1914–1918).* Zurich: Chronos Verlag, 1999.

Lafargue, Paul. *The Right to Be Lazy*, translated by Harriet E. Lothorp. Terre Haute, IN: Standard Publishing, 1904.

Layard, Austen Henry. *A Popular Account of Discoveries at Nineveh.* New York: JC Debry, 1855.

Lukach, Harry Charles. *The Fringe of the East: A Journey through Past and Present Provinces of Turkey.* London: Macmillan, 1913.

Lyon, W. A. *Kurds, Arabs and Britons: The Memoir of Col. W. A. Lyon in Kurdistan, 1918–1945*, edited by D. K. Fieldhouse. London: Tauris, 2002.

Malek, Alia. *The Home That Was Our Country: A Memoir of Syria.* New York: Nation Books, 2017.

Marsh, George Perkins. *Man and Nature: Or, Physical Geography as Modified by Human Action.* London: Sampson Low, Son and Marston, 1864.

Maurin, Amédée. *Invasion de sauterelles: Histoire. Anatomie. Marche. Moeurs. Reproduction. Ravages. Leur importance en Agriculture. Moyens de destruction.* Paris: Callamel, 1866.

Midhat Paşa. *Tabsıra-ı İbret.* Istanbul: Hilal Matbaası, 1325 [1909].

Mikha'il, Dunya. *Fi Suq al-Sabaya.* Milan: Manshurat al-Mutawassit, 2017.

Miller, David Hunter. *My Diary at the Conference of Paris, with Documents.* New York: Appeal Printing Company, 1924.

Mina, Hanna. *Baqaya al-Suwar.* Beirut: Dar al-Adab, 1978.

The Missionary Herald, Containing the Proceedings of the American Board of Commissioners for Foreign Missions with a View of Other Benevolent Operations for the Year 1866. Vol. 62. Cambridge, MA: Riverside Press, 1866.

Morgenthau, Henry. *Ambassador Morgenthau's Story.* Garden City, NY: Doubleday, 1918.

United States Diplomacy on the Bosphorous: The Diaries of Ambassador Morgenthau, 1913–1916. Edited by Ara Sarafian. Princeton and London: Gomidas Institute, 2004.

Mudiriyat al-Zira'a. *Al-Jarad wa Ahamm Turuq Abadatihi*. Damascus: Matba'at al-Jumhuriyya al-Suriyya, 1936.

Mustawfi, Hamdallah. *Nuzhat al-Qulub*. Leiden: Brill, 1915.

Nazif, Süleyman. *El-Cezire Mektupları*. Cairo: Matbaa-yı İctihad, 1906.

Niebuhr, Carsten. *Description de l'Arabie faite sur des observations propres et des avis recueillis dans les lieux memes*. Amsterdam: S. J. Baalde, 1774.

Nogales, Rafael de. *Four Years beneath the Crescent*, translated by Mona Lee. London: Sterndale Classics, 2003.

Odian, Yervant. *Accursed Years: My Exile and Return from Der Zor, 1914–1919*, translated by Ara Stepan Melkonian. London: Gomidas Institute, 2009.

Oppenheim, Max von. *Tel Halaf: A New Culture in Oldest Mesopotamia*, translated by Gerald Wheeler. London: G.P. Putnam's Sons, 1931.

Vom Mittelmeer zum Persischen Golf durch den Haurän, die Syrische Wüste und Mesopotamien. Vol. 2. Berlin: D. Reimer (E. Vohsen), 1899.

Öztoprak, İzzet, Cafer Güler, Murat Karataş, and Güneş Şahin, eds. *Cumhuriyet'in XV Yılında Türkiye*. Ankara: Atatürk Araştırma Merkezi, 2014.

Panian, Karnig. *Goodbye, Antoura: A Memoir of the Armenian Genocide*, edited by Aram Goudsouzian and translated by Simon Beugekian, with an introduction by Keith Watenpaugh. Stanford, CA: Stanford University Press, 2015.

Parrot, Friedrich. *Journey to Ararat*, translated by W. D. Cooley. London: Longman, Brown, Green, and Longmans, 1845.

Parry, Oswald. *Six Months in a Syrian Monastery, Being the Record of a Visit to the Head Quarters of the Syrian Church in Mesopotamia, with Some Account of the Yazidis or Devil Worshippers of Mosul and El Jilwah, Their Sacred Book*. London: Horace Cox, 1895.

Peters, John Punnet. *Nippur or Explorations and Adventures on the Euphrates: The Narrative of the University of Pennsylvania Expedition to Babylonia in the Years 1888–1890*. Vol. 1. New York: G. P. Putnam's Sons, 1898.

Pliny the Elder. *The Natural History*, translated by John Bostock. London: Fleet Street, 1855.

Qadri, Ahmad. *Muddhakirati 'an al-Thawra al-'Arabiyya al-Kubra*. Damascus: Manshurat Wizarat al-Thaqafa, 1993.

Ramzi, 'Uthman and Salim Hana. *Al-Jazira wa Rijalatuha: Ta'rif al-Jazira ila Ibna' Suriyya wa al-Bilad al-'Arabiyya*. Qamishli: Matba'at al-Khabur, undated.

Rao, Y. Ramachandra. *A Preliminary List of Insect Pests of 'Iraq*. Iraq and Bombay: Department of Agriculture, 1921.

The Desert Locust in India. New Delhi: Indian Council of Agricultural Research, 1960.

Rayyis, Najib al-. *Al-A'mal al-Mukhtara: Ya Zalam al-Sijn*. Vol. 1. London: Riad El-Rayyes Books, 1994.

Riley, Charles V. *The Locust Plague in the United States: Being More Particularly a Treatise on the Rocky Mountain Locust or So-Called Grasshopper as It*

Occurs East of the Rocky Mountains, with Practical Recommendations for Its Destruction. Chicago: Rand, McNally, 1877.

Rooke, Horace George D. *Note on Locusts in 'Iraq and the Control Measures Adopted*. Baghdad: Government Press, 1930.

Sachau, Eduard. *In Syrien und Mesopotamien*. Leipzig: F. A. Brockhaus, 1883.

Salih, Yasin al-Hajj. *Al-Thawra al-Mustahila: Al-Thawra, al-Harb al-Ahaliyya, wa al-Harb al-ʿAmma fi Suriya*. Beirut: Al-Muʾassasa al-ʿArabiyya lil-Dirasat wa al-Nashr, 2017.

Sanger, Richard H. *Where the Jordan Flows*. Washington, DC: The Middle East Institute, 1963.

Sarafian, Ara, ed. *United States Official Records on the Armenian Genocide 1915–1917*. London: Gomidas Institute, 2004.

Skaf, Rafeq. "Le Criquet marocain au Proche-Orient et sa grégarisation sous l'influence de l'homme." *Bulletin de la Société d'écologie* 3 (1972): 247–325.

Stafford, Ronald Sempill. *The Tragedy of the Assyrians*. London: George Allen & Unwin, 1935.

Suphi Paşa, Abdüllatif. "Suphi Paşa hazretleri'nin Layihası'ndan." In *Yeni Türk Edebiyatı Antolojisi I, 1839–1865*, edited by Mehmet Kaplan, İnci Enginün, and Birol Emil. Istanbul: Edebiyat Fakültesi Matbaası,1974.

Süreyya, Mehmed. *Haşerat-ı Ziraiye Mecmuası*. Istanbul: Osmanlı Ziraat ve Ticaret Gazetesi Matbaası, 1332 [1914].

Svazlian, Verjiné. *The Armenian Genocide: Testimonies of the Eyewitness Survivors*. Yerevan: National Academy of Sciences of the Republic of Armenia, 2011.

Sykes, Mark. *Dar-ul-Islam: A Record of a Journey Through Ten of the Asiatic Provinces of Turkey*. London: Bickers & Son, 1904.

"Journeys in North Mesopotamia." *The Geographical Journal* 30.4 (1907): 384–398.

The Caliph's Last Heritage: A Short History of the Turkish Empire. London: MacMillan, 1915.

Through Five Turkish Provinces. London: Bickers & Son, 1900.

Tamer, Zakariya. *Al-Jarad fi al-Madina: Qissa*. Illustrated by Hijazi. Beirut: Dar al-Fata al-ʿArabi, 1977.

Taylor, John George. "Journal of a Tour in Armenia, Kurdistan, and Upper Mesopotamia, with Notes of Researches in the Deyrsim Dagh, in 1866." *Journal of the Royal Geographical Society of London* 38 (1868): 281–361.

Tepeyran, Ebubekir Hazim. *Hatıralar*. Istanbul: Pera Turizm ve Ticaret, 1998.

Toynbee, Arnold. *Armenian Atrocities, the Murder of a Nation*. London: Hodder & Stoughton, 1915.

ʿUjayli, ʿAbd al-Salam al-. *Ahadith al-ʿAshiyat*. Damascus: Matabiʿ Wizarat al-Thaqafa wa al-Irshad al-Qawmi, 1965.

Uvarov, Boris. "Ecology of the Moroccan Locust in Iraq and Syria and the Prevention of Its Outbreaks." *Bulletin of Entomological Research* 24 (September 1933): 407–418.

Grasshoppers and Locusts: A Handbook of General Acridology. Vol. 2. London: Centre for Overseas Pest Research, 1977.

"Studies in Orthoptera of Turkey, Iraq and Syria." *Eos: Revista Española de Entomologia* 10.1–2 (August 1934): 21–123.

Whiting, John D. "Jerusalem's Locust Plague: Being a Description of the Recent Locust Influx into Palestine, and Comparing Same with Ancient Locust Invasions as Narrated in the Old World's History Book, the Bible." *National Geographic* 28.6 (December 1915): 511–550.

Wiedemann, M. "Ibrahim Pascha Glück und Ende: Eine Episode aus den Kurdenkämpfen in Klein-Asien." *Asien* 8.3 (1908): 34–37.

Wiedemann, M. "Ibrahim Pascha Glück und Ende: Eine Episode aus den Kurdenkämpfen in Klein-Asien." *Asien* 8.4 (1909): 52–54.

Wigram, William Ainger and Edgar T. A. Wigram. *The Cradle of Mankind: Life in Eastern Kurdistan*. London: A&C Black, 1922.

Wilder, Laura Ingalls. *On the Banks of Plum Creek*. New York: HarperCollins, 2016.

Wright, Richard. *Native Son*. New York: Harper & Row, 1940.

Yazbek, Samar. *A Woman in the Crossfire: Diaries of the Syrian Revolution*, translated by Max Weiss. London: Haus Publishing, 2011.

Yotnakhparian, Levon. *Crows of the Desert*, translated by Victoria Parian and edited by Levon Parian and Ishkhan Jinbashian. Tujunga, CA: Parian Photographic Design, 2012.

Zakariyya, Ahmad Wasfi. "Dhikriyyati ʿan Wadi al-Furat qabl Khamsa wa Arbaʿin ʿAman." In *Rihla ila al-Furat wa Maqalat Ukhra*, 24–53. Damascus: Dar wa Muʾassasat Raslan lil-Tabaʿa wa al-Nashr wa al-Tawziʿa, 2008.

Zaza, Nureddin. *Bir Kürt Olarak Yaşamım*. Istanbul: Peri, 2000.

Zohary, Michael. *Geobotanical Foundations of the Middle East*. Vol. 1. Stuttgart: Gustav Fischer Verlag, 1973.

SECONDARY SOURCES

Ababsa, Myriam. "Contre-réforme agraire et conflits fonciers en Jâzira syrienne (2000–2005)." *Revue des mondes musulmans et de la Méditerranée* 115/116 (2006): 211–230.

Raqqa: Territoires et pratiques sociales d'une ville syrienne. Beirut: Institut Français du Proche-Orient, 2009.

"Syria's Food Security: From Self-Sufficiency to Hunger as a Weapon." In *Syria: From National Independence to Proxy War*, edited by Linda Mater and Ali Kadri, 247–268. Cham: Palgrave Macmillan, 2019.

"The End of a World: Drought and Agrarian Transformation in Northeast Syria (2007–2010)." In *Syria from Reform to Revolt*. Vol. 1. Edited by Raymond Hinnebusch and Tina Zintl, 199–222. Syracuse, NY: Syracuse University Press, 2015.

Ababsa-al-Husseini, Myriam. "Mise en valeur agricole et contrôle politique de la vallée de l'Euphrate (1865–1946): Étude des relations état, nomades, et citadins dans le caza de Raqqa." *Bulletin d'études orientales* 53/54 (2001–2002): 459–487.

Abou-El-Haj, Rifaʿat ʿAli. *Formation of the Modern State: The Ottoman Empire, Sixteenth to Eighteenth Centuries*. Albany: State University of New York Press, 1991.

Abrahamyan, Victoria. "Armenian Refugees, State Formation and Identity Construction in Mandatory Syria." PhD diss., University of Neuchâtel, 2022.

Abu Bakr, Ahmad ʿUthman. *Akrad al-Milli Ibrahim Pasha*. Baghdad: Matbaʿat Dar al-Jahiz, 1973.

Abu Manneh, Butrus. "The Establishment and Dismantling of the Province of Syria, 1865–1888." In *Problems of the Middle East in Historical Perspective: Essays in Honour of Albert Hourani*, edited by John P. Spagnolo, 7–26. Reading, UK: Ithaca Press, 1992.

Adalet, Begüm. *Hotels and Highways: The Construction of Modernization Theory in Cold War Turkey*. Stanford, CA: Stanford University Press, 2018.

Adamiak, Patrick. "To the Edge of the Desert: Caucasian Refugees, Civilization, and Settlement on the Ottoman Frontier, 1866–1918." PhD diss., University of California San Diego, 2018.

Adanır, Fikret and Oktay Özel, eds. *1915: Siyaset, Tehcir, Soykırım*. Istanbul: Tarih Vakfı Yurt Yayınları, 2015.

Adelman, Jeremy. *Frontier Development: Land, Labour, and Capital in the Wheatlands of Argentina and Canada, 1890–1914*. New York: Clarendon Press, 1994.

Adelman, Jeremy and Stephen Aron. "From Borderlands to Borders: Empires, Nation-States, and the Peoples in Between in North American History." *American Historical Review* 104.3 (1999): 814–841.

Advani, Rohan. "Sowing Scarcity: Syria's Wheat Regime from Self-Sufficiency to Import-Dependency." *Jadaliyya*. February 16, 2021.

Ágoston, Gábor. "A Flexible Empire: Authority and Its Limits on the Ottoman Frontiers." In *Ottoman Borderlands: Issues, Personalities, and Political Changes*, edited by Kemal H. Karpat and Robert W. Zens, 15–32. Madison: University of Wisconsin Press, 2003.

Agresta, Abigail. *The Keys to Bread and Wine: Faith, Nature, and Infrastructure*. Ithaca, NY: Cornell University Press, 2022.

Ajl, Max. "The Political Economy of Thermidor in Syria: National and International Dimensions." In *Syria: From National Independence to Proxy War*, edited by Linda Mater and Ali Kadri, 209–246. Cham: Palgrave Macmillan, 2019.

Akçam, Taner. *Killing Orders: Talat Pasha's Telegrams and the Armenian Genocide*. New York: Palgrave Macmillan, 2018.

The Young Turks' Crimes Against Humanity: The Armenian Genocide and Ethnic Cleansing in the Ottoman Empire. Princeton, NJ: Princeton University Press, 2012.

Akgül, Önder Eren. "Osmanlı'da Çevre, İklim ve Insanlar." *Toplumsal Tarih* 312 (December 2019): 24.

Akın, Yiğit. *When the War Came Home: The Ottomans' Great War and the Devastation of an Empire*. Stanford, CA: Stanford University Press, 2018.

Aksakal, Mustafa. *The Ottoman Road to War in 1914: The Ottoman Empire and the First World War*. New York: Cambridge University Press, 2008.

Alff, Kristen. "Levantine Joint-Stock Companies, Trans-Mediterranean Partnerships, and Nineteenth-Century Capitalist Development." *Comparative Studies in Society and History* 60.1 (2018): 150–177.

Altuğ, Seda. "Culture of Dispossession in the Late Ottoman Empire and Early Turkish Republic: Land, Ethnoreligious Difference, and Violence." In *Reverberations: Violence Across Time and Space*, edited by Yael Navaro, Zerrin Özlem Biner, Alice von Bieberstein, and Seda Altuğ, 83–113. Philadelphia: University of Pennsylvania Press, 2021.

"Sectarianism in the Syrian Jazira: Community, Land, and Violence in the Memories of World War I and the French Mandate (1915–1939)." PhD diss., University of Utrecht, 2011.

"The Turkish–Syrian Border and Politics of Difference in Turkey and Syria (1921–1939)." In *Syria: Borders, Boundaries, and the State*, edited by Matthieu Cimino, 47–73. Cham: Palgrave Macmillan, 2020.

Altuğ, Seda and Benjamin White. "Frontières et pouvoir d'état: La frontière turco-syrienne dans les années 1920 et 1930." *Vingtième siècle: Revue d'histoire* 103 (2009): 91–104.

Amara, Ahmad. "Civilizational Exceptions: Ottoman Law and Governance in Late Ottoman Palestine." *Law and History Review* 36.4 (November 2018): 915–941.

Amrith, Sunil. *Crossing the Bay of Bengal: The Furies of Nature and the Fortunes of Migrants*. Cambridge, MA: Harvard University Press, 2013.

Anderson, Bridget. "The Politics of Pests: Immigration and the Invasive Other." *Social Research: An International Quarterly* 84.1 (2017): 7–28.

Anderson, Scott. *Fractured Lands: How the Arab World Came Apart*. New York: Anchor Books, 2017.

Antrim, Zayde. *Mapping the Middle East*. London: Reaktion Books, 2018.

Routes and Realms: The Power of Place in the Early Islamic World. New York: Oxford University Press, 2012.

Appuhn, Karl. "Ecologies of Beef: Eighteenth-Century Epizootics and the Environmental History of Early Modern Europe." *Environmental History* 15.2 (2010): 268–287.

Aras, Ramazan. "Naqshbandi Sufis and Their Conception of Place, Time and Fear on the Turkish–Syrian Border and Borderland." *Middle Eastern Studies* 55.1 (2019): 44–59.

The Wall: The Making and Unmaking of the Turkish–Syrian Border. Cham: Palgrave Macmillan, 2020.

Arsan, Andrew. *Interlopers of Empire: The Lebanese Diaspora in Colonial French West Africa*. New York: Oxford University Press, 2014.

Asad, Nasir al-Din al-. *Muqadimma li-dirasat al-jarad fi turath al-ʿarabi*. Amman: Dar al-Fath lil-Dirasat wa al-Nashr, 2008.

Aslan, Senem. *Nation-Building in Turkey and Morocco: Governing Kurdish and Berber Dissent*. New York: Cambridge University Press, 2015.

Asseraf, Arthur. *Electric News in Colonial Algeria*. New York: Oxford University Press, 2019.

Astourian, Steven. "The Silence of the Land: Agrarian Relations, Ethnicity, and Power." In *A Question of Genocide: Armenians and Turks at the End of the Ottoman Empire*, edited by Ronald Grigor Suny, Fatma Müge Göçek, and Norman M. Naimark, 55–81. New York: Oxford University Press, 2011.

Ateş, Sabri. *The Ottoman–Iranian Borderlands: Making a Boundary, 1843–1914*. New York: Cambridge University Press, 2013.

Avcı, Yasemin. "The Application of *Tanzimat* in the Desert: The Bedouins and the Creation of a New Town in Southern Palestine (1860–1914)." *Middle Eastern Studies* 45.6 (2009): 969–983.

Aydın, Cemil. *The Idea of the Muslim World: A Global Intellectual History*. Cambridge, MA: Harvard University Press, 2017.

The Politics of Anti-Westernism in Asia: Visions of World Order in Pan-Islamic and Pan-Asian Thought. New York: Columbia University Press, 2007.

Aydın,Suavi, Kudret Emiroğlu, Oktay Özel, and Süha Ünsal. *Mardin – Aşiret – Cemaat – Devlet*. Istanbul: Tarih Vakfı, 2000.

Aydın, Ayşegül and Cem Emrence. *Zones of Rebellion: Kurdish Insurgents and the Turkish State*. Ithaca, NY: Cornell University Press, 2015.

ʿAzzawi, ʿAbbas al-. *Tarikh al-ʿIraq bayn al-Ihtilalayn*. Vol. 8. Qum: Intisharat al-Maktaba al-Haydariyya, 2004.

Badem, Candan. *The Ottoman Crimean War (1853–1856)*. Leiden: Brill, 2010.

Bailony, Reem. "From Mandate Borders to Diaspora: Rashaya's Transnational Suffering and the Making of Lebanon in 1925." *Arab Studies Journal* 26.2 (2018): 44–73.

"Transnationalism and the Syrian Migrant Public: The Case of the 1925 Syrian Revolt." *Mashriq and Mahjar: Journal of Middle East Migration Studies* 1.1 (Spring 2013): 8–29.

Bâli, Aslı. "Sykes–Picot and 'Artificial' States." *American Journal of International Law Unbound* 110 (2016): 115–119.

Balistreri, Alexander. "Revisiting *Millî*: Borders and the Making of the Turkish Nation State." In *Regimes of Mobility: Borders and State Formation in the Middle East, 1918–1946*, edited by Jordi Tejel and Ramazan Hakkı Öztan, 29–58. Edinburgh: Edinburgh University Press, 2022.

Banko, Lauren. *The Invention of Palestinian Citizenship, 1918–1947*. Edinburgh: Edinburgh University Press, 2016.

Barak, On. *Powering Empire: How Coal Made the Middle East and Sparked Global Carbonization*. Oakland: University of California Press, 2020.

Barakat, Nora. "Making 'Tribes' in the Late Ottoman Empire." *International Journal of Middle East Studies* 53.3 (2021): 482–487.

"Marginal Actors? The Role of Bedouin in the Ottoman Administration of Animals as Property in the District of Salt, 1870–1912." *Journal of the Economic and Social History of the Orient* 58.1–2 (April 2015): 105–134.

Barr, James. *A Line in the Sand: The Anglo-French Struggle for the Middle East, 1914–1918*. New York: Norton, 2012.

Barut, Muhammad Jamal. *Al-Takawwun al-Tarikhi al-Hadith lil-Jazira al-Suriyya: Asʾila wa Ishkaliyat min al-Badwana ila al-ʿUmran al-Hadari*. Beirut: Al-Markaz al-ʿArabi lil-Abhath wa Dirasat al-Siyasat, 2013.

Bashkin, Orit. *Impossible Exodus: Iraqi Jews in Israel*. Stanford, CA: Stanford University Press, 2017.

Batatu, Hanna. *Syria's Peasantry, the Descendants of Its Lesser Rural Notables, and Their Politics*. Princeton, NJ: Princeton University Press, 1999.

The Old Social Classes and the Revolutionary Movements of Iraq: A Study of Iraq's Old Landed and Commercial Classes and of Its Communists, Baʿthists, and Free Officers. Princeton, NJ: Princeton University Press, 1978.

Baumgartner, Alice. *South to Freedom: Runaway Slaves to Mexico and the Road to the Civil War*. New York: Basic Books, 2020.
Bawalsa, Nadim. *Transnational Palestine: Migration and the Right of Return before 1948*. Stanford, CA: Stanford University Press, 2022.
Beckert, Sven. *Empire of Cotton: A Global History*. New York: Knopf, 2014.
Bein, Alex. "The Jewish Parasite: Notes on the Semantics of the Jewish Problem, with Special Reference to Germany." *The Leo Baeck Institute Year Book* 9.1 (January 1964): 3–40.
Bello, David. "Consider the Qing Locust." *East Asian Science, Technology, and Medicine* 48 (2018): 49–80.
Bennett, Julian. "The 'Aynali Martini': The Ottoman Army's First Modern Rifle." *Anatolica* 44 (2018): 229–255.
Berberian, Houri. *Armenians and the Connected Revolutions in the Russian, Iranian, and Ottoman Worlds*. Oakland: University of California Press, 2019.
Berg, Peter. "The Post-Environmental Directions of Bioregionalism." In *The Biosphere and Bioregionalism: Essential Writings of Peter Berg*, edited by Cheryll Glotfelty and Eve Quesnel, 64–69. New York: Routledge, Taylor & Francis, 2015.
Bernhardsson, Magnus Thorkell. *Reclaiming a Plundered Past: Archaeology and Nation Building in Modern Iraq*. Austin: University of Texas Press, 2005.
Bet-Shlimon, Arbella. *City of Black Gold: Oil, Ethnicity, and the Making of Modern Kirkuk*. Stanford, CA: Stanford University Press, 2019.
Beysanoğlu, Şevket. "Ziya Gökalp'ın Yaşam Öyküsü." *Ziya Gökalp* 1.2 (April 1975): 326–336.
Bhattacharyya, Debjani. *Empire and Ecology in the Bengal Delta: The Making of Calcutta*. New York: Cambridge University Press, 2018.
Biehler, Dawn Day. *Pests in the City: Flies, Bedbugs, Cockroaches, and Rats*. Seattle: University of Washington Press, 2013.
Biner, Zerrin Özlem. *States of Dispossession: Violence and Precarious Coexistence in Southeast Turkey*. Philadelphia: University of Pennsylvania Press, 2019.
Bivar, Venus. *Organic Resistance: The Struggle over Industrial Farming in Postwar France*. Chapel Hill: University of North Carolina Press, 2018.
Black, Megan. *The Global Interior: Mineral Frontiers and American Power*. Cambridge, MA: Harvard University Press, 2018.
Blackhawk, Ned. *Violence over the Land: Indians and Empires in the American West*. Cambridge, MA: Harvard University Press, 2006.
Blanc, Jacob, Frederico Freitas, and Zephyr Frank, eds. *Big Water: The Making of Borderlands Between Brazil, Argentina, and Paraguay*. Tucson: University of Arizona Press, 2018.
Bloxham, Donald. *The Great Game of Genocide: Imperialism, Nationalism, and the Destruction of the Ottoman Armenians*. Oxford: Oxford University Press, 2005.
Blumi, Isa. *Ottoman Refugees, 1878–1939: Migration in a Post-Imperial World*. London: Bloomsbury, 2013.

Bolaños, Isacar A. "The Ottomans during the Global Crises of Cholera and Plague: The View from Iraq and the Gulf." *International Journal of Middle East Studies* 51.4 (2019): 603–620.

Boratav, Pertev Naili. *100 Soruda Türk Folkloru*. Istanbul: Gerçek Yayınevi, 1984.

Bostan, İdris. "Zor Sancağı'nın Imar ve Islahı ile Alakalı Üç Layiha." *Osmanlı Araştırmaları* 6 (1986): 163–220.

Bosworth, Albert Brian. *Conquest and Empire: The Reign of Alexander the Great*. New York: Cambridge University Press, 1993.

Bozdoğan, Sibel. *Modernism and Nation Building: Turkish Architectural Culture in the Early Republic*. Seattle: University of Washington Press, 2001.

Brady, Lisa M. "Life in the DMZ: Turning Diplomatic Failure into an Environmental Success." *Diplomatic History* 32.4 (2008): 585–611.

Braudel, Fernand. *La Méditerranée et le monde méditerranéen à l'époque de Philippe II*. Paris: Armand Colin, 1982.

Brown, Kate. *A Biography of No Place: From Ethnic Borderland to Soviet Heartland*. Cambridge, MA: Harvard University Press, 2004.

Dispatches from Dystopia: Histories of Places Not Yet Forgotten. Chicago: University of Chicago Press, 2015.

Brown, Wendy. *Walled States, Waning Sovereignty*. New York: Zone Books, 2010.

Brumett, Palmira. *Image and Imperialism in the Ottoman Revolutionary Press, 1901–1911*. Albany: State University of New York Press, 2000.

Budak, Mustafa. *Misâk-ı Millî'den Lozan'a: Idealden Gerçege Türk Dış Politikası*. Istanbul: Küre Yayınları, 2008.

Büssow, Johann. "Negotiating the Future of a Bedouin Polity in Mandatory Syria: Political Dynamics of the Sbaʿa-ʿAbada During the 1930s." *Nomadic Peoples* 15.1 (2011): 70–95.

Çağlar, Ayşe and Nina Glick Schiller. *Migrants and City-Making: Dispossession, Displacement, and Urban Regeneration*. Durham, NC: Duke University Press, 2018.

Camp, Stephanie. *Closer to Freedom: Enslaved Women and Everyday Resistance in the Plantation South*. Chapel Hill: University of North Carolina Press, 2004.

Campos, Michelle. *Ottoman Brothers: Muslims, Christians, and Jews in Early Twentieth-Century Palestine*. Stanford, CA: Stanford University Press, 2011.

Can, Lâle. *Spiritual Subjects: Central Asian Pilgrims and the Ottoman Hajj at the End of Empire*. Stanford, CA: Stanford University Press, 2020.

Çavuş, Ömer. "Osmanlı Devleti'nde Çekirgelere Karşı Mücadele." *Osmanlı Bilimi Araştırmaları* 13.2 (2017): 59–76.

Çelik, Semih. "Humans in *Animalscapes*: Reconstructing Vermin–Human Interactions in Rural Anatolia and Mesopotamia (ca. 1600–1850)." *Diyar* 3.1 (2022): 49–66.

Çetinsaya, Gökhan. *Ottoman Administration of Iraq, 1890–1908*. London: Routledge, 2006.

Ceylan, Ebubekir. *The Ottoman Origins of Modern Iraq: Political Reform, Modernization and Development in the Nineteenth-Century Middle East*. London: Tauris, 2011.

Chalabian, Antranig. *Revolutionary Figures: Mihran Damadian, Hambardzum Boyadjian, Serod Aghbiur, Hrair-Dzhoghkk, Gevork Chavush, Sebastastatsi Murad, Nikol Duman*, translated by Arra S. Avakian. USA: Antranig Chalabian, 1994.

Chambers, Richard L. "The Education of a Nineteenth-Century Ottoman Alim, Ahmed Cevdet Pasha." *International Journal of Middle East Studies* 4.4 (October 1973): 440–464.

Châtel, Francesca de. "The Role of Drought and Climate Change in the Syrian Uprising: Untangling the Triggers of the Revolution." *Middle Eastern Studies* 50.4 (2014): 521–535.

Chatty, Dawn. "Bedouin Tribes in Contemporary Syria: Alternative Perceptions of Authority, Management, and Control." In *Tribes and States in a Changing Middle East*, edited by Uzi Rabi, 145–170. New York: Oxford University Press, 2016.

Displacement and Dispossession in the Modern Middle East. Cambridge: Cambridge University Press, 2010.

Syria: The Making and Unmaking of a Refuge State. London: Hurst & Company, 2017.

Chaudhuri, Kirti N. *Trade and Civilisation in the Indian Ocean: An Economic History from the Rise of Islam to 1750*. New York: Cambridge University Press, 1985.

Chiang, Connie Y. *Nature Behind Barbed Wire: An Environmental History of Japanese American Incarceration*. New York: Oxford University Press, 2018.

Christmas, Sakura. "Japanese Imperialism and Environmental Disease on a Soy Frontier, 1890–1940." *The Journal of Asian Studies* 78.4 (2019): 809–836.

Çiçek, M. Talha. *Negotiating Empire in the Middle East: Ottomans and Arab Nomads in the Modern Era, 1840–1914*. New York: Cambridge University Press, 2021.

Coates, Peter. *American Perceptions of Immigrant and Invasive Species*. Berkeley: University of California Press, 2007.

"The Muskrat's New Frontier: The Rise and Fall of an American Animal Empire in Britain." *Environmental History* 25.2 (April 2020): 207–236.

Coen, Deborah R. *Climate in Motion: Science, Empire, and the Problem of Scale*. Chicago: University of Chicago Press, 2018.

Cole, Camille Lyans. "Nafia for the Tigris: The Privy Purse and the Infrastructure of Development in Late Ottoman Iraq, 1882–1914." *History of Science* (2021): 1–23.

"Precarious Empires: A Social and Environmental History of Steam Navigation on the Tigris." *Journal of Social History* 50.1 (2016): 74–101.

Copeland, Marion. *Cockroach*. London: Reaktion, 2003.

Cora, Yaşar Tolga. "The Market as a Means of Post-Violence Recovery: Armenians and Oriental Carpets in the Late Ottoman Empire (c. 1890s–1910s)." *International Review of Social History* 66.2 (2021): 217–241.

Cora, Yaşar Tolga, Dzovinar Derderian, and Ali Sipahi, eds. *The Ottoman East in the Nineteenth Century: Societies, Identities and Politics*. London: IB Tauris, 2016.

Courtois, Sebastián de. *The Forgotten Genocide: Eastern Christians, The Last Arameans*, translated by Vincent Aurora. Piscataway, NJ: Gorgias Press, 2004.

Craib, Raymond B. *Cartographic Mexico: A History of State Fixations and Fugitive Landscapes*. Durham, NC: Duke University Press, 2004.

Cronon, William. *Changes in the Land: Indians, Colonists, and the Ecology of New England*. 20th anniversary ed. New York: Hill and Wang, 2003.

Nature's Metropolis: Chicago and the Great West. New York: Norton, 1991.

Crosby, Alfred. *Ecological Imperialism: The Biological Expansion of Europe, 900–1900*. New York: Cambridge University Press, 1986.

Cushman, Gregory T. *Guano and the Opening of the Pacific World: A Global Ecological History*. New York: Cambridge University Press, 2013.

Dağlıoğlu, Emre Can, ed. *Arapların 1915'i: Soykırım, Kimlik, Coğrafya*. Istanbul: İletişim, 2021.

Dağyeli, Jeanine Elif. "The Fight against Heaven-Sent Insects: Dealing with Locust Plagues in the Emirate of Bukhara." *Environment & History* 26.1 (February 2020): 79–104.

Danforth, Nicholas. *The Remaking of Republican Turkey: Memory and Modernity since the Fall of the Ottoman Empire*. New York: Cambridge University Press, 2021.

Daniel, Pete. *Toxic Drift: Pesticides and Health in the Post-World War II South*. Baton Rouge: Louisiana State University Press, 2005.

Daoudy, Marwa. *The Origins of the Syrian Conflict: Climate Change and Human Security*. New York: Cambridge University Press, 2020.

Davis, Bradley Camp. *Imperial Bandits: Outlaws and Rebels in the China–Vietnam Borderlands*. Seattle: University of Washington Press, 2017.

Davis, Diana K. *Resurrecting the Granary of Rome: Environmental History and French Colonial Expansion in North Africa*. Athens: Ohio University Press, 2007.

The Arid Lands: History, Power, Knowledge. Cambridge, MA: MIT Press, 2016.

Davis, Mike. *Late Victorian Holocausts: El Niño, Famines, and the Making of the Third World*. London: Verso, 2001.

Davison, Roderic. *Reform in the Ottoman Empire, 1856–1876*. Princeton, NJ: Princeton University Press, 1963.

De León, Jason. *Land of Open Graves: Living and Dying on the Migrant Trail*. Oakland: University of California Press, 2015.

Demuth, Bathsheba. *The Floating Coast: An Environmental History of the Bering Strait*. New York: WW Norton, 2019.

Der Matossian, Bedross. *Shattered Dreams of Revolution: From Liberty to Violence in the Late Ottoman Empire*. Stanford, CA: Stanford University Press, 2014.

The Horrors of Adana: Revolution and Violence in the Early Twentieth Century. Stanford, CA: Stanford University Press, 2022.

Deringil, Selim. "'The Armenian Question Is Finally Closed': Mass Conversions of Armenians in Anatolia during the Hamidian Massacres of 1895–1897." *Comparative Studies in Society and History* 51.2 (2009): 344–371.

The Well-Protected Domains: Ideology and the Legitimation of Power in the Ottoman Empire, 1876–1909. London: Tauris, 1998.

"'They Live in a State of Nomadism and Savagery': The Late Ottoman Empire and the Post-Colonial Debate." *Comparative Studies in Society and History* 45.2 (2003): 311–342.

Derr, Jennifer L. *The Lived Nile: Environment, Disease, and Material Colonial Economy in Egypt*. Stanford, CA: Stanford University Press, 2019.

Diken, Şeyhmus. *Ahım Var Diyarbakır*. Istanbul: Aras Yayıncılık, 2017.

Dodge, Toby. *Inventing Iraq: The Failure of Nation Building and a History Denied*. New York: Columbia University Press, 2003.

Dolbee, Samuel. "Borders, Disease, and Territoriality in the Post-Ottoman Middle East." In *Regimes of Mobility: Borders and State Formation in the Middle East, 1918–1946*, edited by Jordi Tejel and Ramazan Hakkı Öztan, 205–227. Edinburgh: Edinburgh University Press, 2022.

"Empire on the Edge: Desert, Nomads, and the Making of an Ottoman Provincial Border." *American Historical Review* 127.1 (March 2022): 129–158.

"The Desert at the End of Empire: An Environmental History of the Armenian Genocide." *Past & Present* 247 (May 2020): 197–233.

Donabed, Sargon George. *Reforging a Forgotten History: Iraq and the Assyrians in the 20th Century*. Edinburgh: Edinburgh University Press, 2015.

Donoghue, Michael E. *Borderland on the Isthmus: Race, Culture, and the Struggle for the Canal Zone*. Durham, NC: Duke University Press, 2014.

Dubnov, Arie and Laura Robson, eds. *Partitions: A Transnational History of Territorial Separatism*. Stanford, CA: Stanford University Press, 2019.

Duffy, Andrea A. *Nomad's Land: Pastoralism and French Environmental Policy in the Nineteenth-Century Mediterranean World*. Lincoln: University of Nebraska Press, 2019.

Dündar, Fuat. *Modern Türkiye'nin Şifresi: İttihat ve Terakki'nin Etnisite Mühendisliği (1913–1918)*. Istanbul: İletişim Yayınları, 2008.

Edwell, Peter M. *Between Rome and Persia: The Middle Euphrates, Mesopotamia and Palmyra under Roman Control*. London: Routledge, 2008.

Eger, A. Asa. *The Islamic–Byzantine Frontier: Interaction and Exchange Among Muslim and Christian Communities*. London: Tauris, 2015.

Ekinci, Mehmet R. "Osmanlı Devleti Döneminde Milli Aşireti XVIII.–XIX. YY." PhD diss., Fırat Üniversitesi, 2017.

Ekmekçioğlu, Lerna. *Recovering Armenia: The Limits of Belonging in Post-Genocide Turkey*. Stanford, CA: Stanford University Press, 2016.

Eldem, Edhem. "A Shameful Debate? A Critical Reassessment of the 'Torossian Debate.'" *Journal of Genocide Research* 19.2 (2017): 258–273.

Elden, Stuart. *Terror and Territoriality: The Spatial Extent of Sovereignty*. Minneapolis: University of Minnesota Press, 2009.

Ellis, Matthew H. "Over the Borderline? Rethinking Territoriality at the Margins of Empire and Nation in the Modern Middle East (Part I)." *History Compass* 13.8 (2015): 411-422.

"Over the Borderline? Rethinking Territoriality at the Margins of Empire and Nation in the Modern Middle East (Part II)." *History Compass* 13.8 (2015): 423-434.

Desert Borderland: The Making of Modern Egypt and Libya. Stanford, CA: Stanford University Press, 2018.

Emrence, Cem. *Remapping the Ottoman Middle East: Modernity, Imperial Bureaucracy, and the Islamic State*. New York: Tauris, 2012.

Ertem, Özge. "Eating the Last Seed: Famine, Empire, Survival and Order in Ottoman Anatolia in the Late 19th Century." PhD diss., European University Institute, 2012.

Fahmy, Khaled. *All the Pasha's Men: Mehmed Ali, His Army, and the Making of Modern Egypt*. Cambridge: Cambridge University Press, 1997.

Fahrenthold, Stacy D. *Between the Ottomans and the Entente: The First World War in the Syrian and Lebanese Diaspora, 1908–1925*. New York: Oxford University Press, 2019.

Faroqhi, Suraiya, ed. *Animals and People in the Ottoman Empire*. Istanbul: Eren, 2010.

Fattah, Hala. *The Politics of Regional Trade in Iraq, Arabia, and the Gulf, 1745–1900*. Albany: State University of New York Press, 1997.

Fattah, Hala and Candan Badem, "The Sultan and the Rebel: Saʿdun Al-Mansur's Revolt in the Muntafiq, c. 1891–1911," *International Journal of Middle East Studies* 45.4 (2013): 677–693.

Ferllini, Roxana and Alexandra Croft. "The Case of an Armenian Mass Grave." *Journal of Human Rights* 8.3 (August 2009): 229–244.

Few, Martha. "Killing Locusts in Colonial Guatemala." In *Centering Animals in Latin American History*, edited by Zeb Tortorici and Martha Few, 62–92. Durham, NC: Duke University Press, 2013.

Fiege, Mark. *Irrigated Eden: The Making of an Agricultural Landscape in the American West*. Seattle: University of Washington Press, 1999.

Finkel, Caroline. *Osman's Dream: The Story of the Ottoman Empire, 1300–1923*. New York: Basic Books, 2007.

Fisher, H. J. "The Eastern Maghrib and the Central Sudan." In *The Cambridge History of Africa*, edited by Roland Oliver, vol. 3, 232–330. New York: Cambridge University Press, 1977.

Fissell, Mary. "Imagining Vermin in Early Modern England." *History Workshop Journal* 47 (1999): 1–29.

Fleming, Laurence. *The Entokil Man: The Life of Harold Maxwell-Lefroy*. London: Dexter Haven, 2015.

Fletcher, Robert. *British Imperialism and "The Tribal Question": Desert Administration and Nomadic Societies in the Middle East, 1919–1936*. Oxford: Oxford University Press, 2015.

"The Locust, the Empire, and the Museum." *Evolve* 31 (2017): 46–53.

Forestier-Peyrat, Etienne. "Fighting Locusts Together: Pest Control and the Birth of Soviet Developmental Aid, 1920–1939." *Global Environment* 7 (2014): 292–327.

Foster, Zachary. "The 1915 Locust Attack in Syria and Palestine and Its Role in the Famine during the First World War." *Middle Eastern Studies* 51.3 (2014): 1–25.

Fratantuono, Ella. "Producing Ottomans: Internal Colonization and Social Engineering in Ottoman Immigrant Settlement." *Journal of Genocide Research* 21.1 (2019): 1–24.

Fromkin, David. *A Peace to End All Peace: Creating the Modern Middle East: 1914–1922*. London: Andre Deutsch, 1989.

Fuccaro, Nelida. *The Other Kurds: Yazidis in Colonial Iraq*. London: Tauris, 1999.

Gaunt, David. "Failed Identity and the Assyrian Genocide." In *Shatterzone of Empires: Coexistence and Violence in the German, Habsburg, Russian, and Ottoman Borderlands*, edited by Omer Bartov and Eric D. Weitz, 317–333. Bloomington: Indiana University Press, 2013.

Massacres, Resistance, Protectors: Muslim–Christian Relations in Eastern Anatolia during World War I. Piscataway, NJ: Gorgias Press, 2006.

"Two Documents on the 1895 Massacres in the Province of Diyarbekir: A Discussion." *Études arméniennes contemporaines* 10 (2018): 187–201.

Gaunt, David, Naures Atto, and Soner O. Barthoma, eds. *Let Them Not Return: Sayfo – The Genocide against the Assyrian, Syriac, and Chaldean Christians in the Ottoman Empire*. New York: Bergahn Books, 2017.

Gelvin, James. *Divided Loyalties: Nationalism and Mass Politics in Syria at the Close of Empire*. Berkeley: University of California Press, 1998.

Ghazarian, Matthew. "Ghost Rations: Empire, Ecology, and Community in the Ottoman East, 1839–94." PhD diss., Columbia University, 2021.

Ghosh, Amitav. "Shared Sorrows: Indians and Armenians in the Prison Camps of Ras al-ʿAin, 1916–1918," unpublished paper.

The Great Derangement: Climate Change and the Unthinkable. Chicago: University of Chicago Press, 2016.

Giesen, James. *Boll Weevil Blues: Cotton, Myth, and Power in the American South*. Chicago: University of Chicago Press, 2011.

Gil-Har, Yitzhak. "French Policy in Syria and Zionism: Proposal for a Zionist Settlement." *Middle Eastern Studies* 30.1 (Jan 1994): 155–165.

Gingeras, Ryan. *Sorrowful Shores: Violence, Ethnicity, and the End of the Ottoman Empire, 1912–1923*. New York: Oxford University Press, 2009.

Ginio, Eyal. *The Ottoman Culture of Defeat: The Balkan Wars and Their Aftermath*. New York: Oxford University Press, 2016.

Göçek, Fatma Müge. *Denial of Violence: Ottoman Past, Turkish Present, and Collective Violence Against Armenians, 1789–2009*. Oxford: Oxford University Press, 2015.

Göçen, Hüseyin. "Sığırcık Suyu Şeyhleri." *Toplumsal Tarih* 298 (October 2018): 38–48.

Goodwin, Barry K. and Thomas J. Greenes. "Tsarist Russia and the World Wheat Market." *Explorations in Economic History* 35 (1998): 405–430.

Gonzalez, Johnhenry. *Maroon Nation: A History of Revolutionary Haiti*. New Haven, CT: Yale University Press, 2019.
Goswami, Manu. *Producing India: From Colonial Economy to National Space*. Chicago: University of Chicago Press, 2004.
Grandin, Greg. *The End of the Myth: From the Frontier to the Border Wall in the Mind of America*. New York: Metropolitan Books, 2019.
Gratien, Chris. "The Ottoman Quagmire: Malaria, Swamps, and Settlement in the Late Ottoman Mediterranean." *International Journal of Middle East Studies* 49.4 (2017): 583–604.
"The Rice Debates: Political Ecology in the Ottoman Parliament." In *Seeds of Power: Explorations in Ottoman Environmental History*, edited by Onur İnal and Yavuz Köse. Winwick: White Horse Press, 2019.
The Unsettled Plain: An Environmental History of the Late Ottoman Frontier. Stanford, CA: Stanford University Press, 2022.
Greene, Jack P. and Philip D. Morgan, eds. *Atlantic History: A Critical Appraisal*. New York: Oxford University Press, 2009.
Grehan, James. "The Legend of the Samarmar: Parades and Communal Identity in Syrian Towns, c. 1500–1800." *Past & Present* 204 (2009): 89–125.
Güldoğan, Vedat. *Diyarbakır Tarihi*. Ankara: Kripto, 2011.
Gündoğdu, Cihangir. "The State and the Stray Dogs in Late Ottoman Istanbul: From Unruly Subjects to Servile Friends." *Middle Eastern Studies* 54.4 (2018): 555–574.
Güngör, Salahattin. "Bir Canlı Tarih Konuşuyor." *Resimli Tarih* 43.4 (July 1953): 2444–2445.
Gürel, Perin E. *The Limits of Westernization: A Cultural History of America in Turkey*. New York: Columbia University Press, 2017.
Gürün, Kamuran. *The Armenian File: The Myth of Innocence Exposed*. Nicosia: K. Rustem & Brother, 1985.
Gutman, David. *The Politics of Armenian Migration to North America, 1885–1915*. Edinburgh: Edinburgh University Press, 2019.
Hadjian, Avedis. *Secret Nation: The Hidden Armenians of Turkey*. London: Tauris, 2018.
Haines, Daniel. *Rivers Divided: Indus Basin Waters in the Making of India and Pakistan*. New York: Oxford University Press, 2017.
Haji, Golan. "La hidad ʿala mawt al-hashara." *Jadaliyya*. July 30, 2018.
Halaçoğlu, Yusuf. *Facts on the Relocation of Armenians, 1914–1918*. Ankara: Turkish Historical Society Printing House, 2002.
Hamalainen, Pekka. *The Comanche Empire*. New Haven, CT: Yale University Press, 2008.
Hamdan, Ali Nehme. "Breaker of Barriers? Notes on the Geopolitics of the Islamic State in Iraq and Sham." *Geopolitics* 21.3 (2016): 605–627.
Hamed-Troyansky, Vladimir. "Becoming Armenian: Religious Conversions in the Late Imperial South Caucasus." *Comparative Studies in Society and History* 63.1 (2021): 242–272.
"Circassian Refugees and the Making of Amman, 1878–1914." *International Journal of Middle East Studies* 49.4 (2017): 605–623.

"Imperial Refuge: Resettlement of Muslims from Russia in the Ottoman Empire, 1860-1914." PhD diss., Stanford University, 2018.

Hanioğlu, M. Şükrü. *The Young Turks in Opposition*. New York: Oxford University Press, 1995.

Hannoyer, Jean. "Politique des notables en Syrie: La naissance d'une ville (Deyr al-Zôr, 1850–1921)." *Bulletin d'études orientales* 41/42 (1989–1990): 113–142.

Hanoosh, Yasmeen. *The Chaldeans: Politics and Identity in Iraq and the American Diaspora*. London: Tauris, 2019.

Hanssen, Jens. *Fin de Siècle Beirut: The Making of an Ottoman Provincial Capital*. Oxford: Clarendon Press, 2005.

Harootunian, Harry. *The Unspoken as Heritage: The Armenian Genocide and Its Unaccounted Lives*. Durham, NC: Duke University Press, 2019.

Harraway, Donna J. *When Species Meet*. Minneapolis: University of Minnesota Press, 2008.

Harvey, David. *Spaces of Global Capitalism: Towards a Theory of Uneven Development*. New York: Verso, 2006.

Haugen, Gary A. and Victor Boutros. *The Locust Effect: Why the End of Poverty Requires the End of Violence*. New York: Oxford University Press, 2014.

Heimpel, Wolfgang. "Moroccan Locusts in Qattunan." *Revue d'Assyriologie et d'archéologie orientale* 90.2 (1996): 101–120.

Heyd, Uriel. *Foundations of Turkish Nationalism: The Life and Teachings of Ziya Gökalp*. London: Luzac & Company and the Harvill Press, 1950.

Hiltermann, Joost. *A Poisonous Affair: America, Iraq, and the Gassing of Halabja*. New York: Cambridge University Press, 2007.

Hinnebusch, Raymond. *Peasant and Bureaucracy in Ba'thist Syria: The Political Economy of Rural Development*. Boulder, CO: Westview Press, 1989.

Hızlı, Kasım. "Çekirge İstilasına Çözüm Sığırcık Kuşu." *Yedikıta: Tarih ve Kültür Dergisi* 70 (June 2014): 50–54.

Hole, Frank. "Drivers of Unsustainable Land Use in the Semi-Arid Khabur River Basin, Syria." *Geographical Research* 47.1 (March 2009): 4–14.

Hopkins, Benjamin. *Ruling the Savage Periphery: Frontier Governance and the Making of the Modern World*. Cambridge, MA: Harvard University Press, 2020.

Hourani, Albert. "A Vision of History: An Examination of Professor Toynbee's Ideas." In *A Vision of History: Near Eastern and Other Essays*, 1–34. Beirut: Khayats, 1961.

Husain, Faisal. "In the Bellies of the Marshes: Water and Power in the Countryside of Ottoman Baghdad." *Environmental History* 19 (October 2014): 638–664.

Rivers of the Sultan: The Tigris and the Euphrates in the Ottoman Empire. New York: Oxford University Press, 2021.

Husri, Khaldun Sati al-. "The Assyrian Affair of 1933 (I)." *International Journal of Middle East Studies* 5.2 (April 1974): 161–176.

"The Assyrian Affair of 1933 (II)." *International Journal of Middle East Studies* 5.3 (June 1974): 344–360.

Hütteroth, Wolf-Dieter. "Settlement Destruction in the Gezira Between the 16th and 19th Century." In *The Syrian Land in the 18th and 19th Century*, edited by Thomas Philipp, 285–294. Stuttgart: Franz Steiner Verlag, 1992.
Ibrahim, Muhammad ʿAli. *Amir Umaraʾ Kurdistan – Mir Mirani Kurdistan: Ibrahim Basha al-Milli, 1845–1908*. Erbil: Al-Akadamiyya al-Kurdiyya, 2009.
Iğsız, Aslı. *Humanism in Ruins: Entangled Legacies of the Greek–Turkish Population Exchange*. Stanford, CA: Stanford University Press, 2018.
İnal, Onur and Yavuz Köse, eds. *Seeds of Power: Explorations in Ottoman Environmental History*. Winwick: White Horse Press, 2019.
İnalcık, Halil, ed. *An Economic and Social History of the Ottoman Empire, 1300–1914*. New York: Cambridge University Press, 1994.
Ingold, Tim. *Lines: A Brief History*. New York: Routledge, 2016.
Issawi, Charles. *The Economic History of Turkey, 1800–1914*. Chicago: University of Chicago Press, 1980.
Jaafar, Hadi H. and Eckart Woertz. "Agriculture as a Funding Source of ISIS: A GIS and Remote Sensing Analysis." *Food Policy* 64 (October 2016): 14–25.
Jakes, Aaron. *Egypt's Occupation: Colonial Economism and the Crises of Capitalism*. Stanford, CA: Stanford University Press, 2020.
Jennings, Ronald C. *Christians and Muslims in Ottoman Cyprus and the Mediterranean World, 1571–1640*. New York: New York University Press, 1993.
"The Locust Problem in Cyprus." *Bulletin of the School of Oriental and African Studies* 51.2 (1988): 279–313.
Jinks, Rebecca. "'Marks Hard to Erase': Reclamation of 'Absorbed' Armenian Women, 1919–1927." *American Historical Review* 123.1 (February 2018): 86–123.
Johnson, Walter. *River of Dark Dreams: Slavery and Empire in the Cotton Kingdom*. Cambridge, MA: Belknap Press of Harvard University Press, 2014.
Jones, Susan. "Becoming a Pest: Prairie Dog Ecology and the Human Economy in the Euroamerican West." *Environmental History* 4.4 (October 1999): 531–552.
Jongerden, Joost. "Elite Encounters of a Violent Kind: Milli Ibrahim Paşa, Ziya Gökalp and Political Struggle in Diyarbekir at the Turn of the 20th Century." In *Social Relations in Ottoman Diyarbekir, 1870–1915*, edited by Joost Jongerden and Jelle Verheij, 55–84. Leiden: Brill, 2012.
The Settlement Issue in Turkey and the Kurds. Leiden: Brill, 2007.
Jongerden, Joost and Jelle Verheij, eds. *Social Relations in Ottoman Diyarbekir, 1870–1915*. Leiden: Brill, 2012.
Jwaideh, Albertine and J. W. Cox. "The Black Slaves of Turkish Arabia during the 19th Century." *Slavery & Abolition* 9.3 (1988): 45–59.
Kafadar, Cemal. *Between Two Worlds: The Construction of the Ottoman State*. Berkeley: University of California Press, 1995.
Kendine ait bir Roma: Diyar-ı Rum'da Kültürel Coğrafya ve Kimlik Üzerine. Istanbul: Metis, 2017.
Kaiser, Hilmar. *At the Crossroads of Der Zor: Death, Survival, and Humanitarian Resistance in Aleppo, 1915–1917*. London: Gomidas Institute, 2002.

The Extermination of Armenians in the Diarbekir Region. Istanbul: Istanbul Bilgi University, 2014.

Karaman, Oktay. "Diyarbakır Valisi Hatunoğlu Kurt İsmail Paşa'nın Diyarbakır'daki Aşiretleri Islah ve İskan Çalısması (1868–1875)." *History Studies: International Journal of History* 4 (2012): 227–249.

Karpat, Kemal H. "The Land Regime, Social Structure, and Modernization in the Ottoman Empire." In *Beginnings of Modernization in the Middle East: The Nineteenth Century*, edited by William R. Polk and Richard L. Chambers, 69–90. Chicago: University of Chicago Press, 1968.

Kasaba, Reşat. *A Moveable Empire: Ottoman Nomads, Migrants, and Refugees*. Seattle: University of Washington Press, 2009.

The Ottoman Empire and the World Economy in the Nineteenth Century. Albany: State University of New York, 1988.

Katz, Cindi and Neil Smith. "An Interview with Edward Said." *Environment and Planning D: Society and Space* 21 (2003): 635–651.

Kayalı, Hasan. *Arabs and Young Turks: Ottomanism, Arabism, and Islamism in the Ottoman Empire, 1908–1918*. Berkeley: University of California Press, 1997.

Imperial Resilience: The Great War's End, Ottoman Longevity, and Incidental Nations. Oakland: University of California Press, 2021.

Kaynar, İhsan Seddar. "Kıbrıs'ın Çekirge Sorunu ve İktisadi Felaketi (1878–1930)." *Kebikeç* 46 (2018): 121–154.

Kelley, Colin P., Shahrzad Mohtadi, Mark A. Cane, Richard Seager, and Yochanan Kushnir. "Climate Change in the Fertile Crescent and Implications of the Recent Syrian Drought." *Proceedings of the National Academy of Sciences* 112.11 (March 17, 2015): 3241–3246.

Kennedy, Dane. *The Last Blank Spaces: Exploring Africa and Australia*. Cambridge, MA: Harvard University Press, 2013.

Kévorkian, Raymond. "Earth, Fire, or Water: Or How to Make the Armenian Corpses Disappear." In *Destruction and Human Remains: Disposal and Concealment in Genocide and Mass Violence*, edited by Élisabeth Anstett and Jean-Marc Dreyfus, 89–116. Manchester: Manchester University Press, 2014.

The Armenian Genocide: A Complete History. London: Tauris, 2011.

Khater, Akram. *Inventing Home: Emigration, Gender, and the Middle Class in Lebanon, 1870–1920*. Berkeley: University of California Pres, 2001.

Khazanov, Anatoly M. *Nomads and the Outside World*, translated by Julia Crookenden. Madison: University of Wisconsin Press, 1993.

Khidr, Tha᾽ir Hamid Muhammad. *Tarikh al-Muhammad al-Jarba wa Qabilat al-Shammar al-ʿArabiyya fi Iqlim Najd wa al-Jazira*. Beirut: Dar al-ʿArabiyya lil Mawsuʿat, 2002.

Khoury, Dina Rizk. *State and Provincial Society in the Ottoman Empire: Mosul, 1540–1834*. Cambridge: Cambridge University Press, 1997.

Khoury, Philip S. *Syria and the French Mandate: The Politics of Arab Nationalism, 1920–1945*. Princeton, NJ: Princeton University Press, 1987.

"The Tribal Shaykh, French Tribal Policy, and the Nationalist Movement in Syria between Two World Wars." *Middle Eastern Studies* 18.2 (1982): 180–193.

Khuri-Makdisi, Ilham. *The Eastern Mediterranean and the Making of Global Radicalism, 1860–1914*. Berkeley: University of California Press, 2010.

Kieser, Hans-Lukas. *Talaat Pasha: Father of Modern Turkey, Architect of Genocide*. Princeton, NJ: Princeton University Press, 2018.

Kıran, Eyüp. *Kürt Milan Aşiret Konfederasyonu: Ekolojik, Toplumsal ve Siyasal bir İnceleme*. Istanbul: Elma Yayınları, 2003.

Kiyotaki, Keiko. *Ottoman Land Reform in the Province of Baghdad*. Leiden: Brill, 2019.

Klein, Janet. "State, Tribe, Dynasty, and the Contest over Diyarbekir at the Turn of the 20th Century." In *Social Relations in Ottoman Diyarbekir*, edited by Joost Jongerden and Jelle Verheij, 147–178. Leiden: Brill, 2012.

The Margins of Empire: Kurdish Militias in the Ottoman Tribal Zone. Stanford, CA: Stanford University Press, 2011.

Kriwaczek, Paul. *Babylon: Mesopotamia and the Birth of Civilization*. New York: St. Martin's Press, 2010.

Kurmuş, Orhan. "The Cotton Famine and Its Effects on the Ottoman Empire." In *The Ottoman Empire and the World-Economy*, edited by Huri İslamoğlu-İnan, 160–169. New York: Cambridge University Press, 1987.

Kurt, Ümit Selim. *The Armenians of Aintab: The Economics of Genocide in an Ottoman Province*. Cambridge, MA: Harvard University Press, 2021.

Kurtiç, Ekin. "Sediment in Reservoirs: A History of Dams and Forestry in Turkey." In *Transforming Socio-Natures in Turkey: Landscapes, State and Environmental Movements*, edited by Onur İnal and Ethemcan Turhan, 90–111. London: Routledge, 2020.

LaDow, Beth. *The Medicine Line: Life and Death on a North American Borderland*. New York: Routledge, 2001.

Lampe, John R. and Marvin R. Jackson. *Balkan Economic History, 1550–1950: From Imperial Borderland to Developing Nations*. Bloomington: Indiana University Press, 1982.

Lange, Katharina. "Shawāyā: Economic Melange, Pure Origins? Outsiders' and Insiders' Accounts of Tribal Identity in Northern Syria." In *Shifts and Drifts in Nomad–Sedentary Relations*, edited by Stefan Leder and Bernhard Streck, 99–122. Wiesbaden: Dr. Ludwig Reichert Verlag, 2005.

Latchininsky, Alexandre V. "Moroccan Locust *Dociostaurus maroccanus* (Thunberg, 1815): A Faunistic Rarity or an Important Economic Pest?" *Journal of Insect Conservation* 2 (1998): 167–178.

Leal, Claudia. *Landscapes of Freedom: Building a Postemancipation Society in the Rainforests of Western Colombia*. Tucson: University of Arizona Press, 2018.

Leick, Gwendolyn. *Mesopotamia: The Invention of the City*. New York: Penguin Press, 2001.

Lewis, Martin W. and Karen E. Wigen. *The Myth of Continents: A Critique of Metageography*. Berkeley: University of California Press, 1997.

Lewis, Norman. *Nomads and Settlers in Syria and Jordan, 1800–1980*. New York: Cambridge University Press, 1987.

Lindner, Rudi Paul. *Nomads and Ottomans in Medieval Anatolia*. Bloomington: Indiana University Press, 1983.

Lockman, Zachary. *Contending Visions of the Middle East: The History and Politics of Orientalism*. New York: Cambridge University Press, 2004.

Field Notes: The Making of Middle East Studies in the United States. Stanford, CA: Stanford University Press, 2016.

Lockwood, Jeffrey. *Locust: The Devastating Rise and Mysterious Disappearance of the Insect That Shaped the American Frontier*. New York: Basic Books, 2004.

Lorenz, Fredrick Walter. "The 'Second Egypt': Cretan Refugees, Agricultural Development, and Frontier Expansion in Ottoman Cyrenacia, 1897–1904." *International Journal of Middle East Studies* 53.1 (2021): 89–105.

Low, Michael Christopher. *Imperial Mecca: Ottoman Arabia and the Indian Ocean Hajj*. New York: Columbia University Press, 2020.

Lowry, Heath. *The Story Behind Ambassador Morgenthau's Story*. Istanbul: The ISIS Press, 1990.

Lydon, Ghislaine. *On Trans-Saharan Trails: Islamic Law, Trade Networks, and Cross-Cultural Exchange in Nineteenth Century West Africa*. New York: Cambridge University Press, 2009.

MacKeen, Dawn. *The Hundred-Year Walk: An Armenian Odyssey*. New York: Houghton Mifflin Harcourt, 2016.

Mavhunga, Clapperton Chakanetsa. "Vermin Beings: On Pestiferous Animals and Human Game." *Social Text* 29.1 (2011): 151–176.

Maier, Charles. *Once within Borders: Territories of Power, Wealth, and Belonging since 1500*. Cambridge, MA: Belknap Press of Harvard University Press, 2016.

Makdisi, Ussama. *Culture of Sectarianism: Community, History, and Violence in Nineteenth-Century Ottoman Lebanon*. Berkeley: University of California Press, 2000.

"Ottoman Orientalism." *American Historical Review* 107.3 (June 2002): 768–796.

The Age of Coexistence: The Ecumenical Frame and the Making of the Modern Arab World. Oakland: University of California Press, 2019.

Maksudyan, Nazan. *Ottoman Children and Youth during World War I*. Syracuse, NY: Syracuse University Press, 2019.

Marcus, Aliza. *Blood and Belief: The PKK and the Kurdish Fight for Independence*. New York: New York University Press, 2007.

Marcus, Harold. *The Middle East on the Eve of Modernity: Aleppo in the Eighteenth Century*. New York: Columbia University Press, 1989.

Marrus, Michael R. *The Unwanted: European Refugees in the Twentieth Century*. New York: Oxford University Press, 1985.

Martin, Laura J. "Proving Grounds: Ecological Fieldwork in the Pacific and the Materialization of Ecosystems." *Environmental History* 23.3 (July 2018): 567–592.

Marx, Karl. *Capital: A Critique of Political Economy*, translated by Ben Fowkes. Vol. 1. New York: Penguins, 1990.

Masalha, Nur. *Expulsion of the Palestinians: The Concept of "Transfer" in Zionist Political Thought, 1882–1948*. Washington, DC: Institute for Palestine Studies, 1992.

Masters, Bruce. *The Arabs of the Ottoman Empire, 1516–1918: A Social and Cultural History*. New York: Cambridge University Press, 2013.

The Origins of Western Economic Dominance in the Middle East: Mercantilism and the Islamic Economy in Aleppo, 1600–1750. New York: New York University Press, 1988.

McCarthy, Justin. *The Ottoman Peoples and the End of Empire*. London: Arnold, 2001.

McDowall, David. *A Modern History of the Kurds*. London: IB Tauris, 2004.

McMeekin, Sean. *The Berlin–Baghdad Express: The Ottoman Empire and Germany's Bid for World Power*. Cambridge, MA: Harvard University Press, 2010.

McNeill, John Robert. *Mosquito Empires: Ecology and War in the Greater Caribbean, 1620–1914*. Cambridge: Cambridge University Press, 2010.

McWilliams, James E. *American Pests: The Losing War on Insects from Colonial Times to DDT*. New York: Columbia University Press, 2008.

Meiton, Fredrik. *Electrical Palestine: Capital and Technology from Empire to Nation*. Oakland: University of California Press, 2019.

Melillo, Edward D. *The Butterfly Effect: Insects and the Making of the Modern World*. New York: Knopf, 2020.

Mendoza, Mary. "Treacherous Terrain: Racial Exclusion and Environmental Control at the U.S.–Mexico Border." *Environmental History* 23 (2018): 117–126.

Middleton, Nick. *Deserts: A Very Short Introduction*. New York: Oxford University Press, 2009.

Mikhail, Alan. "Foreword: Ottoman and Nature." In *Seeds of Power: Explorations in Ottoman Environmental History*, edited by Onur İnal and Yavuz Köse, vii–ix. Winwick: White Horse Press, 2019.

"Middle East Environmental History: The Fallow Between the Fields." In *Water on Sand: Environmental Histories of the Middle East and North Africa*, edited by Alan Mikhail, 1–25. New York: Oxford University Press, 2013.

Nature and Empire in Ottoman Egypt: An Environmental History. New York: Oxford University Press, 2011.

The Animal in Ottoman Egypt. New York: Oxford University Press, 2013.

Under Osman's Tree: The Ottoman Empire, Egypt, and Environmental History. Chicago: University of Chicago Press, 2017.

Miller, Owen. "Rethinking Violence in the Sasun Mountains (1893–1894)." *Études arméniennes contemporaries* 10 (2018): 97–123.

Minawi, Mostafa. "Beyond Rhetoric: Reassessing Bedouin–Ottoman Relations along the Route of the Hijaz Telegraph Line at the End of the Nineteenth Century." *Journal of the Economic and Social History of the Orient* 58 (2015): 75–104.

The Ottoman Scramble for Africa: Empire and Diplomacy in the Sahara and the Hijaz. Stanford, CA: Stanford University Press, 2016.

Mitchell, Timothy. *Carbon Democracy: Political Power in the Age of Oil*. New York: Verso, 2011.

Rule of Experts: Egypt, Techno-politics, Modernity. Berkeley: University of California Press, 2003.

Mitman, Gregg. *Breathing Space: How Allergies Shape Our Lives and Landscapes*. New Haven, CT: Yale University Press, 2007.

Mohamed, Jama. "'The Evils of Locust Bait': Popular Nationalism during the 1945 Anti-Locust Control Rebellion in Colonial Somaliland." *Past & Present* 174 (2002): 184–216.

Moon, David. *The Plough That Broke the Steppes: Agriculture and Environment on Russia's Grasslands, 1700–1914*. New York: Oxford University Press, 2013.

Moore, Jason W. *Capitalism in the Web of Life: Ecology and the Accumulation of Capital*. New York: Verso, 2016.

Moughalian, Sato. *Feast of Ashes: The Life and Art of David Ohannessian*. Stanford, CA: Redwood Press, 2019.

Mouradian, Khatchig. "The Meskeneh Concentration Camp, 1915–1917: A Case Study of Power, Collaboration, and Humanitarian Resistance during the Armenian Genocide." *Journal of the Society for Armenian Studies* 24 (2015): 44–55.

The Resistance Network: The Armenian Genocide and Humanitarianism in Ottoman Syria, 1915–1918. East Lansing: Michigan State University Press, 2021.

Nash, Linda. *Inescapable Ecologies: A History of Environment, Disease, and Knowledge*. Berkeley: University of California Press, 2006.

Needham, Andrew. *Power Lines: Phoenix and the Making of the Modern Southwest*. Princeton, NJ: Princeton University Press, 2014.

Neep, Daniel. *Occupying Syria under the French Mandate: Insurgency, Space and State Formation*. New York: Cambridge University Press, 2012.

Nieuwenhuis, Tom. *Politics and Society in Early Modern Iraq: Mamluk Pashas, Tribal Shayks, and Local Rule Between 1802 and 1831*. The Hague: Nijhoff Publishers, 1981.

Nixon, Rob. *Slow Violence and the Environmentalism of the Poor*. Cambridge, MA: Harvard University Press, 2011.

Olsson, Tore. *Agrarian Crossings: Reformers and the Remaking of the U.S. and Mexican Countryside*. Princeton, NJ: Princeton University Press, 2017.

Owen, Roger. *Cotton and the Egyptian Economy, 1820–1914: A Study in Trade and Development*. Oxford: Clarendon Press, 1969.

The Middle East and the World Economy, 1800–1914. New York: Tauris, 1981.

Özbilge, Nevcihan. *Çekirgeler, Kürtler ve Devlet: Erken Cumhuriyet Dönemine Yeniden Bakmak*. Istanbul: Tarih Vakfı Yurt Yayınları, 2020.

Özer, Sevilay. *Anadolu'da Görülen Çekirge İstilaları ve Halk Üzerindeki Etkisi (1914–1945)*. Ankara: Türk Tarih Kurumu, 2016.

Özok-Gündoğan, Nilay. "A Peripheral Approach to the 1908 Revolution in the Ottoman Empire: Land Disputes in Peasant Petitions in Post-Revolutionary Diyarbekir." In *Social Relations in Ottoman Diyarbekir, 1870–1915*, edited by Joost Jongerden and Jelle Verheij, 179–215. Leiden: Brill, 2012.

"Ruling the Periphery, Governing the Land: The Making of the Modern Ottoman State in Kurdistan, 1840–70." *Comparative Studies of South Asia, Africa and the Middle East* 34.1 (2014): 160–175.

Öztan, Ramazan Hakkı. "Republic of Conspiracies: Cross-Border Plots and the Making of Modern Turkey." *Journal of Contemporary History* 56.1 (January 2021): 55–76.

"The Great Depression and the Making of Turkish–Syrian Border, 1921–1939." *International Journal of Middle East Studies* 52.2 (May 2020): 311–326.

"Tools of Revolution: Global Military Surplus, Arms Dealers and Smugglers in the Late Ottoman Balkans, 1878–1908." *Past & Present* 237 (2017): 167–195.

Pamuk, Şevket. *Uneven Centuries: Economic Development of Turkey Since 1820*. Princeton, NJ: Princeton University Press, 2018.

Peçe, Uğur Zekeriya. "The Conscription of Greek-Ottomans into the Sultan's Army, 1908–1912." *International Journal of Middle East Studies* 52.3 (2020): 433–448.

Pedersen, Susan. *The Guardians: The League of Nations and the Crisis of Empire*. Oxford: Oxford University Press, 2015.

Pehlivan, Zozan. "Abandoned Villages in Diyarbekir Province at the End of the 'Little Ice Age.'" In *The Ottoman East: Transregionalism, Fluid Identities and Local Politics in the 19th and 20th Centuries*, edited by Yaşar Tolga Cora, Dzovinar Derderian, and Ali Sipahi, 223–246. London: Tauris, 2016.

"El Niño and the Nomads: Global Climate, Local Environment, and the Crisis of Pastoralism in Late Ottoman Kurdistan." *Journal of the Economic and Social History of the Orient* 63.3 (2020): 316–356.

Peloquin, Claude. "Locust Swarms and the Spatial Techno-Politics of the French Resistance in World War II." *Geoforum* 49 (2013): 103–113.

Perry, Jackson. "'Conquered by the Sparrows': Avian Invasions in French North Africa, circa 1871–1920." *Environmental History* 25.2 (2020): 310–334.

Petersen, Andrew. *The Medieval and Ottoman Hajj Route in Jordan: An Archaeological and Historical Survey*. Oxford: Oxbow Books, 2012.

Petrov, Milen V. "Everyday Forms of Compliance: Subaltern Commentaries on Ottoman Reform, 1864–1868," *Comparative Studies in Society and History* 46.4 (2004): 730–759.

"Midhat Paşa and the Vilayet of Danube, 1864–1868." PhD diss., Princeton University, 2006.

Philipp, Thomas, ed. *The Syrian Land in the 18th and 19th Century*. Stuttgart: Franz Steiner Verlag, 1992.

Philliou, Christine. *Biography of an Empire: Governing Ottomans in an Age of Revolution*. Berkeley: University of California Press, 2011.

Turkey: A Past Against History. Oakland: University of California Press, 2021.

Pitts, Graham Auman. "A Hungry Population Stops Thinking about Resistance: Class, Famine, and Lebanon's World War I Legacy." *Journal of the Ottoman and Turkish Studies Association* 7.2 (Fall 2020): 217–236.

"The Ecology of Migration: Remittances in World War I Mount Lebanon." *Arab Studies Journal* 26.2 (2018): 102–129.

Powell, Eve Troutt. *A Different Shade of Colonialism: Egypt, Great Britain, and the Mastery of Sudan*. Berkeley: University of California Press, 2003.

Pritchard, Sara B. *Confluence: The Nature of Technology and the Remaking of the Rhône*. Cambridge, MA: Harvard University Press, 2011.

Provence, Michael. *The Great Syrian Revolt and the Rise of Arab Nationalism*. Austin: University of Texas Press, 2005.

The Last Ottoman Generation and the Making of the Modern Middle East. New York: Cambridge University Press, 2017.

Purdy, Jedediah. *After Nature: A Politics for the Anthropocene*. Cambridge, MA: Harvard University Press, 2015.

Pursley, Sarah. *Familiar Futures: Time, Selfhood, and Sovereignty in Iraq*. Stanford, CA: Stanford University Press, 2019.

"'Lines Drawn on an Empty Map': Iraq's Borders and the Legend of the Artificial State." *Jadaliyya*. June 2, 2015.

Qattan, Najwa al-. "When Mothers Ate Their Children: Wartime Memory and the Language of Food in Syria and Lebanon." *International Journal of Middle East Studies* 46.4 (2014): 719–36.

Quataert, Donald. "The Age of Reforms," in *An Economic and Social History of the Ottoman Empire, 1300–1914*, edited by Halil İnalcık, 856–861. New York: Cambridge University Press, 1994.

The Ottoman Empire, 1700–1922. New York: Cambridge University Press, 2000.

Radding, Cynthia. *Wandering Peoples: Colonialism, Ethnic Spaces, and Ecological Frontiers in Northwestern Mexico, 1700–1850*. Durham, N.C.: Duke University Press, 1997.

Radner, Karen. "Fressen und gefressen werden: Heuschrecken als Katastrophe und Delikatesse im Alten Vorderen Orient." *Die Welte des Orients* 34 (2004): 7–24.

Raffles, Hugh. *Insectopedia*. New York: Vintage Books, 2011.

Reeves, Madeleine. *Border Work: Spatial Lives of the State in Rural Central Asia*. Ithaca, NY: Cornell University Press.

Robinson, Chase. *Empire and Elites after the Muslim Conquest: The Transformation of Northern Mesopotamia*. New York: Cambridge University Press, 2000.

Robson, Laura. *States of Separation: Transfer, Partition, and the Making of the Modern Middle East*. Oakland: University of California Press, 2017.

The Politics of Mass Violence in the Middle East. New York: Oxford University Press, 2020.

Rodrigue, Aron. *French Jews, Turkish Jews: The Alliance Israelite Universelle and the Politics of Jewish Schooling in Turkey, 1860–1925*. Bloomington: Indiana University Press, 1990.

Rogan, Eugene. "Aşiret Mektebi: Abdülhamid II's School for Tribes (1892–1907)." *International Journal of Middle East Studies* 28.1 (1996): 83–107.

Frontiers of the State in the Late Ottoman Empire. New York: Cambridge University Press, 1999.

The Fall of the Ottomans: The Great War in the Middle East. New York: Basic Books, 2015.

Rogers, Naomi. "Germs with Legs: Flies, Disease, and the New Public Health." *Bulletin of the History of Medicine* 63.4 (1989): 599–617.

Rothschild, N. Harry. "Sovereignty, Virtue, and Disaster Management: Chief Minister Yao Chong's Proactive Handling of the Locust Plague of 715–16." *Environmental History* 17.4 (October 2012): 783–812.

Roy, Rohan Deb. *Malarial Subjects: Empire, Medicine and Nonhumans in British India, 1820–1909*. Cambridge: Cambridge University Press, 2017.

Russell, Edmund. *War and Nature: Fighting Humans and Insects with Chemicals from World War I to Silent Spring*. New York: Cambridge University Press, 2001.

Saggs, Henry William Frederick. *The Might That Was Assyria*. London: Sidgwick and Jackson, 1984.

Sahlins, Peter. *Boundaries: The Making of France and Spain in the Pyrenees*. Berkeley: University of California Press, 1989.

"Natural Frontiers Revisited: France's Boundaries Since the Seventeenth Century." *American Historical Review* 95.5 (December 1990): 1423–1451.

Said, Edward. *Culture and Imperialism*. New York: Vintage Books, 1994.

Sajadian, China. "Debts of Displacement: Syrian Refugee Farmworkers at the Lebanese-Syrian Border." PhD diss., City University of New York-Graduate Center, 2022.

Sakatni, Mehdi. "From Camel to Truck? Automobiles and the Pastoralist Nomadism of Syrian Tribes during the French Mandate (1920–1946)." *Comparative Studies of South Asia, Africa and the Middle East* 39.1 (2019): 159–169.

Saliba, Najib E. "The Achievements of Midhat Pasha as Governor of the Province of Syria, 1878–1880." *International Journal of Middle East Studies* 9 (1978): 307–323.

Salzmann, Ariel. *Tocqueville in the Ottoman Empire: Rival Paths to the Modern State*. Leiden: Brill, 2004.

Sanchez-Bayo, Francisco and Kris A. G. Wyckhuys. "Worldwide Decline of the Entomofauna: A Review of Its Drivers." *Biological Conservation* 232 (April 2019): 8–27.

Saraçoğlu, M. Safa. *Nineteenth-Century Local Governance in Ottoman Bulgaria: Politics in Provincial Councils*. Edinburgh: Edinburgh University Press, 2017.

Saraiva, Tiago. *Fascist Pigs: Technoscientific Organisms and the History of Fascism*. Cambridge, MA: MIT Press, 2016.

Sarmani, Hanifa ʿUmar Faruq. *Tarikh Iqlim al-Jazira al-Furatiyya khilal al-ʿAsr al-Marwani al-Umawi, 64–132 H/684–750 M*. Beirut: Dar al-Nawadir al-Lubnaniyya, 2014.

Satia, Priya. *Spies in Arabia: The Great War and the Cultural Foundations of Britain's Covert Empire in the Middle East*. New York: Oxford University Press, 2008.

Scaramelli, Caterina. *How to Make a Wetland: Water and Moral Ecology in Turkey*. Stanford, CA: Stanford University Press, 2021.

Schayegh, Cyrus. *The Middle East and the Making of the Modern World*. Cambridge, MA: Harvard University Press, 2017.

Schendel, Willem van. "Geographies of Knowing, Geographies of Ignorance: Jumping Scale in Southeast Asia." *Environment and Planning D: Society and Space* 20.6 (2002): 647–668.

The Bengal Borderland: Beyond State and Nation in South Asia. London: Anthem Press, 2005.

Schiebinger, Londa L. *Plants and Empire: Colonial Bioprospecting in the Atlantic World*. Cambridge, MA: Harvard University Press, 2004.

Scott, James C. *Against the Grain: A Deep History of the Earliest States*. New Haven, CT: Yale University Press, 2017.

The Art of Not Being Governed: An Anarchist History of Upland Southeast Asia. New Haven, CT: Yale University Press, 2009.

Seeley, Samantha. *Race, Removal and the Right to Remain: Migration and the Making of the United States*. Williamsburg, VA: Omohundro Institute of Early American History and Culture; Chapel Hill: University of North Carolina Press, 2021.

Seikaly, Sherene. *Men of Capital: Scarcity and Economy in Mandate Palestine*. Stanford, CA: Stanford University Press, 2017.

Seitz, Jack. "Scientizing the Steppe: Agronomists, Nomads, and the Settler Colony on the Kazakh Steppe, 1881–1928." PhD diss., Iowa State University, 2019.

Semerdjian, Elyse. "Bone Memory: The Necrogeography of the Armenian Genocide in Dayr al-Zur, Syria." *Human Remains and Violence* 4.1 (2018): 56–75.

Serra, Gianluca. "Over-Grazing and Desertification in the Syrian Steppe Are the Root Causes of War." *The Ecologist*, June 5, 2015.

Serres, Michel. *Le Parasite*. Paris: B. Grasset, 1980.

Shafir, Nir. "Moral Revolutions: The Politics of Piety in the Ottoman Empire." *Comparative Studies in Society and History* 61.3 (2019): 595–623.

Shah, Sonia. *The Next Great Migration: The Beauty and Terror of Life on the Move*. London: Bloomsbury, 2020.

Shahin, Mazin Muhammad Fayz al-. *Ahdath Khalida fi Tarikh Dayr al-Zur*. Dayr al-Zur: Mazin Muhammad Fayz al-Shahin, 2007.

Shaw, Stanford and Ezel Kural Shaw. *Reform, Revolution, and Republic: The Rise of Modern Turkey, 1808–1875*. New York: Cambridge University Press, 1977.

Shemmassian, Vahram L. "The League of Nations and the Reclamation of Armenian Genocide Survivors." In *Looking Back, Moving Forward: Confronting the Armenian Genocide*, edited by Richard Hovannisian, 81–112. New Brunswick, NJ: Transaction Publishers, 2003.

Shields, Sarah. *Fezzes in the River: Identity Politics and European Diplomacy in the Middle East on the Eve of World War II*. New York: Oxford University Press, 2011.

Mosul before Iraq: Like Bees Making Five-Sided Cells. Albany: State University of New York Press, 2000.

"Sheep, Nomads and Merchants in Nineteenth-Century Mosul." *Journal of Social History* 25.4 (1992): 773–789.

Shlaim, Avi. *Israel and Palestine: Reappraisals, Revisions, Refutations*. New York: Verso, 2009.

Slobodkin, Yan. "Empire of Hunger: Famine and the French Colonial State, 1867–1945." PhD diss., Stanford University, 2018.

Sluglett, Peter. *Britain in Iraq: Contriving King and Country*. London: Tauris, 2007.

"The Resilience of a Frontier: Ottoman and Iraqi Claims to Kuwait, 1871–1990." *The International History Review* 24.4 (2002): 783–816.

Sollberger, E. "Mr. Taylor in Chaldaea." *Anatolian Studies* 22 (1972): 129–139.

St. John, Rachel. *Line in the Sand: A History of the Western U.S.–Mexico Border*. Princeton, NJ: Princeton University Press, 2011.

Stocker, Laura. "The 'Camel Dispute': Cross-Border Mobility and Tribal Conflicts in the Iraqi-Syrian Borderland, 1929–34." In *Regimes of Mobility: Borders and State Formation in the Middle East, 1918–1946*, edited by Jordi Tejel and Ramazan Hakkı Öztan, 319–350. Edinburgh: Edinburgh University Press, 2022.

Stoetzer, Bettina. "Ruderal Ecologies: Rethinking Nature, Migration, and the Urban Landscape in Berlin." *Cultural Anthropology* 33.2 (2018): 295–323.

Subramanian, Divya. "Legislating the Labor Force: Sedentarization and Development in India and the United States, 1870–1915." *Comparative Studies in Society and History* 61.4 (2019): 835–863.

Suny, Ronald Grigor. *"They Can Live in the Desert but Nowhere Else": A History of the Armenian Genocide*. Princeton, NJ: Princeton University Press, 2015.

Suny, Ronald Grigor, Fatma Müge Göçek, and Norman M. Naimark, eds. *A Question of Genocide: Armenians and Turks at the End of the Ottoman Empire*. New York: Oxford University Press, 2011.

Sutter, Paul. "Nature's Agents or Agents of Empire? Entomological Workers and Environmental Change during the Construction of the Panama Canal." *Isis* 98 (2007): 724–754.

"The World with Us: The State of American Environmental History." *Journal of American History* 100.1 (June 2013): 94–119.

Tabak, Faruk. *The Waning of the Mediterranean, 1550–1870: A Geohistorical Approach*. Baltimore, MD: Johns Hopkins University Press, 2008.

Tachjian, Vahé. *Daily Life in the Abyss: Genocide Diaries, 1915–1918*. New York: Bergahn, 2017.

"Gender, Nationalism, Exclusion: The Reintegration Process of Female Survivors of the Armenian Genocide." *Nations and Nationalism* 15.1 (2009): 60–80.

La France en Cilicie et en Haute-Mésopotamie: Aux confins de la Turquie, de la Syrie et de l'Irak. Paris: Karthala, 2004.

Tagliacozzo, Eric. *Secret Trades, Porous Borders: Smuggling and States along a Southeast Asian Frontier, 1865–1915*. New Haven, CT: Yale University Press, 2005.

Tamari, Salim. *Year of the Locust: A Soldier's Diary and the Erasure of Palestine's Ottoman Past*. Berkeley: University of California Press, 2011.

Tanielian, Melanie S. *The Charity of War: Famine, Humanitarian Aid, and World War I in the Middle East*. Stanford, CA: Stanford University Press, 2018.

Tauber, Eliezer. *The Formation of Modern Syria and Iraq*. Portland, OR: Frank Cass, 1995.

"The Struggle for Dayr al-Zur: The Determination of Borders between Syria and Iraq." *International Journal of Middle East Studies* 23 (1991): 361–385.

Tejel, Jordi. "'Des femmes contre des moutons': Franchissements féminins de la frontière turco-syrienne (1929–1944)." *20 & 21 Revue d'Histoire* 146 (2020): 35–47.

Syria's Kurds: History, Politics and Society, translated by Emily Welle and Jane Welle. London: Routledge, 2009.

"The Last Ottoman Rogues: The Kurdish–Armenian Alliance in Syria and the New State System in the Interwar Middle East." In *Age of Rogues: Rebels, Revolutionaries and Racketeers at the Frontiers of Empires*, edited by Ramazan Hakkı Öztan and Alp Yenen, 355–382. Edinburgh: Edinburgh University Press, 2021.

Tejel Gorgas, Jordi. *Le mouvement kurde de Turquie en exil: Continuités et discontinuités du nationalisme kurde sous le mandat français en Syrie et au Liban (1925–1946)*. Bern: Peter Lang, 2007.

"The Kurdish Cultural Movement in Mandatory Syria and Lebanon: An Unfinished Project of 'National Renaissance,' 1932–46." *Iranian Studies* 47.5 (2014): 839–855.

"The Terrier Plan and the Emergence of a Kurdish Policy under the French Mandate in Syria, 1926–1936." *International Journal of Kurdish Studies* 21.1–2 (2007): 93–108.

"Un territoire de marge en haute Djézireh syrienne (1921–1940)." *Études rurales* 186 (2010): 61–76.

Tekin, Kenan. "Reforming Categories of Science and Religion in the Late Ottoman Empire." PhD diss., Columbia University, 2016.

Thistle, John. *Resettling the Range: Animals, Ecologies, and Human Communities in British Columbia*. Vancouver: University of British Columbia Press, 2015.

Thomas, Lewis V. *A Study of Naima*, edited by Norman Itzkowitz. New York: New York University Press, 1972.

Thompson, Elizabeth. *Colonial Citizens: Republican Rights, Paternal Privilege, and Gender in French Syria and Lebanon*. New York: Columbia University Press, 2000.

Ticktin, Miriam. "Invasive Others: Toward a Contaminated World." *Social Research: An International Quarterly* 84.1 (Spring 2017): xxi–xxxiv.

Toksöz, Meltem. *Nomads, Migrants and Cotton in the Eastern Mediterranean: The Making of the Adana-Mersin Region, 1850–1908*. Leiden: Brill, 2010.

Treitel, Corinna. *Eating Nature in Modern Germany: Food, Agriculture and Environment, c. 1870–2000*. New York: Cambridge University Press, 2017.

Truett, Samuel. *Fugitive Landscapes: The Forgotten History of the U.S.–Mexico Borderlands*. New Haven, CT: Yale University Press, 2006.

Tsing, Anna. *The Mushroom at the End of the World: On the Possibility of Life in Capitalist Ruins*. Princeton, NJ: Princeton University Press, 2015.

Tural, Erkan. "Osmanlı Orman ve Maadin ve Ziraat Nezareti'nde Büroktraik Reform (1908–1914)." *Tarih Dergisi* 48.2 (2008): 85–115.

Turna, Nalan. *19. YY'den 20. YY'ye Osmanlı Topraklarında Seyahat, Göç ve Asayış Belgeleri: Mürür Tezkereleri*. Istanbul: Kaknus Yayınları, 2013.

Üngör, Uğur Ümit. *The Making of Modern Turkey: Nation and State in Eastern Anatolia, 1913–1950*. New York: Oxford University Press, 2012.

Üngör, Uğur Ümit and Mehmet Polatel. *Confiscation and Destruction: The Young Turk Seizure of Armenian Property*. London: Continuum, 2011.

Vail, David D. *Chemical Lands: Pesticides, Aerial Spraying, and Health in North America's Grasslands since 1945*. Tuscaloosa: University of Alabama Press, 2018.

Van Bruinessen, Martin. *Agha, Shaikh and State: The Social and Political Structures of Kurdistan*. London: Zed Books, 1992.

Van der Watt, L. M. "'To Kill the Locusts, but Not Destroy the Farmers': Officials, Farmers and the Plagues of Pharaoh, c. 1920–1935." *South African Historical Journal* 62.2 (2010): 356–383.

Varlık, Nükhet. *Plague and Empire in the Early Modern Mediterranean World: The Ottoman Experience, 1347–1600*. New York: Cambridge University Press, 2015.

Velud, Christian. "Une expérience d'administration régionale en Syrie Durant le mandat français: conquête, colonization et mise en valeur de la Gazira, 1920–1936." PhD diss., Université Lumière Lyon 2, 1991.

Verheij, Jelle. "Diyarbekir and the Armenian Crisis of 1895." In *Social Relations in Ottoman Diyarbekir, 1870–1915*, edited by Joost Jongerden and Jelle Verheij, 85–145. Leiden: Brill, 2012.

Wadewitz, Lissa K. *The Nature of Borders: Salmon, Boundaries, and Bandits on the Salish Sea*. Seattle: University of Washington Press, 2012.

Walker, Brett. *The Lost Wolves of Japan*. Seattle: University of Washington Press, 2008.

Toxic Archipelago: A History of Industrial Disease in Japan. Seattle: University of Washington Press, 2010.

Waloff, Nadejda and George Basil Popov. "Sir Boris Uvarov (1889–1970): The Father of Acridology." *Annual Review of Entomology* 35 (1990): 1–26.

Wang, Jessica. *Mad Dogs and Other New Yorkers: Rabies, Medicine, and Society in an American Metropolis, 1840–1920*. Baltimore, MD: Johns Hopkins University Press, 2019.

Watenpaugh, Heghnar Zeitlian. *The Missing Pages: The Modern Life of a Medieval Manuscript from Genocide to Justice*. Stanford, CA: Stanford University Press, 2020.

Watenpaugh, Keith David. *Being Modern in the Middle East: Nationalism, Colonialism, and the Arab Middle Class*. Princeton, NJ: Princeton University Press, 2006.

Bread from Stones: The Middle East and the Making of Modern Humanitarianism. Oakland: University of California Press, 2015.

"The League of Nations' Rescue of Armenian Genocide Survivors and the Making of Modern Humanitarianism, 1920–1927." *American Historical Review* 115.5 (2010): 1315–1339.

Webb, Peter. *Imagining the Arabs: Arab Identity and the Rise of Islam*. Edinburgh: Edinburgh University Press, 2016.

Weisiger, Marsha. *Dreaming of Sheep in Navajo Country*. Seattle: University of Washington Press, 2011.

Weitzberg, Keren. *We Do Not Have Borders: Greater Somalia and the Predicaments of Belonging in Kenya*. Athens: Ohio University Press, 2017.

Weizman, Eyal and Fazal Sheikh. *The Conflict Shoreline: Colonization as Climate Change in the Negev Desert*. Göttingen: Steidl, 2015.

White, Benjamin. "A Grudging Rescue: France, the Armenians of Cilicia, and the History of Humanitarian Evacuations." *Humanity* 10.1 (2019): 1–27.

"Humans and Animals in a Refugee Camp: Baquba, Iraq, 1918–20." *Journal of Refugee Studies* 32.2 (2019): 216–236.

"Refugees and the Definition of Syria." *Past & Present* 235 (May 2017): 141–178.

The Emergence of Minorities in the Middle East: The Politics of Community in French Mandate Syria. Edinburgh: Edinburgh University Press, 2012.

White, Richard. *Railroaded: The Transcontinentals and the Making of Modern America*. New York: Norton, 2011.

The Organic Machine. New York: Hill and Wang, 1995.

White, Sam. *The Climate of Rebellion in the Early Modern Ottoman Empire*. New York: Cambridge University Press, 2011.

Wickman, Thomas. *Snowshoe Country: An Environmental and Cultural History of Winter in the Early American Northeast*. New York: Cambridge University Press, 2018.

Widell, Magnus. "Historical Evidence for Climate Instability and Environmental Catastrophes in Northern Syria and the Jazira: The Chronicle of Michael the Syrian." *Environment and History* 13.1 (February 2007): 47–70.

Wilkinson, Tony James and D. J. Tucker. *Settlement Development in the North Jazira, Iraq: A Study of the Archaeological Landscape*. Baghdad: British School of Archaeology in Iraq and Department of Antiquities and Heritage, 1995.

Williams, Elizabeth. "Cultivating Empires: Environment, Expertise, and Scientific Agriculture in Late Ottoman and French Mandate Syria." PhD diss., Georgetown University, 2015.

Williamson, John Frederick. "A Political History of the Shammar Jarba Tribe of Al-Jazira: 1800–1958." PhD diss., Indiana University, 1975.

Willoughby, Bam. "Opposing a Spectacle of Blackness: Arap Bacı, Bacı Kalfa, Dadi, and the Invention of African Presence in Turkey." *Lateral* 10.1 (2021).

Wilson, Robert. *Seeking Refuge: Birds and Landscapes of the Pacific Flyway*. Seattle: University of Washington Press, 2010.

Wingham, Zavier. "*Arap Bacı'nın Ara Muhaveresi*: Under the Shadow of the Ottoman Empire and Its Study." *YILLIK: Annual of Istanbul Studies* 3 (2021): 177–183.

Winter, Stefan. "Alep et l'émirat du desert (*çöl beyliği*) au XVIIe–XVIIIe siècle." In *Aleppo and Its Hinterland in the Ottoman Period / Alep et sa province à l'époque ottoman*, edited by Stefan Winter and Malfalda Ade, 86–108. Leiden: Brill, 2019.

"The Other '*Nahdah*': The Bedirxans, the Millîs, and the Tribal Roots of Kurdish Nationalism in Syria." *Istituto per l'Oriente* 86 (2006): 461–474.

Wolf, Eric. *Europe and the People without History*. Berkeley: University of California Press, 1997.

Wolff, Larry. *The Idea of Galicia: History and Fantasy in Habsburg Political Culture*. Stanford, CA: Stanford University Press, 2010.

Worboys, Michael. "Imperial Entomology: Boris P. Uvarov and Locusts, c. 1920–c. 1950." *The British Journal for the History of Science* 55 (2022): 27–51.

Worster, Donald. "Transformations of the Earth: Toward an Agroecological Perspective in History." *The Journal of American History* 76.4 (March 1990): 1087–1106.

Yadırgı, Veli. *The Political Economy of the Kurds in Turkey: From the Ottoman Empire to the Turkish Republic*. New York: Cambridge University Press, 2017.

Yassin-Kassab, Robin and Leila al-Shami. *Burning Country: Syrians in Revolution and War*. London: Pluto Press, 2016.

Yaycıoğlu, Ali. *Partners of the Empire: The Crisis of the Ottoman Order in the Age of Revolutions*. Stanford, CA: Stanford University Press, 2016.

Yıldız, Kerem. *The Kurds in Syria: The Forgotten People*. London: Pluto Press, 2005.

Yılmaz, İlkay. "Governing the Armenian Question through Passports in the Late Ottoman Empire (1876–1908)." *Journal of Historical Sociology* 32.4 (2019): 388–403.

Yılmaz, Seçil. "Love in the Time of Syphilis: Medicine and Sex in the Ottoman Empire, 1860–1922." PhD diss., City University of New York-Graduate Center, 2016.

Yosmaoğlu, İpek. *Blood Ties: Religion, Violence, and the Politics of Nationhood in Ottoman Macedonia, 1878–1908*. Ithaca, NY: Cornell University Press, 2014.

Zahra, Tara. *Kidnapped Souls: National Indifference and the Battle for Children in the Bohemian Lands, 1900–1948*. Ithaca, NY: Cornell University Press, 2008.

The Lost Children: Reconstructing Europe's Families after World War II. Cambridge, MA: Harvard University Press, 2011.

Zakariyya, Ahmad Wasfi. *ʿAshaʾir al-Sham*. Vol. 2. Damascus: Dar al-Fikir, 1983.

Zandi-Sayek, Sibel. *Ottoman Izmir: The Rise of a Cosmopolitan Port, 1840–1880*. Minneapolis: University of Minnesota Press, 2012.

Zerubavel, Yael. *Desert in the Promised Land*. Stanford, CA: Stanford University Press, 2019.

Zhang, Ling. *The River, the Plain, and the State: An Environmental Drama in Northern Song China, 1048–1128*. New York: Cambridge University Press, 2016.

Zubaida, Sami. "Contested Nations: Iraq and the Assyrians." *Nations and Nationalism* 6.3 (2000): 363–382.

Zürcher, Erik Jan. *The Young Turk Legacy and Nation Building: From the Ottoman Empire to Atatürk's Turkey*. London: Tauris, 2010.

Index